PERFORMANCE ENHANCEMENTS IN A FREQUENCY HOPPING GSM NETWORK

Performance Enhancements in a Frequency Hopping GSM Network

by

Thomas Toftegaard Nielsen
Aalborg University, CPK
ERICSSON Telebit

and

Jeroen Wigard
Aalborg University, CPK
NOKIA Networks

SPRINGER SCIENCE+BUSINESS MEDIA, LLC

A C.I.P. Catalogue record for this book is available from the Library of Congress.

ISBN 978-1-4419-4976-9 ISBN 978-0-306-47313-5 (eBook)
DOI 10.1007/978-0-306-47313-5

Printed on acid-free paper

Originally published by Kluwer Academic Publishers, Boston in 2000
Softcover reprint of the hardcover 1st edition 2000

Contents

Preface

Mobile communications has during the last couple of years undergone an explosive progress in terms of number of subscribers as well in the effort put into related research. The subscriber increase has lead to requirements concerning better network quality and higher network capacity in order for the operators to be able to handle the requests. During the last 5 years a substantial amount of resources has therefore been put into enhancement to existing mobile radio systems like GSM.

This book provides a detailed description on how to enhance the BSS part of a GSM network using frequency hopping. The intention is to present a newly developed method for modelling a frequency hopping GSM network as well as to show the performance gains of different capacity enhancements. Everything is done within the scope of enhancing the performance of a frequency hopping GSM network.

One of the main issues in this book is to describe a new way of designing radio system performance enhancement features by using detailed computer network modelling. It has been done by combining link level and system level simulations to be able to achieve a high resolution in time. The link simulator developed and exploited provides a link performance model of the slow associated control channel (SACCH) as well as the full rate traffic channel (TCH/FS) in GSM. The network simulator, able to model the BSS part of the GSM network, is described and used extensively. Effects like cell structure, handover and power control algorithms, discontinuous transmission, traffic distribution, radio propagation and other network functionality's are modelled. In the book a model of the gain from frequency hopping is described and used for link as well as for system level calculations. Correspondingly the book treats the issue of measuring network quality in a frequency hopping network using simulations as well as real data. Alternative ways of exploiting frequency hopping using MAIO-management are also proposed.

The second major issue of the book, concrete capacity and quality enhancements, are documented throughout several of the chapters. Investigations of the how to improve the capacity along with the implications when combining downlink power control and discontinuous transmission in a frequency hopping GSM network are described. Also, a handover algorithm for GSM is studied for the frequency hopping GSM environment. Several handover enhancements are proposed and investigated. A detailed study, using a handover algorithm that enables reuse partitioning, of how to increase the cell capacity is also treated. This principle is based on a combination of the IUO reuse partitioning algorithm and frequency hopping. Various models are developed to investigate the absolute capacity potential. Several enhancements are furthermore proposed, increasing the cell capacity even further. The last performance enhancing subject treated takes a more practical view. It concerns frequency planning in a frequency hopping GSM network. Initially a newly developed method of how to do frequency planning of frequency hopping networks is described. This method includes the gain from frequency hopping directly in the allocation process. Concerning the same issue an improved method for graphical visualisation of a frequency plan for frequency hopping networks has been developed and is also described.

In general Chapter 1 through 6 provides an overview of the overall subject and the methods used to treat frequency hopping in GSM. In Chapter 7 the issue of downlink power control and DTX is treated in combination with frequency hopping. Chapter 8 deals with various existing and newly proposed handover algorithms. In Chapter 9 the network capacity is enhanced by combining reuse partitioning and frequency hopping. The book concludes treating the issue of frequency planning of frequency hopping GSM networks.

The book is written in such a way that it should be possible to read each of the design chapters (4 through 10) individually if only a certain subject is of interest. In general throughout the book many references have been used. Frequently more than one reference is used at a time. The idea of this is to provide the reader with easy access to related literature. In that way the book can be used as a work of reference. The literature list is included in the end of the book.

The book is intended for everyone interested in mobile radio communication systems. In general a high level of practical relevance relates to all the subjects treated, making the book especially relevant for network infrastructure manufacturers and network operators.

Acknowledgements

The material in this book originates from the research conducted as part of the two Ph.D. theses we have finalized in the spring of 1999, at Aalborg University, Center for PersonKommunikation (CPK), Denmark

We would like to thank our supervisors Jørgen Bach Andersen and Preben E. Mogensen for assigning us to this very interesting project and providing us with relationships to the industry, which has been of great help during the project. Several different participants have been involved in the project, where all have been more or less directly involved in the conducted research and the chosen subjects. The Danish GSM-900/1800 network operator SONOFON has sponsored Thomas Toftegaard Nielsen, while the Finish telecommunications vendor NOKIA Telecommunications has sponsored Jeroen Wigard.

Initially two completely separate Ph.D. projects were initiated with the same objective, namely capacity enhancements of the radio resources in a GSM system. However, quickly it became obvious that if possible, there would be a unique possibility of research teamwork between the research institution Center for PersonKommunikation (CPK) at Aalborg University, SONOFON and NOKIA, if the research was carried out jointly. Besides exploiting the experience from both SONOFON and NOKIA the possibility of developing complex network features in close co-operation with CPK, having NOKIA implementing the specific feature in their system and finally enable real live tests in the SONOFON network has been significant for the work carried out. Furthermore, from SONOFON, the many practical everyday comments have initiated very up-to-date and relevant research. The interaction of all involved parties reflecting this close co-operation is shown in Figure 1.

We would like to acknowledge SONOFON as well as NOKIA for their financial support, which has made the research possible. We are also grateful for the many discussions and proposals of new ideas, which have influenced quite a lot on the themes treated.

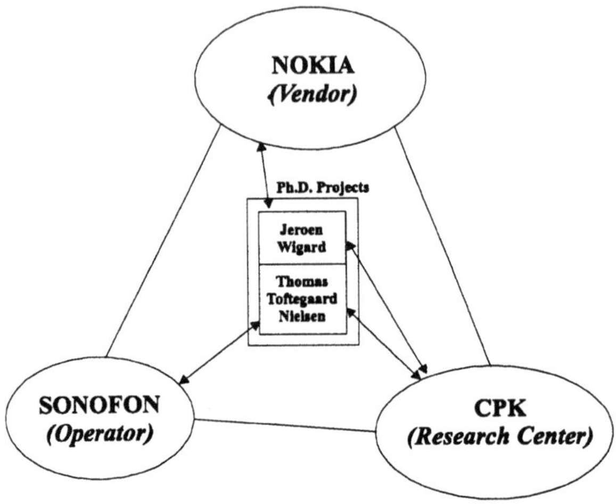

Figure 1. The co-operation between CPK, SONOFON and NOKIA

During the entire process we have had terrific colleagues with whom we have been able to discuss relevant as well as irrelevant research issues. The pleasant working environment at the cellular system group (CSG) at CPK and the capacity planning group at SONOFON has been greatly appreciated. In particular it should be said that a great part of the work has been done in co-operation with CSG at CPK, to which we are both formally employed. The development of the software tool CAPACITY, presenting a new way of modelling a mobile communications network, has been done in a close co-operation with Per Henrik Michaelsen for which we thank him.

Finally we acknowledge with gratitude the support of our girlfriends, Berit and Anita, who have provided constant encouragement during the writing of this book. They have both been impressively patient with our sometimes absent state of mind and extreme working schedule.

We hope this book will help explain the functionality of frequency hopping in GSM as well as the potential advantages and problems associated with frequency hopping. Correspondingly we would greatly appreciate if the proposed performance improvements will inspire further studies to enhance GSM in the future to come.

Jeroen Wigard
Thomas Toftegaard Nielsen

January 2000

Chapter 1

INTRODUCTION

The ability to exchange information becomes more and more important in today's society. This is reflected in the effort that is being put into research of the different telecommunication fields. Due to an enormous progress in the field of semiconductors, telecommunication today is relatively cheap with examples such as the telegraph, the telephone, digital mobile telephones, the Internet and various satellite communication systems being some of the landmarks in the development of electronic communication systems.

1. EVOLUTION OF DIGITAL SYSTEMS

The common drive in the research of communication systems,[1] is the need for faster exchange of information, i.e. the need for exchange of increasing amounts of information per time unit. Some of the goals of research within this field are to make communication less expensive and more efficient. The efficiency involves, among other parameters, the ability to be mobile while communicating. This has over the last couple of years become possible, enabling a commercial success of digital speech systems. The need for capacity has therefore increased enormously. Some digital examples are the European GSM-900 and GSM-1800, which have both become world-wide spread. These types of systems are generally referred to as 2nd generation Personal Communication Service systems, PCS [158]. In recent years Europe has witnessed a massive growth in mobile communications, where some northern European countries, such as Finland, have experienced penetration rates of more than 55 % [93]. On the way towards new 3rd generation mobile systems, such as the European Universal Mobile Telecommunications System (UMTS), Wireless Local Area Networks (WLAN) and Mobile Broadband Systems (MBS), allowing broadband data transmission (see Figure 2), it is therefore necessary to deal with the capacity problem of the existing 2nd generation systems. Along with the increasing

[1] In this book the word 'system' is used as a reference to a complete digital communication system. This includes everything in the link from end-user to end-user.

capacity requirements, requirements to the network performance increase the need for enhanced 2nd generation systems even further.

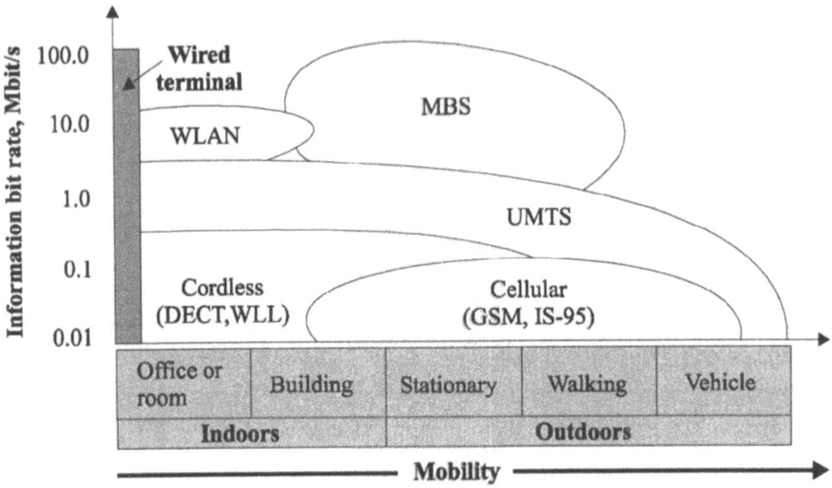

Figure 2. Different wireless communication systems, illustrating the available information bit rate for each system as a function of the mobility of the user [46].

In the case of 3rd generation systems a large effort is currently put into specifying, designing and standardising the individual systems. For 2nd generation systems this has previously been done, and therefore a large effort is now put into research/development of features and methods to optimise the performance of the individual system. Such features are in particular related to network algorithms of the 2nd generation system, but can in many cases correspondingly be used in 3rd generation systems. The two different research "branches" in the area of digital wireless communication are shown in Figure 3.

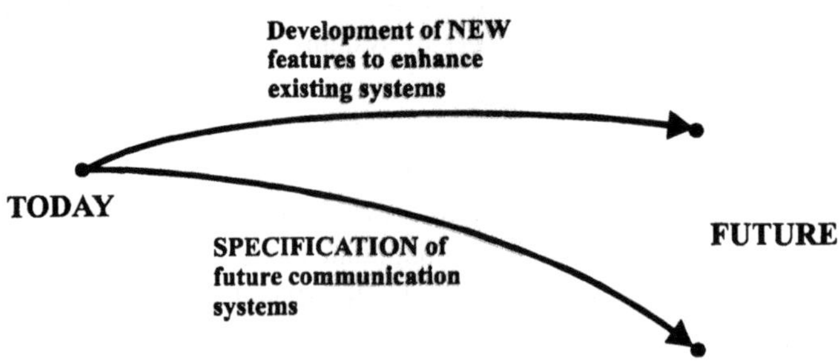

Figure 3. The two different kind of research going on right now: improving existing 2nd generation systems and specifying new 3rd generation systems.

This book deals with improvements to 2nd generation systems. In order to be able to design relevant performance features it has been chosen to limit the description to deal with one specific system and GSM has been chosen.

The term 'performance' is quite complex and linked closely to different network parameters such as capacity and quality. Part of this book treats the search for features/algorithms to enhance the network performance of the radio communication part of a GSM network, the BSS network. In order to clarify this statement the term performance, as defined in relation to this book, is described in the following.

2. PERFORMANCE OF A MOBILE NETWORK

Depending on the target group quite different parameters are of interest in determining the system performance. The two parties of interest, as considered here, are the network operator and the mobile user. Therefore the performance parameters considered are also seen from this perspective. The most essential parameters are:

Capacity - The GSM network operators have only a limited number of channels at their disposal. This means that the system capacity must be optimised with a fixed number of channels. The network should be designed to meet the requirements of capacity arising from the users. Intensive research on how to do this is continuously being conducted using different approaches. Different methods on how to enhance the GSM network capacity are proposed throughout this book, making capacity one of the most important parameters.

Quality – Throughout this book the term quality reflects the experience by the system end-user. Different quality measures are used, like the signal to interference ratio (*C/I*), the bit error rate (BER), the frame erasure rate (FER), the dropped call rate and the number of blocked calls. In some situations these quality measures are correlated, while in others some are more independent of each other. All of them are not necessarily correlated to the subjective quality experienced by the mobile user. Many different factors decide the network quality, making it another key parameter in this book.

Coverage - In order to have a high performance cellular mobile network, a certain level of coverage has to be provided. Since such a network is assumed, coverage becomes a secondary issue.

Cost - The impact of the price of a commercial communication system may never be underestimated. The mobiles, as well as the base stations including the rest of the cellular network infrastructure, have to be relatively cheap to ensure commercial success. However, in this book the cost issue is not treated any further.

New Services – The effect of offering new services to the subscribers becomes more important for network operators. Having lowered the air-time price as much as possible, one parameter that can be used to ensure the proper revenue is new services. By inventing new services that does not necessarily require a large amount of capacity, an income based on the functionality of the service can be generated. Since such new services can be operator specific, the individual operator can differentiate itself from the competing operators and thereby get more customers. Currently one of the new services for GSM is enhanced data-rates to introduce Internet applications from GSM. This subject is briefly introduced in Chapter 2.

Complexity/Flexibility - From the network operators' point of view a low network complexity is highly desirable. This is related to what is generally referred to as network maintenance and for a great part taken care of by the Operating and Maintenance Centre (OMC). From the radio network engineers point of view, especially frequency and parameter planning is important. A fairly large part of this book is concerned with the network complexity when considering the problem of frequency planning.

The six performance parameters introduced above all have some kind of an influence on each other making the overall network performance evaluation quite complex. Examples of such influences are the trade off between quality and capacity or between quality/capacity and cost.

Another way of illustrating this parameter interaction is by looking at the typical life cycle of the highest priority of these parameters, if specified by the operator. This is shown in Figure 4.

Coverage ⟶ Capacity ⟶ Quality

Figure 4. The shift in primary performance parameter as a function of time.

Initially, during the roll-out of the network (stage 1), the primary aim is to provide as much coverage as possible in order to offer mobile telephony to as many people as possible. With increasing coverage the need for capacity becomes more and more important (stage 2). Furthermore, the various national operators compete about the customers, which increases the capacity requirements even further since the individual operator typically will do quite inventive things to get new customers. A very effective way to accelerate the movement towards stage 2 is when operators starts selling mobile phones at a very low cost (e.g. at the price of 1 DKr. as was the case in Denmark in 1997). At stage 2 the idea is typically to provide a satisfactorily quality, while increasing the network capacity substantially. After a certain period of time the number of potential new customers becomes smaller and the primary aim becomes to improve the network quality (stage 3). Of course the quality has to be as good as possible during all stages, but now it becomes the primary aim.

3. THE AIM OF THE BOOK

Research aiming at developing new system specific network features, has traditionally been a complex matter since the only way to investigate if an idea is good or bad, is by trying it out in a real network. Many features can only be evaluated properly by collecting statistics from a clustered (large) part of the network. Furthermore, an important point of live network trials is that the operator will typically not allow quality degradations in the operating network. Therefore, the investigated ideas have to be of little risk to the network. This minimises the steps taken in each trial and correspondingly in the research conducted. Another important thing is the time put into the implementation of individual network features (by the vendor as well as the operator) which can easily be quite extensive.

For the reasons described, it would be desirable to have a computer model of the BSS part of the network, in which the proposed features could be tried out and enhanced further. Such a model can be used for computer aided network design (CAND), i.e. for designing and optimising the network.

Besides developing new performance enhancement network features to increase the network quality as well as capacity, the aim has therefore also been to develop a computer aided network design model for the BSS part of the GSM network. The model should have a resolution in time as well as in system level detail so high that the tendencies found in the output from the model can be related to real networks. The level of detail should correspondingly allow new features to be implemented in a way that allows a realistic implementation afterwards in the live network. The underlying idea of the book is therefore to:

1. Describe the effect of frequency hopping in a GSM BSS system using a designed computer network model

2. Document new network capacity and quality enhancement features for frequency hopping GSM BSS networks.

Chapter 2

PERFORMANCE ENHANCING STRATEGIES AND EVALUATION METHODS

Inventions of new radio network features have in the past not always been unproblematic. Due to the extensive trial demands when considering complete networks the development has traditionally been done by looking at smaller parts of the entire system. With some features this approach is however not a good solution since the desired/undesired effect may only be observed if treating a larger part of the network.

The approach used to develop new network features is introduced in this chapter. Initially, in Section 1, some effort has been put into giving an overview of existing radio performance enhancement principles. Among these are the features exploited in this book. Since the number of features is almost unlimited, only the ones we believe to be the most important strategies are described. More detailed information on the methods and strategies can be found in the literature referred to throughout the section. In Section 2 the computer aided design model is briefly presented to give an idea of how it works and how it has been designed. To emphasise that other methods than simulations have been used during the study, Section 3 and 4 have been written. Here it is described how, in some simplified cases, classic traffic theory is used for comparison. Finally, to verify the correlation between the real world and the simulation results, live network field trials have been carried out if possible.

1. RADIO PERFORMANCE ENHANCEMENTS

Fundamentally different approaches to improve the performance of a cellular mobile radio systems exists. One graphical way of illustrating these is shown in Figure 5. The different strategies are illustrated as a super highway, where it is up to the individual radio network designer/planner to choose which one of the possible roads to take [17]. Of course the choice is not limited to one single strategy.

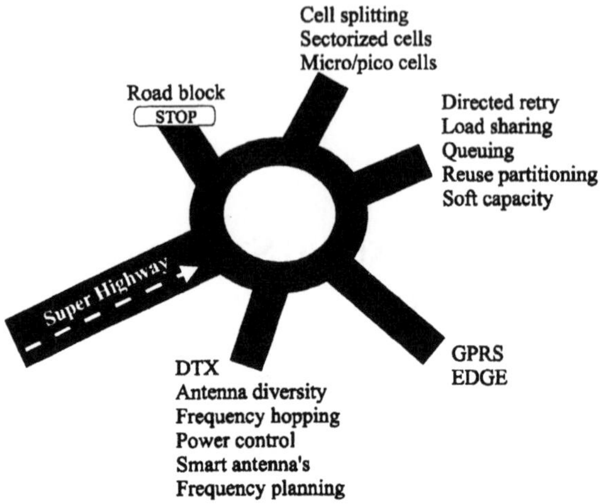

Figure 5. A super-highway illustrating the four major strategic ways to improve the performance of a cellular mobile communication network.

The first method is the engineering of the network infrastructure. Here the enhancement methods are described by cell splitting, the use of sectorized cells and the implementation of micro and pico cells as an addition to an existing digital cellular network. Another way is the dynamic arrangement of available radio channels, such as directed retry, load sharing, queuing, reuse partitioning and soft capacity algorithms. The third method is to employ technologies to reduce the influence from interference between frequencies so as to enable higher possible frequency reuse. Employing features like discontinuous transmission, antenna diversity, frequency hopping, power control and/or smart antennas can do this. Also the practical issue of doing frequency planning falls in this category. The last major way treats the issue of implementing new services in the network. Several different methods can be used. Here GPRS and EDGE have been chosen.

Other ways of increasing cellular network performance than the ones shown in Figure 5 exists, however these strategies present a general overview of what we believe are the most dominating ones.

1.1 Engineering of the Network Infrastructure

As seen in Figure 5 three methods to increase the network performance by modifying the network infrastructure are described; cell splitting, sectorization and hierarchical cell structures.

1.1.1 Cell Splitting

The basic idea of cell splitting is to divide a cell into smaller cells as the amount of traffic grows [129]. By adjusting the neighbour parameters, the frequency reuse can be retained [113]. Cell splitting is probably the method, which has the highest potential in terms of capacity gain, but is also a very expensive solution for the network operator as it requires more base station sites. For areas with a very high traffic load, where small cells are required in order to achieve adequate capacity, it is good practice to use a hierarchical network structure consisting of an overlaying macro cellular layer and an underlying micro cellular layer. More about this is found in Section 1.1.3. Cell splitting is not treated any further.

1.1.2 Sectorized Cells

Antenna sectorization is commonly used for GSM macro cellular base stations. Using antenna sectorization, several cells can be served by one base station. The offered capacity per base station is increased this way making it an economically attractive solution for the network operator. However, sectorization requires a larger effective frequency reuse distance. For an omni directional base station configuration a reuse of 7/7 is nominal for GSM, whereas for a three sectorized configuration is commonly referred to have a nominal reuse of 4/12 [66]. The capacity increase of a three sectorized base station is a factor of 1.5-1.75 [66]. Base station configurations with three 120° sectors are most commonly used, but also six sectorized 60° configurations have been studied [110]. From a capacity point of view, a very narrow sector beam is not exclusively positive due to the change in effective frequency reuse distance and an increased handover rate. Most of the research conducted throughout the book is based on sectorization of sites.

1.1.3 Hierarchical or Multi Layer Cell Structures

Another promising solution to satisfy the clustered high teletraffic demand is the small-cells approach [221]. The idea is very simple. The network uses a number of small-size cells, referred to as micro and pico cells, to serve demands in the high-density areas. These small cells can be combined with the relatively large cells, the macro cells, as can be seen in Figure 6.

The macro cells are used primarily for coverage. The micro and pico cells are intended to provide services in residential areas, offices, public places and streets. The physical size of these low power base stations is very small, and they can therefore be mounted on street lampposts or nearly anywhere within a building.

The idea is that the base stations are close to their mobile users and are typically joined to their control centres either by high capacity point-to-point radio links or optical fibers. Almost every existing digital cellular network operator has some kind of a macro-cell based network in order to provide satisfactorily coverage. At the same time more and more operators are required combine these with micro cells to increase the network capacity [132].

A pure micro cellular system, i.e. a system without macro cells, suffers from high costs and complexity. Such a system may involve a large number of base stations and a conversation may need several handovers. Another practical drawback of the pure micro cell system, at least initially, is the lack of wide area continuous radio coverage.

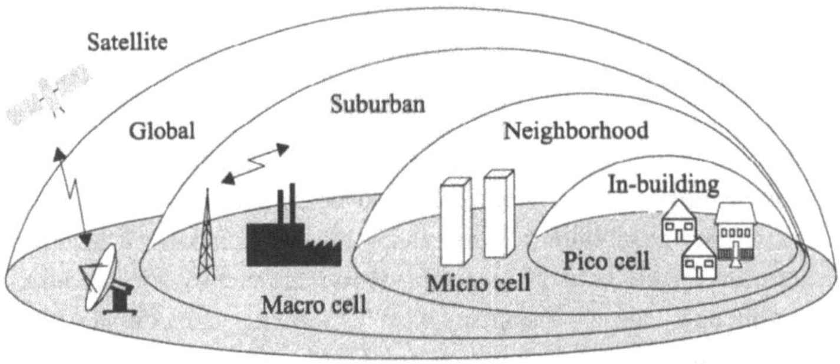

Figure 6. The different types of cells generally used in describing continuous coverage ranging from satellite communication to small pico cells [46].

With hierarchical cells the network complexity quickly increases. A clear benefit can be seen in co-ordinating the macro and micro cellular layer. Handovers can be arranged in such a way that different priorities are given to different mobiles. Such priorities could e.g. be based on parameters as speed and location. The micro cells could handle the slowly moving traffic, while the macro cells carries the users of high speed. For those who can receive strong enough signals from both types of cells, macro cells can serve as a traffic overflow handler, when all the channels in the micro cells are fully occupied. By doing this the trunking efficiency, and thereby the network capacity, is increased.

Throughout the book a cellular network environment consisting of macro cells has in general been assumed.

1.2 Radio Interface Channel Allocation Techniques

Various methods for allocating the mobile station to the best suitable radio channel exist. Some of the techniques for GSM, such as directed retry, load sharing, queuing, soft capacity and reuse partitioning, are introduced in this section.

1.2.1 Directed Retry

When an initial connection is attempted, the serving cell will be selected by the mobile station, based on the received signal strength. When attached to the network, the mobile station can, for congestion reasons, be requested to make a re-selection of the serving cell. When this happens during the call-setup it is called a directed retry.

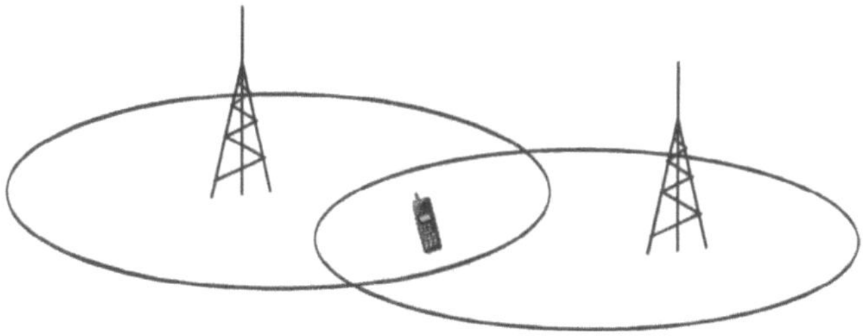

Figure 7. Example showing the overlapping service areas of neighbouring cells.

This way the blocking in the network can be decreased. The directed retry feature can only be used when the mobile station has more than one serving cell candidate. Thus, in order for the feature to be efficient, it is required that neighbouring cells have a large overlapping serving area, see Figure 7, which in theory makes the feature more powerful with large frequency reuses. Directed retry is not treated any further.

1.2.2 Load Sharing

For mobiles in dedicated mode, carrying out a conversation, the overlapping service areas can again be exploited. This functionality is referred to as load sharing or traffic reason handover. Different ways of implementing the functionality can be imagined for GSM. Load sharing is treated in the handover study carried out in Chapter 8 and is therefore not treated further here.

1.2.3 Queuing

To be able to load a network to a higher level, i.e. increase the capacity, queuing on both incoming calls and handovers can be introduced. A lot of research, on how queuing of handovers affect the capacity of cellular networks, has been carried out [73]. Some network operators use queuing on both handovers and new calls. By using queuing, the blocking can be decreased at the price of a delay. The queue should be configured (queue holding time and length) so the blocking probability is reduced to a minimum, while the time delays are acceptable. Furthermore, queuing can, with benefit in certain situations, be combined with directed retry. More about queuing in GSM is found in Chapter 8.

1.2.4 Reuse Partitioning

The basic idea of reuse partitioning is to enlarge the possibility of using the same frequency more often in a certain number of cells, i.e. to decrease the frequency reuse. By changing the normal frequency allocation strategy and allowing different reuse patterns for different frequencies, an overall tighter frequency reuse of the available frequencies can be achieved. Mobile stations close to the base station can use the overlay network, while mobile stations close to the border of the cell, stay on the underlayer, see Figure 8. The frequency reuse of the overlay can be smaller than the reuse of the underlayer, since the mobile stations on the overlayer are likely to have better quality than the ones on the underlayer. As an example, using sectorized sites, a reuse factor of 1/3 could be used on the overlayer, while the reuse at the underlayer correspondingly could be 3/9.

Several different principles exist for reuse partitioning, such as the concentric cell[2] and the intelligent underlay/overlay[3] (IUO). Both these concepts use two layers, one with the relatively loose frequency reuse and the other with a much tighter reuse scheme. IUO is based on an evaluation of the *C/I* experienced by the mobile stations, whereas the concentric cell concept use the concept of inner and outer zones. It is based on the idea that a higher measured power level automatically implies a higher *C/I* ratio. In reality this philosophy means that the inner zone can serve only mobiles close to the site [94]. Chapter 9 is devoted to reuse partitioning by IUO for frequency hopping GSM networks.

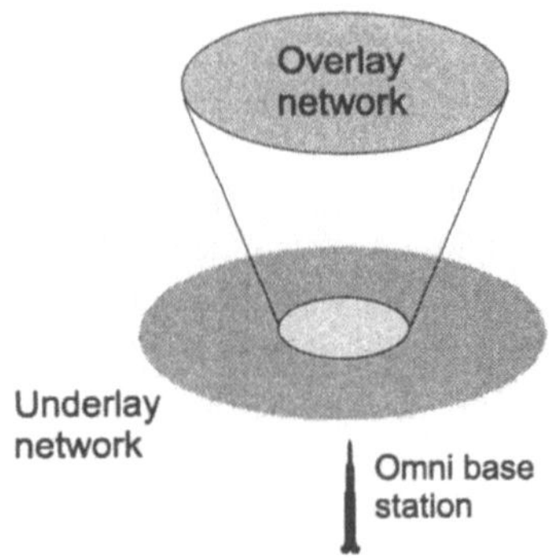

Figure 8. Reuse partitioning, illustrated by an omni directional site with coverage area for both of the frequency groups.

[2] Different vendors have introduced the concentric cell concept, of which Motorola is one.

[3] Nokia and the mobile network operator CSL from Hong Kong have originally introduced the intelligent underlay/overlay concept.

1.2.5 Soft Capacity Algorithms

An alternative to the multi layered cell structure is the principle of soft capacity algorithms. The idea is to allocate frequencies according to a reuse factor smaller than the "normally" acceptable minimum reuse. If the loose "normal" reuse factor is 12, the factor may de decreased to 9 or 10 or even down to 3 or 1. In the channel allocation algorithm a criterion is added, which only allows an allocation if the overall quality (in the cell and the neighbouring cells using the same frequency) is good. This way the network will never be loaded completely, but a very dynamic allocation algorithm adjusted to the actual traffic distribution is achieved. At the same time, the trunking efficiency becomes much better, since more channels are available per cell. The major problem with these soft capacity algorithms is how to design a measure describing the quality of the clustered cell satisfactorily. The soft blocking issue is not treated any further.

1.3 Techniques to Limit the Influence of Interference

The subject of interference limiting techniques is probably one of the hottest of all the performance enhancing strategies described so far. The basic idea of limiting the performance degradation from interference is to provide better quality or to reduce the frequency reuse factor to get more capacity.

1.3.1 Discontinuous Transmission

Discontinuous transmission (DTX) is a powerful and simple way of decreasing the interference in a network. The idea is only to transmit/receive when necessary [151], i.e. when the subscriber is speaking. For speech communications, each person typically does not talk more than 40% of the time, corresponding to a DTX factor of 0.4. Results from simulations have shown a linear proportionality between the DTX factor and the improvement in *C/I* [105] when combined with random frequency hopping. The functionality of DTX in combination with power control and frequency hopping is treated in Chapter 7.

1.3.2 Antenna Diversity

Antenna diversity on the base station is a well-known feature [54,147]. In the uplink, the idea is to let the base station have two or more antennas, where each antennas receives the signal. The antennas have to be configured in a way that the fast fading of the signal received by the different antennas is independent. Two different ways of designing the base station hardware exists for antenna diversity. The traditional design uses space diversity with the antennas being physically separated enough to ensure independent fast fading. Uplink antenna space diversity is of great interest for coverage extension, i.e. for open land terrain. The other possibility, polarisation diversity, where two antennas with orthogonal polarisation are used [49]. Polarisation diversity is especially found interesting for urban and bad urban areas. Antenna diversity at the base station can also be used for the downlink direction. Here the functionality is denoted transmit diversity [142].

For capacity reasons it is also desirable to use downlink antenna diversity at the mobile station. This functionality can be implemented in different ways. The theoretically best way to implement downlink antenna diversity is by using a Maximal-Ratio type of combining diversity [145]. Unfortunately this solution of net very cost-effective, since it requires two parallel RF-receiving paths within the mobile. Different sub-optimum solutions exist of which some are subject to research to be introduced in current and future cellular systems [58]. The issue of antenna diversity is not treated any further.

1.3.3 Frequency Hopping

No description of the feature is given here, since it is the underlying main issue of the book. Here the idea is simply to state that frequency hopping falls into the category of techniques that limits the influence from interference.

1.3.4 Uplink/Downlink Power Control

Dynamic power control can be applied in both the uplink and downlink direction. In the uplink situation, the mobile adjusts the output power. This happens typically when a mobile is close to the serving base station, then the received signal strength is unnecessarily high. This has two advantages; the interference is decreased and the mobile consumes less battery power. In the same situation, the base station can apply downlink power control to decrease the overall level of downlink interference. This interference reduction can be converted to a capacity increase if used in a frequency hopping network. The issue of downlink power control in a frequency hopping GSM network is found in Chapter 7.

1.3.5 Smart Antennas at the Base Station

An even more advanced capacity enhancement method is the use of intelligent or smart antennas. The term 'smart antenna' is used for several different types of antennas, but in this brief description the meaning is simply an antenna, which is more intelligent than a passive antenna.[4]

Typically a grid of narrow overlapping beams, e.g. four or eight, is created by means of a phased array to cover a sector of 120°. A mobile station performing an angular sweep in a cell will make a seamless handover between antenna beams

[4] A passive antenna is an antenna not able to change the radiation pattern or antenna gain.

instead of performing an inter-cell handover between sectors. This type of intelligent antenna is called a 'switched-beam system' [128].

Other more advanced systems are being developed, such as in the TSUNAMI II[5] project [46,148], where a mobile station is tracked by an antenna beam. In such systems idealised simulated capacity increases of as much as 300-400 % have been found [45]. Currently a great effort by different mobile network manufacturers and operators is put into determining the effectiveness of current and future smart antenna systems [146]. Smart antennas are not treated any further.

1.3.6 Frequency Planning

A more practical way of minimising the influence from interference is by allocating the frequencies in the best possible way. This could seem a trivial task, however for many operators the procedure used is far for optimum. A high level of accuracy in the modelling of the environment as well as the radio propagation is required and often not satisfied. Over the last couple of years more and more advanced algorithms for doing automated frequency planning has become available [37]. For a frequency hopping GSM network, the allocation problem increases in complexity. This issue is treated in Chapter 10.

1.4 Data Services for GSM

As described in Chapter 1, the current system evolution of GSM primarily treats the issue of migrating towards the third generation systems like UMTS. This migration consists of a large number of projects including improved voice coding and advanced data transmission services. The goal of these new data transmission services is to offer higher data-rates and make the network more flexible when considering the issue of offering various mobile data services. Two of the most important projects are GPRS and EDGE.

1.4.1 Packet Switched Services - GPRS

The general packet radio service, GPRS, is the first GSM functionality, which resembles a packet switched type of protocol. The purpose of GPRS is to accomplish efficient dynamic spectrum resource sharing between data sources that are bursty in nature. By offering 4 different data-rates per channel as well as multiple timeslots per used, the bit rate offered to the user can ideally vary from around 10 kbit/s to more than 170 kbit/s [60]. The design of GPRS is well integrated into the GSM structure to allow smooth introduction into already existing GSM networks. The time needed to successfully transmit and receive a certain amount of

[5] The TSUNAMI II project is part of the research program Advanced Communication Technologies and Services, ACTS, carried out in the development of 3G platforms.

data decreases, when a lower coding rate is used. However, the total transmit time should also include potential retransmissions, which are more likely to occur in case of little channel coding. Hence there is a channel coding trade off between short nominal transmission time and few retransmissions. The optimal channel coding selection depends on the radio conditions, which in turn depend on the frequency reuse [121]. GPRS is not treated any further.

1.4.2 Packet Switched Services - EDGE

The use of alternative modulation schemes to provide enhanced data rates for global evolution, EDGE, is currently being standardised for GSM by the European Telecommunications Standards Institute (ETSI). The block structure and retransmission protocol is based on the GPRS standard. However, in EDGE not only the coding scheme, as in GPRS, but also the modulation scheme is changed. EDGE utilises 8-PSK. It means that instead of having one bit per symbol 3 bit per symbol is introduced, increasing the effective data-rate. However, from the nature of this modulation scheme the spectrum efficiency is decreased and the network becomes more sensitive to the *C/I*. To compensate for this, a link adaptation algorithm ensures that the most efficient scheme is always used [70]. This is done by adaptively choosing the modulation and coding, which give the maximum throughput, according to the time varying link quality.

1.5 Closing Comments on Performance Enhancements

One reason why it in general is quite complex to analyse the network performance is because almost all of the features have a great influence on each other. Typically the combined effect of several of the features is investigated. Examples are directed retry and queuing, antenna diversity and frequency hopping [145] or DTX and frequency hopping [17]. An almost unlimited number of ways to enhance the capacity of a specific network exist, which makes it very difficult to decide which one(s) of the possible ways to go.

2. COMPUTER AIDED GSM NETWORK DESIGN

As described in Chapter 1, one of the two primary aims of the work conducted has been to develop a model of the BSS part of the GSM network, which has been done as a computer program. In other words, exploiting this program is in our terms equivalent to using a computer aided network design (CAND) approach. The basic principle is described in the following.

2.1 The Simulation Tool

A GSM network simulation tool, CAPACITY, has been developed at Center for PersonKommunikation (CPK) at Aalborg University. CAPACITY is capable of modelling a frequency hopping GSM type of BSS network. It can be used to create simulation results of a dynamically changing network. The program can simulate several of the factors that affect the performance of a GSM system, such as different handover algorithms, power control and discontinuous transmission.

In order to get a complete performance evaluation of a GSM system it is necessary to include the performance of each individual mobile station. This could be done by simulating the link performance of all the mobiles in the system, but would require unrealistically powerful computers to get reliable results within an acceptable amount of time. Therefore it was decided to use statistics from a GSM link simulation tool, to create lookup tables describing the performance of the mobile station under various conditions. This concept is shown in Figure 9.

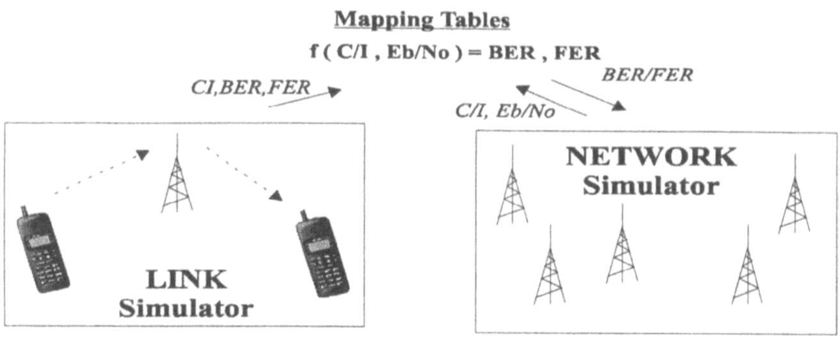

Figure 9. Combining LINK and NETWORK simulations to create CAPACITY.

Chapter 5 is devoted to a description of how to do system level modelling of the BSS part of a GSM network, i.e. the implementation of CAPACITY.

Each of the operations in the GSM transmission path, including a fast fading radio channel with thermal white Gaussian noise, is modelled in the GSM link simulation tool. A block diagram showing the relationship between these operations is given in Figure 10.

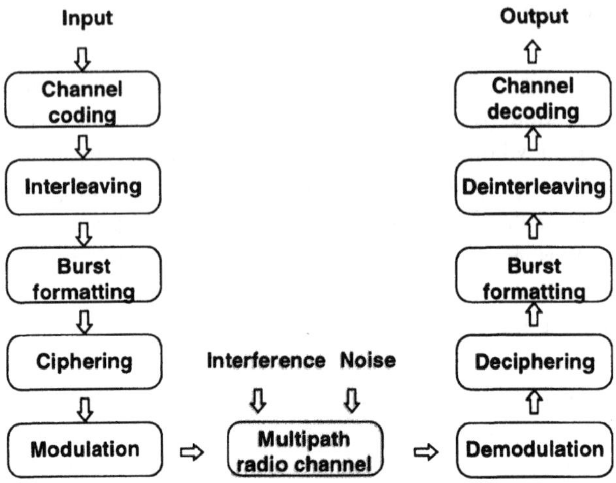

Figure 10. Block structure of the GSM link simulator.

The link simulation tool is capable of modelling two different logical GSM channels, namely the full rate traffic channel (TCH/F) [237] and the slow associated control channel (SACCH) [165]. Chapter 4 is devoted to a detailed description of all the link level issues including the link simulator.

3. CLASSIC TRAFFIC THEORY

Most of the research conducted with the purpose of enhancing the network performance has been related to the invention of features used on system level. In some specific situations it is possible to use classical traffic theory (overflow/queuing etc.) for simple idealised traffic systems. This type of modelling has been used whenever possible.

All the classic traffic calculations carried out throughout this book are based on the general overview provided in [103], in which the theories of *A.K.Erlang* [59] and *E.Brockmeyer* [24] play an important role.

4. NETWORK FIELD TRIALS

Since a close co-operation with the Danish GSM900/1800 operator SONOFON has taken place, the possibility to test some of the features in a real network has also been utilised as much as possible. Of course the limitations of always requiring the network to be running with little or no degradation have been fulfilled. In general however, valuable information has been achieved from the field trials.

Chapter 3

A BRIEF INTRODUCTION TO THE GSM SYSTEM

In GSM the mobiles, when active in the network, are attached to the network which is typically separated into geographical areas, each referred to as cells. Each user should be able to move around in the entire network, between cells, with no disturbance. Such a service is therefore transparent to the user. The user interface to the GSM system is a mobile station, where the different services are brought to the user by the base station, located in the serving cell. This chapter deals with the connection and communication within the GSM system. Not all aspects are described, but the ones that are modelled in the simulations and required in order to understand the subjects of the book, are treated.

Section 1 describes the GSM system structure, while a description of the multiple access scheme is found in Section 2. Section 3 and 4 treats the channel types in GSM as well as the mapping from the logical channels on to the physical channels. The modulation scheme and typical cell architecture used in GSM are briefly introduced in Section 5 and 6. The type of measurement reporting available in GSM is the subject of Section 7, while the last 3 sections deal with frequency hopping, discontinuous transmission (DTX) and the GSM dropped call parameter.

1. GSM SYSTEM STRUCTURE

The GSM network is structured in a hierarchical way. Each mobile station (MS) is interfacing to the base station subsystem (BSS), which contains the base stations (BTS's) and the base station controllers (BSC's). The BTS handles all the radio transmission and reception devices up to and including the base station antenna in one cell. The signal processing of the radio interface is also taken care of by the BTS. The internal structure and organisation of the BSS is handled by the BSC. It is in charge of all radio interface management. It controls the BTS and MS and its main operations include allocation and release of radio channels and mobility management using handovers. The BSS is in charge of providing and managing transmission paths between the mobile station and mobile services switching centre (MSC), which sometimes is referred to as the network and switching sub-system (NSS) [151]. The NSS includes the main switching functions of GSM, as well as the data bases needed for subscriber data and mobility management. The main role of the NSS is to manage the communication between the GSM users and users in other types of telecom networks as well as of users between different BSS systems. A sketch of the network structure is shown in Figure 11.

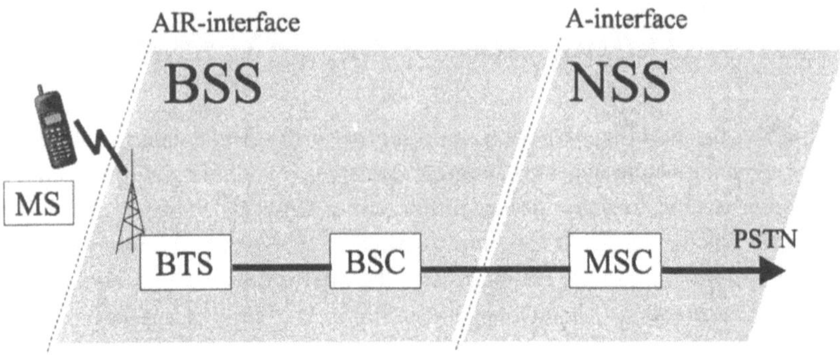

Figure 11. Simplification of the hierarchical network structure in GSM.

This book concentrates on the BSS part of the network. When specifically considering the issue of capacity enhancement features, such features are directly related to the development of algorithms executed in the BSC (see Chapter 7, 8 and 9), while when considering the quality enhancements from frequency planning (Chapter 10) only the BTS's are in principle treated.

2. MULTIPLE ACCESS SCHEME IN GSM

In GSM the multiple access scheme is based on the multi carrier, time division multiple access and frequency division duplex, MC/TDMA/FDD [62] principle. Two frequency bands are defined for GSM-900: the band 890-915 MHz is used for the uplink whereas 935-960 MHz is used for the downlink. These bands are in most countries divided among 2 or 3 operators. Besides the two 900 MHz bands two more bands are available around 1800 MHz, ranging from 1710 MHz to 1785 MHz and from 1805 MHz to 1880 MHz. These two bands are usually divided between 2 to 4 operators. The carrier spacing is in both cases 200 kHz allowing 124 carriers in GSM-900 and 374 carriers in GSM-1800 (if leaving a guard band of 200 kHz at each end of the two sub-bands). Some operators have frequencies available in both the 900 MHz and the 1800 MHz band and are correspondingly referred to as dual band operators.

Each radio frequency is divided into TDMA frames of 4.615 ms. with each TDMA frame subdivided into 8 full timeslots. Each of these timeslots can carry a full rate traffic channel, two half rate traffic channels or one of the control channels.

One timeslot on one frequency is called a slot. The structure in the time and frequency domain is shown in Figure 12.

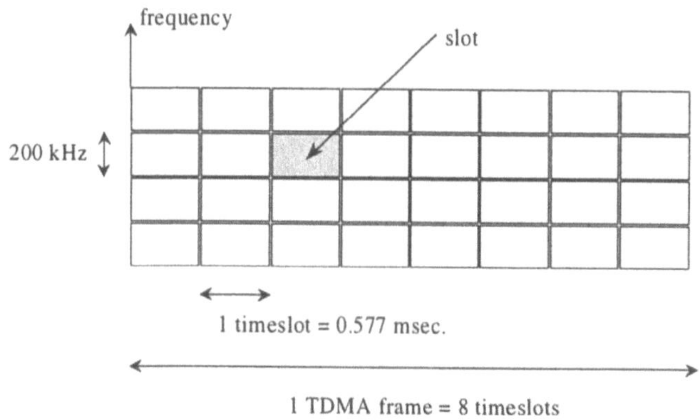

Figure 12. The multiple access scheme used in GSM.

The data transmitted in one slot is denoted a burst. Five different types of bursts exist; the normal burst, the access burst, the frequency correction burst, the synchronisation burst and the dummy burst. The information and format of the individual burst depends on the type of channel it belongs to. This is thoroughly described in the literature in e.g. [151, p. 231]. A short overview is given later in the chapter in Section 4.

3. CHANNEL TYPES IN GSM

Traditionally channels are described on two levels: the physical and the logical level. A physical channel corresponds to one timeslot on one carrier, while a logical channel reflects the specific type of information carried by the physical channel. This means that the different sorts of information will be sent on different logical channels. The logical channels are mapped or multiplexed on the physical channels, which will be described in the next section.

Typically the logical channels are divided in two groups: traffic channels and control channels. The traffic channels are the resources available to the user for either speech or data. A logical traffic channel in GSM is abbreviated by TCH. It can be used for either data or speech. The radio interface must support bi-directional transmission in order to support speech and data communication.

The traffic channels can either be full rate (TCH/F) or half rate (TCH/H) channels. The full rate traffic channel is a 13 kbit/s coded speech/data channel with a raw data rate of 9.6 kbit/s, 4.8 kbit/s or 2.4 kbit/s. The half rate supports both 7 kbit/s, 4.8 kbit/s and 2.4 kbit/s [62].

The other type of channels, the control channels, are the channels used for signalling and controlling. In other words they control the traffic channels. In GSM there are common as well as dedicated control channels. Below a short description of each of the control channels can be found, starting with the common channels:

- **The Frequency Correction Channel (FCCH):** This channel carries information for frequency correction of the mobile station. It simply consists of a row of zeros, which means that the transmitted frequency is equal to the carrier frequency plus a quarter of the bit-frequency.
- **The Synchronisation Channel (SCH):** This channel contains information of the identification of a BTS and frame synchronisation for the mobile station. The SCH must contain two encoded parameters, the BTS identity code and a reduced TDMA frame number.
- **The Broadcast Control Channel (BCCH):** This is probably the most important control channel within the GSM. Often, a mobile station can receive, and potentially be received by, several cells, possibly in different networks or even in different countries. It has then to choose one of them, and some information is required for the choice, like the network to which each cell belongs. This information is broadcast regularly in each cell, to be listened to by all the mobile stations in idle mode. The channel doing this is the downlink unidirectional channel BCCH. Furthermore, in dedicated mode the mobile station listens to the BCCH of the neighbouring cells and monitors the received signal strength from them. That way the mobile knows when and to which cell to make a handover.
- **The Paging Channel (PCH):** If there is an incoming call, or a short message, the network will page the mobile station, using the PCH in all cells in the location area where the mobile station is registered. The PCH contains the mobile station identity number, the IMSI or TMSI.
- **The Random Access Channel (RACH):** The RACH is used initially by the mobile when attempting an access to the network. By making that access, the mobile station is requesting a signalling channel. The reason for the access could be a page response or initiation of a call. Since the distance between the mobile station and base station is unknown, the access burst is as short as possible in order not to interfere with the adjacent time slot.
- **The Access Granted Channel (AGCH):** The AGCH is used for acknowledge of the access attempt sent on the RACH. On this channel the mobile station will assign a signalling channel (SDCCH) to continue the signalling according to the reason for the access.

The dedicated logical control channels are:

- **The Stand-alone Dedicated Control Channel (SDCCH):** This is the channel on which the signalling will take place. It may be used for call set-up, authentication, ciphering or transmission of text messages (short message or cell broadcast). This bi-directional channel is subdivided into eight sub-channels that can handle the signalling needed by one mobile station. Thus, eight calls can be set up simultaneously. At call set-up the mobile station will be assigned a traffic channel after the signalling on the SDCCH is completed.
- **The Slow Associated Control Channel (SACCH):** This channel is used to transfer signalling data while having an ongoing call on a traffic channel. This channel can carry about two messages per second in each direction. It is used for non-urgent procedures. On the downlink the mobile station is informed about what neighbouring cells to measure (for handover purposes). Furthermore, the mobile is also told what output power and timing advance to use. In the uplink direction the mobile station will report the downlink measurements to the BTS.
- **The Fast Associated Control Channel (FACCH):** The FACCH is used when there is a need for higher capacity signalling in parallel with ongoing traffic. The FACCH works in stealing mode, meaning that the transmitting side throws away a 20 ms segment of speech to fill the bursts with signalling information instead. The FACCH is mainly used for handover commands.

4. MAPPING LOGICAL TO PHYSICAL CHANNELS

The different logical channels are mapped into physical channels. The GSM specifications describe which physical channel to use for each logical channel. Several combinations of the different channels are possible. Here only the TCH/F and corresponding control channels are considered. More information on other combinations can be found in [61].

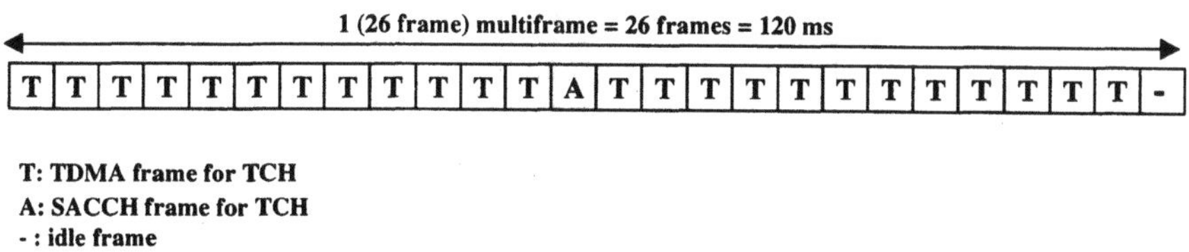

Figure 13. Organisation of the traffic channel and corresponding control channels.

The channel organisation for the TCH/F and the SACCH uses a 26-frame multiframe.[6] It is organised like illustrated in Figure 13, where only one timeslot per TDMA frame is considered.

It is seen from the 26 frames, that 24 are used for traffic, i.e. speech or data. One frame is used for the SACCH channel, while the last one is an idle frame. During this idle frame time interval period a mobile can listen to the other control channels. The complete GSM frame, timeslot and burst structure is seen in Figure 14.

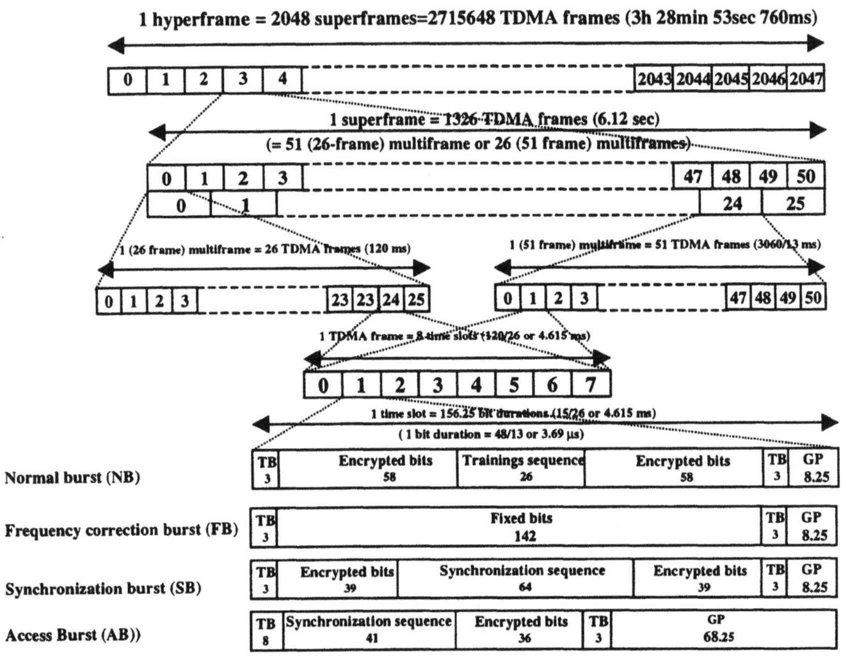

Figure 14. Timeframes, timeslots and bursts in GSM.

5. MODULATION SCHEME IN GSM

The modulation scheme in GSM is Gaussian minimum shift keying (GMSK) with a GBT product of 0.3. The modulation rate is 270.83 kbit/sec [61].

The spectral bandwidth of a GMSK signal with a BT product of 0.3 is only attenuated 10 dB at 100 kHz from the carrier frequency [151]. The choice of a channel separation of 200 kHz [64] in GSM results in a non-negligible overlap between adjacent frequencies. The source of interference can be limited by RF

[6] The term 'frame' is slightly misplaced here, since 4 of such frames are needed to get one complete SACCH frame (= 4 SACCH bursts). Therefore in the book the term SACCH multiframe is used for a period of 480 ms, consisting of 4 of the here mentioned 'frames'.

filtering and careful frequency planning, i.e. by limiting the use of adjacent channel frequencies on the same site and in the same geographical area.

6. TYPICAL CELL ARCHITECTURE

A cellular communication system consists of a number of cells. Various categories of cells are used throughout the literature depending on the size of the individual cell. Macro cells are usually described as cells with a radius between a couple of 100 meters and several tenths of kilometres. Micro cells on the other hand are smaller cells with radii of typically no more than 500 meters, with a coverage area of a few streets or buildings. Other definitions differentiating between different cell sizes exist. Throughout this book macro cells are treated.

In Figure 15 (a) a BTS with 3 cells (a 3 sector site) is depicted. In the same way Figure 15 (b) show a larger part of the network, with hexagonal coverage area of each 3-sector BTS. For simplicity cells are usually depicted as hexagons [28,104]. In reality cells are of course not hexagonal, but have some kind of an irregular shape. In the case of micro cells another type of network grid, referred to as a Manhattan grid, is often used. The network structure of a typical Manhattan grid is shown in Figure 15 (c), where the cluster is split according to buildings and streets.

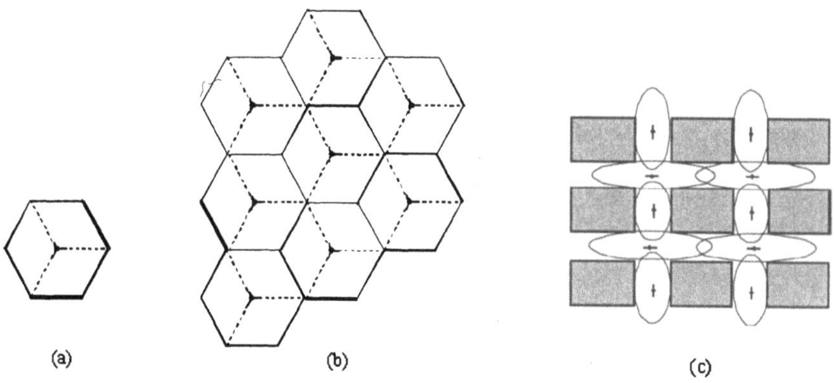

Figure 15. Graphic illustration of one macro cell (a), a macro cell based network cluster of cells (b), and a micro cell based network according to the Manhattan grid including clutters equivalent to buildings (c).

Each cell has been assigned a number of radio channels. Since there is a limited number of available frequencies, frequencies are reused with a certain reuse distance equal to the physical distance between two cells using the same frequency [109]. For the radio network planner an important network parameter is therefore the minimum allowed reuse distance.

Sometimes the term reuse factor is used. The reuse factor indicates the cluster size of cells within which each frequency is used only once. So a cluster size of 7 means that within a group of 7 cells all frequencies are used exactly once, i.e. each frequency is only used in 1/7 of the cells. For a hexagonal cell structure the homogeneous or ideal cluster sizes can take the values K = 1,3,4,7,9....[7] The reuse factor is typically denoted as *x/y*, where *x* is the reuse factor for base stations and *y* the reuse factor for cells. This means that a re-use factor of 3/9, corresponds to a cellular network consisting of 3-sectors sites, and each frequency is only used once within 3 sites, i.e. once per 9 cells. A 3/3 reuse means that each frequency is used once within 3 sites, however in this case for omni directional sites.

7. MEASUREMENT REPORTING IN GSM

The handover and power control algorithms are not specified in the GSM standard, so each individual vendor and network operator can create their own algorithms. However, the parameters that can be used in the algorithms, the radio link measurements, are given by the specifications [65] in order for the mobile stations to meet the same requirements. Descriptions of exactly how the radio link measurements can be used in the power control and handover algorithms are given in Chapter 7 and 9. In the following the available radio link measurements are described.

According to the system specifications, the received signal strength at the receiver must be measured in the complete range from -48 dBm down to the required receiver sensitivity of -110 dBm. The reported parameter, must be an average (over a period of 480 ms) of the received signal level measurement samples in dBm. Furthermore the measured levels must be mapped into a RXLEV value between 0 and 63 before they are reported. The 64 different possible signal levels are shown in Table 1.

Table 1. Quantification of the received signal level into RXLEV.

VALUE	Corresponding Signal Strength Level
RXLEV 0	less than -110 dBm
RXLEV 1	-110 dBm - -109 dBm
RXLEV 2	-109 dBm - -108 dBm
.	.
.	.
RXLEV 62	- 49 dBm - - 48 dBm
RXLEV 63	Larger than - 48 dBm

[7] In practice the frequency reuse schemes are almost never regular or homogeneous.

As well as the received signal strength, the radio link parameter, describing the quality or BER, is specified in the GSM specifications. This estimate of the reported BER should be the average BER obtained *before* channel decoding. The reported parameter is denoted RXQUAL, where RXQUAL, like RXLEV, is quantified into a number of discrete values. 8 such values are defined, as shown in Table 2. The RXLEV as well as the RXQUAL measurement samples are reported once each 480 ms on the SACCH.

Table 2. Quantification of the received BER into RXQUAL.

VALUE	Corresponding BER
RXQUAL 0	Less than 0.2 %
RXQUAL 1	0.2 % - 0.4 %
RXQUAL 2	0.4 % - 0.8 %
RXQUAL 3	0.8 % - 1.6 %
RXQUAL 4	1.6 % - 3.2 %
RXQUAL 5	3.2 % - 6.4 %
RXQUAL 6	6.4 % - 12.8 %
RXQUAL 7	Larger than 12.8 %

The information deciding what neighbouring cells to measure is transmitted by the BTS to the mobile using the SACCH. After the mobile has performed the measurements it reports them back to the BTS. The mobile can, at most, report the measurements from the 6 strongest neighbours, where the signal strengths of the neighbours are measured on their BCCH frequencies. A more detailed description of these neighbour cell measurements can be found in Chapter 8.

In the GSM standard it is specified how the latest 32 samples of RXLEV and RXQUAL must be stored for both uplink and downlink. They can be used as an optional averaging for the handover and power control algorithms.

8. FREQUENCY HOPPING IN GSM

The radio interface of GSM offers the slow frequency hopping functionality. Here frequency hopping is based on the idea that every mobile station transmits its TDMA frames according to a sequence of frequencies specified by the frequency hopping algorithm [62]. A mobile station transmits on a fixed frequency during one timeslot (approx. 577 μs) and then jumps to another frequency before the next TDMA frame. The uplink and downlink frequencies are always duplex frequencies. Two different modes of hopping are specified in GSM: cyclic and pseudo random hopping [62] as shown in Figure 16.

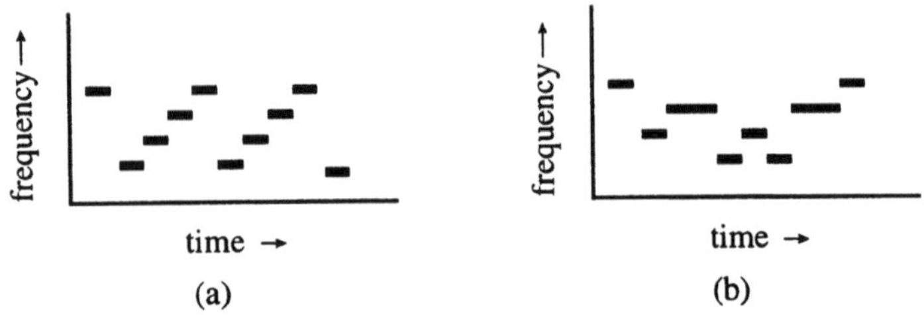

Figure 16. An example of FH in GSM with 4 hopping frequencies: a) cyclic hopping b) pseudo random hopping.

The hopping sequences are predefined in GSM. 64 different sequences are allowed and can each contain up to as many as 64 frequencies. The actual hopping is described by two parameters: the mobile allocation index offset (MAIO) and the hopping sequence number (HSN). The MAIO can take as many values as there are frequencies in the sequence and indicate the initial frequency in the hopping sequence. The HSN can take 64 different values describing the sequence. The frequencies varies pseudo-randomly when the HSN differs from zero. If it is set to zero, the hopping mode is by default cyclic. Figure 17 show 3 hopping sequences with different HSN and MAIO combinations. Also 3 different frequencies are used in the hopping sequences.

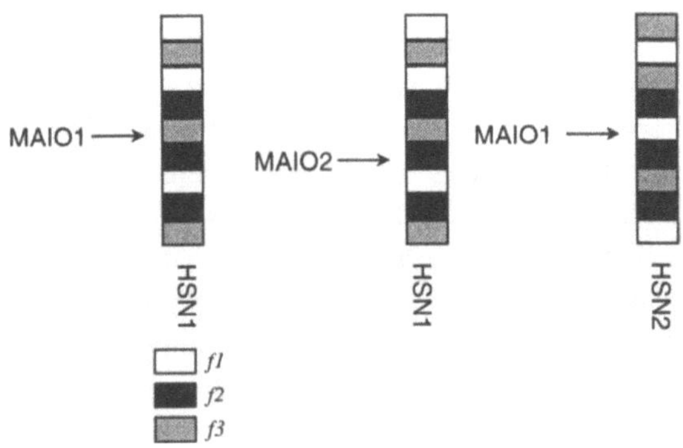

Figure 17. Examples of 3 hopping sequences with different HSN and MAIO combinations.

One of the characteristics of the hopping sequences is that two sequences will never overlap when they have the same HSN, but different MAIO's, i.e. the hopping sequences are orthogonal. Furthermore, it can be derived from general pseudo-random characteristics that two channels having different HSN, but the same frequency list and the same time slot, will interfere in *1/n* of the bursts, where *n* is the number of different frequencies in the hopping sequence. This means that frequency hopping in some sense averages the interference out throughout the network.

Restrictions are applied to some of the control channels. The BCCH channel, which takes up one timeslot on the BCCH carrier, cannot hop since it is used as beacon. Mobile stations access the network using this channel by decoding the base station identification code (BSIC) and the frequency. Therefore everywhere within the cell, at all times, it has to be possible to measure it and it is correspondingly not allowed to participate in the hopping sequence.

Figure 18 show the hopping configurations for baseband and RF hopping.[8] In baseband hopping, every TRX has its own frequency, so when hopping, the mobile hops across the different TRX's. When RF hopping is used, the frequency is changed for each TRX. That way the mobile can stay on the same TRX, while hopping. This implies that the frequency containing the BCCH (timeslot 0 in the figure) cannot hop when RF hopping is used, while with baseband hopping there is no problem (except for timeslot 0). This also means that the hopping length in a baseband hopping system is equal to the number of TRX's, while in a RF hopping system this is not the case.

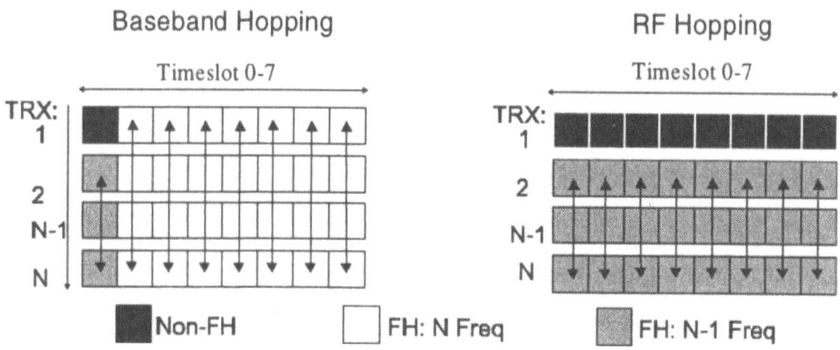

Figure 18. The TRX's to hop across for respectively baseband (left) and RF (right) hopping.

9. DISCONTINUOUS TRANSMISSION IN GSM

The GSM system supports discontinuous transmission (DTX). This feature alters the transmission between speech active phases, with a transmission of one speech frame each 20 ms during which the transmission falls to 12 bursts each 480 ms instead of 100 for the non-DTX mode [151]. This means that interference is reduced in the interval where no speech is transmitted. DTX in GSM is one of the main subjects of Chapter 7.

[8] RF hopping is in the literature also referred to as synthesised hopping.

10. THE DROPPED CALL ALGORITHM

In GSM the network determined terminated of a call is based on a dropped call algorithm. The actual implementation is not specified in the GSM standard and again it is therefore up to the network vendor and operator of how to implement it. The goal of a dropped call algorithm is to remove calls which experience such as bad quality that retaining the connection is meaningless. Here a short description of the how the structure of the algorithm could look like. The algorithm is based on a calculation where the various reasons for experiencing failures in the network are measured and summed. The principle is shown in Equation (1).

```
(failures_on_radio_interface +
failures_from_handovers +
failures_on_Abis_interface +
failures_on_A_interface +                                    (1)
failures_from_LAPD +
failures_on_BTS +
failures_from_user_actions +
failures_on_BSC +
tch_netw_act + tch_act_fail_call)
/total_number_of_calls
```

As can be seen, the calculation involves several parameters of which many are not related to the air-interface but to the remaining part of the GSM network. The parameter `failures_on_radio_interface` is the only parameter directly reflecting failures on the radio interface. In the simulation tool, since we are only modelling the BSS part of the network, we can only get information about this radio related parameter which makes it essential for us. It is the only parameter where simulations and real live measurements can be compared when considering the dropped call performance. One significant comment should furthermore be that `failures_on_radio_interface` is not calculated based on the actual performance of the speech timeslots. It is based on the decoding of the SACCH frames.

Chapter 4

LINK MODELLING AND LINK PERFORMANCE

This chapter concentrates on the GSM link level. Initially the ETSI standardised GSM link is described in Section 1, while the implemented GSM link simulator is dealt with in Section 2. Simulation results showing the influence of frequency hopping on the link performance are found in Section 3, while a method of mapping from *C/I* to bit error rate (BER) and frame erasure rate (FER) is presented in Section 4. The chapter ends with a short summary and conclusions on the GSM link level.

1. THE GSM LINK

Several successive operations have to be performed to convert a speech signal into a radio signal. The reverse operations have to be performed at the receiver end in order to regenerate the speech signal. The operations on the receiver as well as on the transmitting side are shown in Figure 19.

The following operations take place on the transmitting side:

- **Source coding** concerts the analogue speech signal into a digital equivalent.
- **Channel coding** introduces redundancy into the data flow, increasing its rate by adding information calculated from the source data, in order to allow detection or even correction of bit errors that might be introduced during transmission.
- **Interleaving** consists in mixing up the bits of the coded data blocks, so that concatenated bits close to each other in the modulated signal are spread out over several data blocks. Since the error probability of successive bits in the modulated stream is typically highly correlated, and since the channel coding performance is better when errors are decorrelated, interleaving aims at decorrelating errors and their position in the coded blocks.
- **Ciphering** modifies the contents of these blocks through a secret recipe known only by the mobile station and the base station.

- **Burst formatting** adds information to the ciphered data, in order to help synchronisation and equalisation of the received signal. Among others a training sequence is added at this stage.
- **Modulation** transforms the binary signal into an analogue signal at the right frequency. Thereby the signal can be transmitted as radio waves.

The receiver side performs the reverse operations as follows:

- **Demodulation** transforms the radio signal received at the antenna, into a binary signal. More sophisticated demodulators also deliver an estimated probability of correctness for each bit. This concept is referred to as soft decision or soft information.
- **Deciphering** modifies the bits by reversing the ciphering recipe.
- **Deinterleaving** puts the bits of the different bursts back in order to rebuild the original code words.
- **Channel decoding** tries to reconstruct the source information from the output of the demodulator, using the added redundancy to detect or correct possible errors arising from the output of the demodulator.
- **Source decoding** converts the digitally decoded source information into an analogue signal to produce the speech.

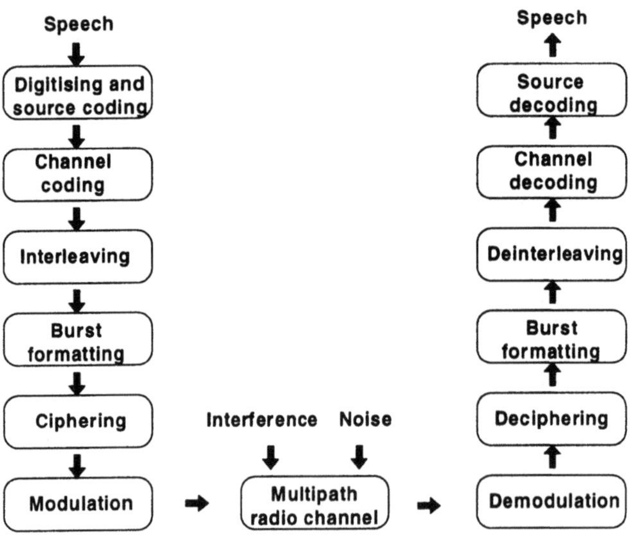

Figure 19. Flow diagram of the link operations according to the ETSI specifications.

Channel coding and interleaving are both essential to achieve a gain from frequency hopping. They are therefore treated in detail in Section 1.1 and 1.2.

1.1 The Channel Coding

Figure 20 show the coding carried out for the TCH/F in GSM, where blocks of 260 information bits are divided into three different classes, class 1a, class 1b and class 2 of which the class 2 bits are uncoded.

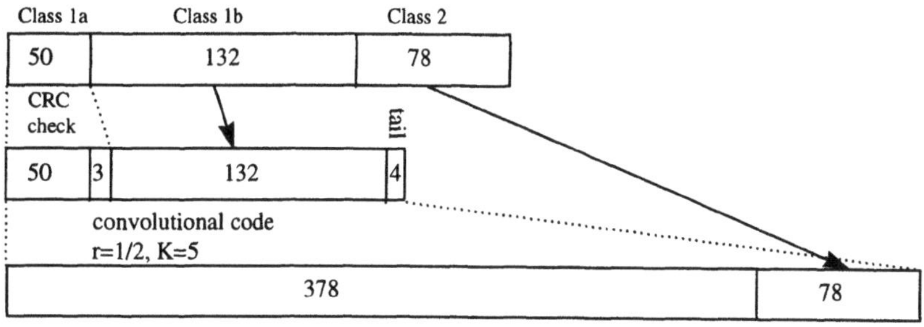

Figure 20. Coding used for the traffic channel (TCH/F) in GSM.

First a 3 bit CRC check is applied to the most important bits, the class 1a bits. After the addition of 4 tail bits all class 1 bits are convolutionally encoded. The convolutional code consists of applying two convolutional codes, whose polynomials are respectively D^4+D^3+1 and D^4+D^3+D+1, as can be seen in Figure 21. This leads to an encoded block length of 456 bits.

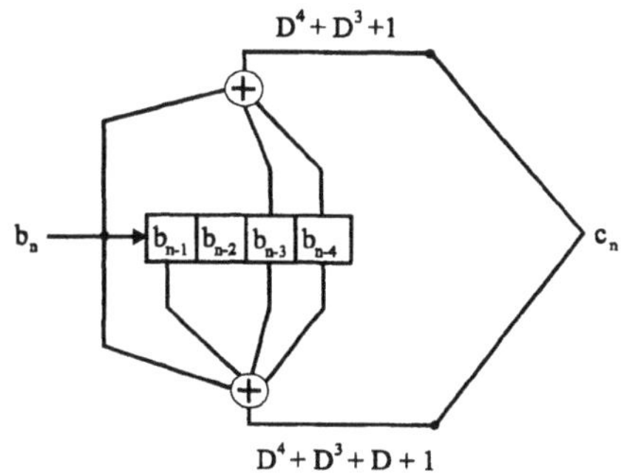

Figure 21. Convolutional coding for the TCH/F channel.

The process taking place for the decoding is the reverse action, where the 378 encoded bits are decoded using a convolutional decoder and a block decoder to retrieve the 182 encoded information bits.

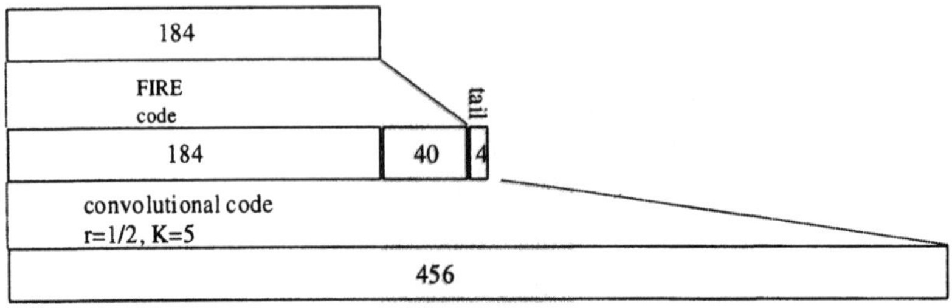

Figure 22. Coding used for the SACCH in GSM.

The coding used for the SACCH is slightly different. It is seen in Figure 22. To a block of 184 information bits a FIRE code of 40 bits is added. Then the addition of 4 tail bits and the same convolutional code as in the TCH/F case follows. The 40 bit FIRE code can correct one clustered group of errors of a length up to 11 [124]. The coded block again consists of 456 bits to fit the same burst format as the TCH/F.

1.2 Interleaving

The interleaving principle of the TCH/F channel can be seen in Figure 23. The coded blocks of 456 bits are divided into 8 groups of 57 bits, each carried by different bursts. A burst therefore contains the contribution of two successive speech blocks A and B. In order to destroy the proximity relations between successive bits, bits from block A use the even positions inside the burst and the bits of block B the odd positions.

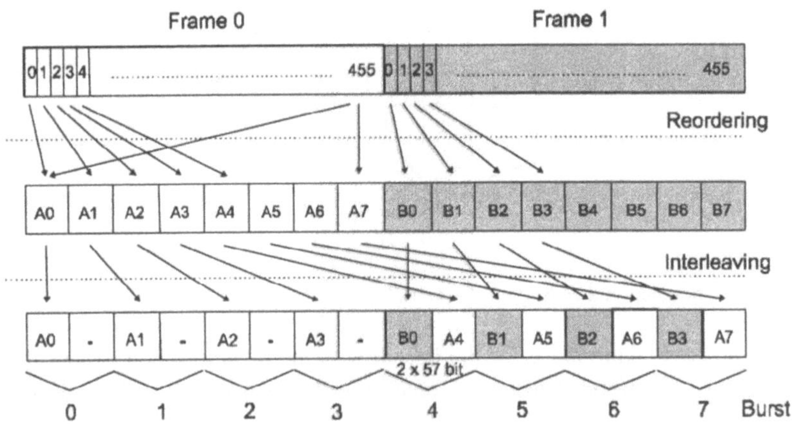

Figure 23. Interleaving of the traffic channel.

The interleaving on the SACCH channel is slightly different. Again an encoded block of 456 bits is divided into 8 blocks of 57 bits, but these 8 blocks are put into 4 bursts, making the interleaving depth equal to 4. The distance (in time) between two consecutive SACCH bursts is a lot greater than the distance between 2 TCH/F bursts, i.e. some spreading has already taken place on the SACCH.

2. THE GSM LINK SIMULATOR

In this section the link simulator is briefly introduced. In Section 2.1 a general overview is given followed by a description of the available output parameters in Section 2.2.

2.1 Structure of the Link Simulator

The link simulation tool is capable of simulating two different GSM channels. The full rate traffic channel (TCH/F) [237] as well as the slow associated control channel (SACCH) [165].

Each of the operations in the GSM transmission path, including a fast fading radio channel and thermal white Gaussian noise, are included as can be seen in Figure 24. At the transmitter side the blocks are implemented as described in the previous section, while for the receiver side a coherent soft decision 16 state Viterbi algorithm (SOVA) [26,112,150] is used.

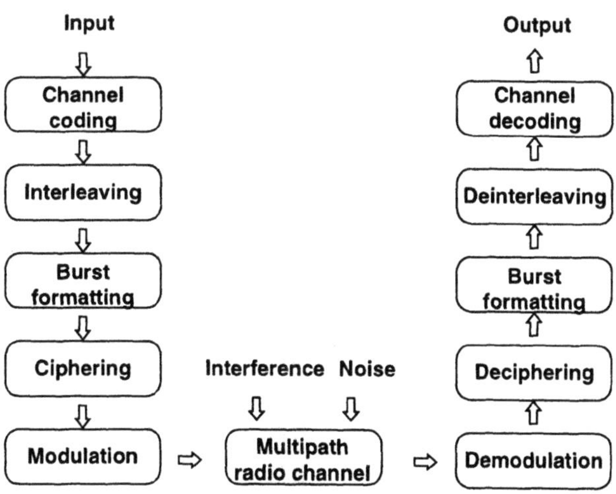

Figure 24. Block structure of the design of the link simulator.

It should be noted that the type of data-receiver used has a strong impact on the link performance when different network features, such as e.g. frequency hopping, are exploited. A soft decision type of receiver adds a gain of several dB's to the performance, compared to the conventional hard decision type of receiver. This is due to the fact that the channel-decoding algorithm (also the Viterbi algorithm) uses an estimate of the probability of having performed a correct decoding as well as the hard decision estimate, a soft decision. Today most new GSM mobile stations use a soft decision type of data-receiver. The difference between the different brands can be found in the way the information is extracted and used.

Different types of channel models can be used. For the simulations in this chapter a typical urban (TU) channel profile has been used [64], with the impulse response shown in Figure 25. This is modelled using a tapped delay-line model with the taps on the places corresponding to the impulse response.

To model the speed of the mobile moving in the typical urban channel, each tab undergoes Rayleigh fading with a variation in time according to the speed. The fading of between each taps is uncorrelated. Other channel models such as Hilly Terrain (HT), Rural Area (RA) or flat fading can be simulated as well. Shadow fading is not included.

Noise or interference can furthermore be added to the channel. The noise used is Gaussian white noise. The interference can be modelled using as many as 9 interfering signals. Each of these interferers transmits bursts filled with random bits. These bursts undergo the same kind of radio channel as the desired user, each with independent fading

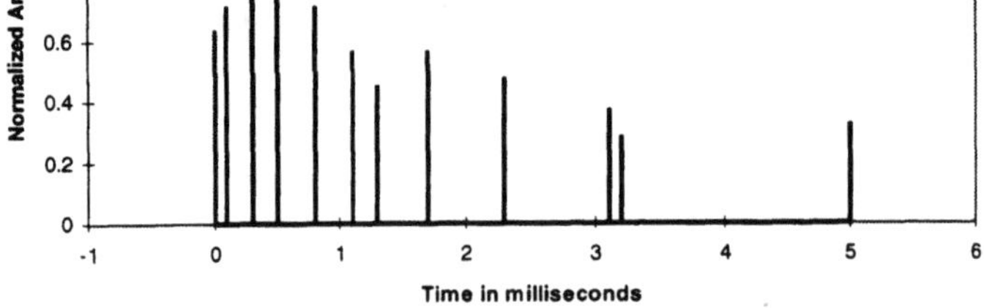

Figure 25. Impulse response of the GSM TU channel model.

2.2 Output Parameters from the Link Simulator

The output of the link simulator consists of the BER and FER as a function of the *C/I* level or noise level (E_b/N_o). The BER is calculated before the decoding, which means that the gain from coding and interleaving is not reflected in this parameter. The BER is of interest since the quality measure in GSM, RXQUAL, is based on the BER as described in Chapter 3.

The FER is the ratio of frame erasures over the total number of frames. A frame erasure occur when the CRC check fails in case of the TCH/F. In case of the SACCH a frame erasure occur, when errors still exist after having applied the convolutional and FIRE decoding. The FER is therefore measured after the deinterleaving and decoding, making it more correlated to the subjective experienced speech quality than the BER.

3. INFLUENCE OF FREQUENCY HOPPING ON THE LINK PERFORMANCE

In this section, the influence of frequency hopping on the link performance is studied. Initially the aim of frequency hopping is described in Section 3.1. In Section 3.2 the settings used in the different simulations to describe frequency hopping are given. Section 3.3 and 0 contain results from the link simulations as well as comparisons to measurements from live tests.

3.1 Aim of Frequency Hopping

As described, the coding and interleaving is superior when bursts belonging to a frame, have been exposed to different channel conditions. These conditions can be changed due to either fading or interference. Correspondingly the advantages of frequency hopping are traditionally described by the two terms, frequency and interference diversity.

The short-term variations of the received radio signal due to the reception of multiple reflections with different phases depends on the speed of the mobile station. Slow moving mobiles can experience a deep fade for quite a while (for several consecutive bursts), whereas fast moving mobiles are typically in a deep fade for a shorter period. It can be said that the bursts of a fast moving mobile experience independent fading, whereas the fading of the consecutive bursts of a slow moving mobile can be very correlated. Frequency hopping provides uncorrelated fading for successive TDMA bursts. By hopping between different frequencies the probability of continuously (for more than one burst in a row) being in a deep multipath fade is decreased. This effect is called frequency diversity. The frequency diversity gain only helps the slow moving mobiles, since the fast moving mobiles already experience uncorrelated fading for successive TDMA bursts.

Figure 26 show the *C/I* of a mobile connection from a GSM network simulation using random FH. The upper plot shows the instantaneous *C/I* (per burst), while the lower plot shows the *C/I* averaged over the interleaving depth of 8 bursts. It can be observed that the large spread in the C/I per burst is reduced significantly by the "averaging" process of coding and interleaving in GSM. The "average" C/I only very seldom goes below 9 dB. By hopping randomly between different frequencies the probability of being interfered by another mobile is averaged out on all the frequencies in the cell. This effect is called interference diversity.

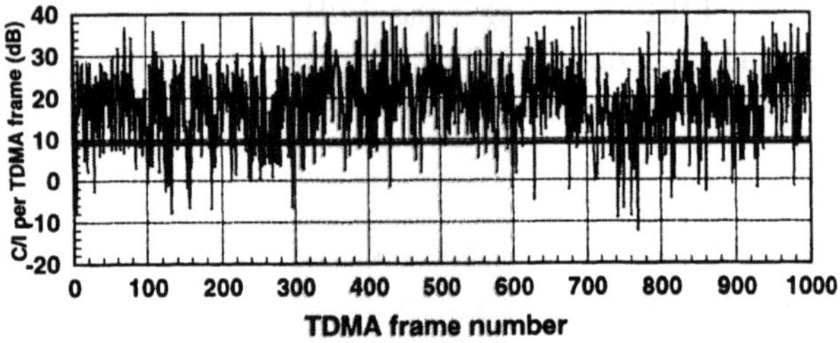

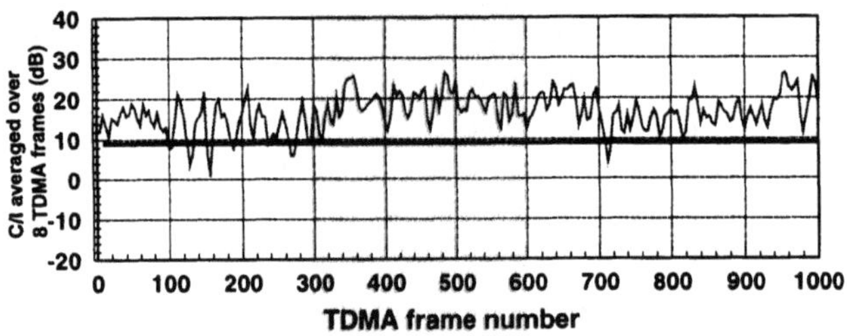

Figure 26. *C/I* samples per burst (upper) and *C/I* averaged over 8 bursts (lower). Random frequency hopping is used.

Without frequency hopping several bursts in a row can be erroneous, due to continuous interference or from being in a fade. The convolutional encoder performs best for random positioned bit errors, which is why reordering and interleaving was originally introduced in the GSM signal transmission flow. However, the reordering and interleaving only improves the coding performance, if the 8 successive bursts carrying the data information of one speech block are exposed to uncorrelated fading and interference. This can be ensured by frequency hopping, as explained above. Random FH leads to a new interference situation for each burst.[9]

Intensive research of the magnitude of the performance gain of both frequency and interference diversity of various frequency hopping systems has been carried out at both link level [249,149] and network level [29,105,183,238]. After the introduction of frequency hopping in existing live networks, numerous performance evaluations have been presented [42,115,177]. It should be noted that the effect of interference diversity is greatly influenced by the use of DTX and RF power control, as will be discussed in Chapter 7.

9 The same effect is achieved with cyclic FH in a frequency plan not using grouped frequency planning.

3.2 Link Simulation Reference Conditions

To understand the performance of the GSM link, some reference link simulations have been made. They concern the full rate speech traffic channels (TCH/F) as well as the slow associated control channel (SACCH) as specified in GSM [61]. Other logical channel types have not been treated. The Typical Urban (TU) channel profile has been used. It represents one of the areas of highest capacity needs, i.e. the places where capacity enhancement techniques are needed. Two different mobile speeds have been simulated, 3 km/h and 50 km/h, which leads to the channel profiles TU3 and TU50.

The link simulations can be divided in two parts; simulations done using the testconditions and simulations using other conditions, here referred to as the non-testconditions [64]. The testconditions are characterised as follows. Noise or interference is included. The noise is additive white Gaussian. One interferer, always interfering on the equivalent frequency as the desired signal, represents the interference. That is the interferer is using exactly the same hopping sequence and this corresponds in reality to all frequencies having the same mean *C/I*. Both the desired and the interfering signals are subject to the same propagation channel and the same speed, but have an independent fading pattern. The difference of the non-testconditions compared to the testconditions is that not all of the frequencies are interfered and that the mean interference level varies for each frequency.

3.3 Link Simulation Results

In this section a summary of the most relevant link simulation results are found. Section 3.3.1 contains the results from the testconditions, while Section 3.3.2 holds the results using the non-testconditions.

3.3.1 Link Simulations using the Testconditions

Specifications of the required performance of individual mobile stations for GSM towards resistance of interference and sensitivity for the testconditions are described in [64]. The required performance in the case of interference can be seen in Table 3 in the case of hopping (FH) and non hopping (NH) for the different channel profiles. The factor α is a value is between 1 and 1.6, depending on the channel condition. These requirements will in the following be denoted as the performance requirements.

Table 3. Reference interference performance at 9 dB *C/I* [64].

	TU3 NH	TU3 FH	TU50NH	TU50 FH
Max. FER	$21\alpha\%$	$3\alpha\%$	$6\alpha\%$	$3\alpha\%$

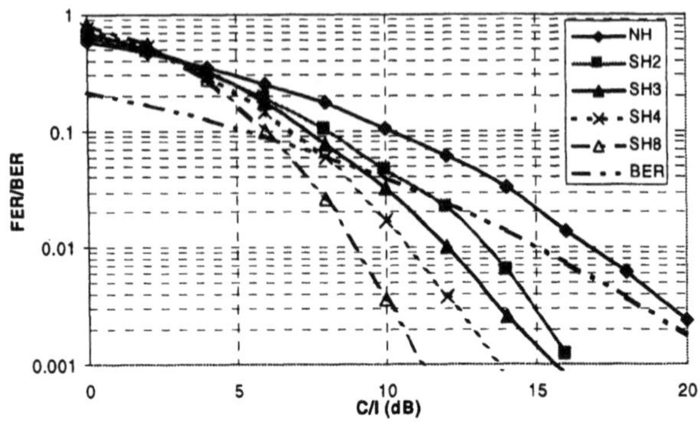

Figure 27. BER and FER for sequential FH and non hopping (NH) with the testconditions, as a function of the *C/I*.

Figure 27 show the FER and BER as function of the *C/I* using the testconditions in the case of sequential and non frequency hopping. The TU3 channel profile has been used. It can be seen that in the NH case a *C/I* of 15 dB is required to get a FER of 2 %. In case of sequential FH over 8 frequencies (SH8), a *C/I* of about only 8 dB is needed. That means, in this reference situation, a gain of about 7 dB is achieved from FH. If compared to the requirements in Table 3, it can be seen that the requirements are easily fulfilled. The BER is the same for all hopping configurations, since the BER is calculated before the decoding with no gain from FH.

The most relevant simulation results for the TCH/F, using the testconditions, are summarised in Table 4 and Table 5 for a channel profile of TU3 and TU50. The following main issues can be observed from the simulations:

- The frequency diversity gain from using cyclic FH over 4 to 8 frequencies is in the order of 5-7 dB in *C/I* for the TU3 profile, and approx. 1-2 dB lower for random FH.
- For the TU50 testconditions the frequency diversity gain from frequency hopping is, as expected, modest around 0.5 - 1.5 dB in *C/I*.
- Frequency hopping ensures decorrelation between successive data bursts. This corresponds to having a link quality, which becomes speed independent. Therefore the absolute performance for TU3 and TU50 is identical.
- The simulation results of the FER agree well with the GSM performance requirements, see Table 3.
- Even for lower number of hopping frequencies as e.g., 2, 3 or 4 a considerable frequency diversity gain is seen for the TU3 channel profile.

Table 4. Simulated FH diversity gain for the TCH/F using the testconditions and the TU3 channel profile.

NUMBER OF FREQUENCIES	*C/I* FOR A FER = 2 %	
Cyclic FH: *C/I*	**Absolute Level [dB]**	**Relative Gain [dB]**
1	15.2	0.0
2	12.1	3.1
3	10.8	4.4
4	9.7	5.5
8	8.2	7.0
Random FH: *C/I*		
1	15.2	0.0
2	13.0	2.2
3	12.0	3.2
4	11.0	4.2
8	10.0	5.2
12	9.2	6.0

When comparing the above results of the TU3 simulations to the results in [184], it is seen that they fit quite well. The difference is less than 1 dB.

Table 5. Simulated FH diversity gain for the TCH/F using the testconditions and the TU50.

NUMBER OF FREQUENCIES	*C/I* FOR A FER = 2 %	
Cyclic FH: C/I	**Absolute Level [dB]**	**Relative Gain [dB]**
1	9.5	0.0
2	8.4	1.1
3	8.4	1.1
4	8.4	1.1
8	8.0	1.5
Random FH: C/I		
1	9.5	0.0
2	8.7	0.8
3	8.0	1.5
4	8.0	1.5
8	8.0	1.5
12	8.0	1.5

As well as the TCH/F traffic channel, the SACCH is also simulated. Figure 28 show the FER on the SACCH as a function of the *C/I*. The mobile speed used to generate these results has been 3 km/h. Only the curves for non hopping and random hopping over 4 and 8 frequencies are shown. It is observed how FH does not give much gain. The reason is that the time period between the different SACCH bursts is longer than is the case of the TCH channels, i.e. the SACCH bursts already have close to independent fading even at low speeds.

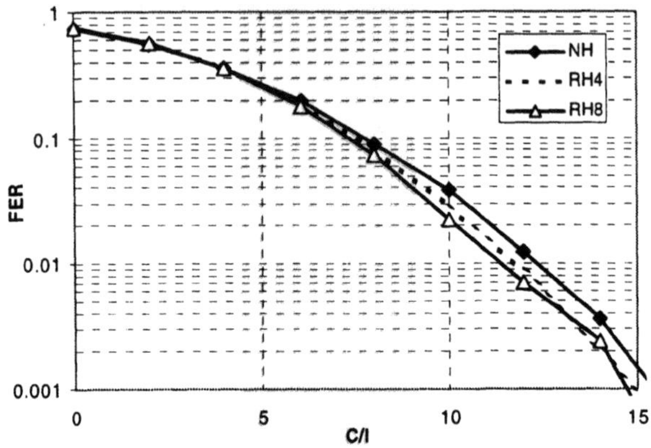

Figure 28. FER on the SACCH using the testconditions and a mobile speed of 3 km/h.

3.3.2 Link Simulation Results using the Non-testconditions

The most essential link simulations using the non-testconditions are shown in Table 6. It should be noted that the *C/I* values in the tables are linear averaged *C/I* values and that the frequencies, which referred to as being not interfered have a *C/I* of 20 dB. For example, it can be seen that with sequential hopping over 2 frequencies (SH2) of which only one is interfered, the FER is equal to 2% when the average *C/I* is equal to 9.3 dB. This means, since the *C/I* on the not-interfered frequency is 20 dB, that the *C/I* on the other frequency is 6.5 dB.

Table 6. *C/I* values for which the FER is equal to 2%, for the non-testconditions.

HOPPING TYPE	1 INTERFERED FREQUENCY	2 INTERFERED FREQUENCIES	3 INTERFERED FREQUENCIES	4 INTERFERED FREQUENCIES
NH	15.2 dB	-	-	-
SH2	9.3 dB	12.1 dB	-	-
SH3	< 0 dB	10.4 dB	10.8 dB	-
SH4	<0 dB	6.3 dB	9.2 dB	9.7 dB
RH2	11.6 dB	13.0 dB	-	-
RH3	9.8 dB	11.5 dB	12.0 dB	-
RH4	6.0 dB	9.9 dB	10.8 dB	11.0 dB

From the results it can be seen, that having some frequencies completely or nearly interference free, quite a large gain can be achieved. When sequential hopping with 3 or 4 frequencies is used and only one frequency is interfered, it is possible to have one very strong interferer on one of the frequencies, while the FER is still satisfactorily good. The general conclusion from these simulations is that not only the mean *C/I*, but also the variance in *C/I* over 8 bursts influences on the FER.

3.4 Performance Comparison to Existing GSM Mobiles

It is relevant to know that the performance of the GSM link simulation tool satisfies the specifications in [64], but an even more interesting thing is to compare our link simulation results to the performance of actual mobile stations. A performance test of 9 different (anonymous) mobile stations operating in a controlled co-channel FH environment has been described in [189]. The test was carried out using the TU3 channel profile with 3 frequencies hopping sequentially. An example of one of the test results is shown in Figure 29, where the performance of the link simulator is included with the others. It can be seen that two mobile stations (MS5 and MS7) are remarkably worse than the others, while it is also shown that the link simulator lies among the mobile stations in the better end of the scale. The accuracy of these measurements is about 1 dB, meaning that the curves can shift about 1 dB relative to each other.

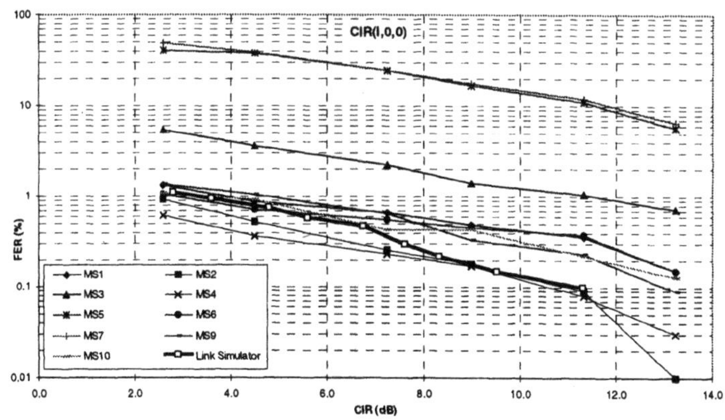

Figure 29. Comparison of the link simulator and the live measurements of cyclic frequency hopping on 3 frequencies with the TU3 profile and with 1 out of 3 frequencies interfered.

4. PREDICTING THE BER/FER WITH FH

In the previous paragraph the quality of a GSM link for a given mean *C/I* (or several mean *C/I* values in case of frequency hopping) was described. The quality was measured using the BER and the FER. Therefore the output indicated the link quality measured over a certain time and averaged over fast fading. The instantaneous *C/I* (per burst) could be quite different from the mean *C/I* (per frame), since both the signal strength of the carrier as well as of the various interferers changes continuously and independently. In this section primarily the instantaneous *C/I* but also the instantaneous E_b/N_0 and the corresponding BER is investigated. Again this is done for both the TCH/F and the SACCH. A frame erasure probability (FEP) is introduced, which indicates the probability of a frame erasure in either traffic or control interleaving frames.

As described in Chapter 1 the GSM link simulation tool is combined with the network level to create CAPACITY. The BER and FER are simulated in the link simulation tool and mapping tables are used as physical layer in CAPACITY. In the following the mapping method is described and evaluated. Throughout this section it is referred to as 'the FER/BER prediction method', even though it is not predicting in the exact meaning of the word, but more calculating the BER and FER from the combination of instantaneous *C/I* values.

4.1 The FER/BER Prediction Method

Using *C/I* values as input, a method using two steps to predict the various FER values of the mapping tables has been developed. The two steps are as follows:

1. First the *C/I* of every burst is translated into a corresponding number of bit errors by a lookup table.
2. Then the average number of errors in 8 following half-bursts (TCH/F) or 4 whole bursts (SACCH) along with the standard deviation of the number of errors, are used to find the FEP using a second lookup table.

The FEP is used directly in the estimation of the FER of a mobile connection. In the following two subsections, the two lookup tables for both the number of errors per burst and the FEP are described.

4.1.1 The Lookup Table for the Number of Errors per Burst

This lookup table is used to find the number of errors in a burst, given a certain *C/I* of that burst. The number of bit errors in each burst depends on:

- Noise.
- Interference. This means co- as well as adjacent channel interference.
- Frequency Selective Fading. At relatively high data rates the channel frequency impulse response is no longer flat across the channel bandwidth. This can introduce inter symbol interference (ISI).

To find the dependency between the number of bit errors and these three factors, link simulations were carried out for various channel conditions. For every simulated burst the power of the carrier, the power of the interference, the noise level and the corresponding number of bit errors are collected. By calculating a mean and a standard deviation from the bit errors, a relation can be derived. In Figure 30 the relation between E_b/N_0 for each individual burst and the number of errors in that burst can be seen. In the rest of this book E_b/N_0 will be neglected, i.e. only interference characteristics are considered.

As just said, the number of errors not only depends on the *C/I* value, but also on the frequency selective fading. Frequency selective fading can introduce errors, since it causes inter symbol interference (ISI). When the desired signal is in a fade, it is most likely that ISI occurs. Therefore not one but four mapping tables have been developed.

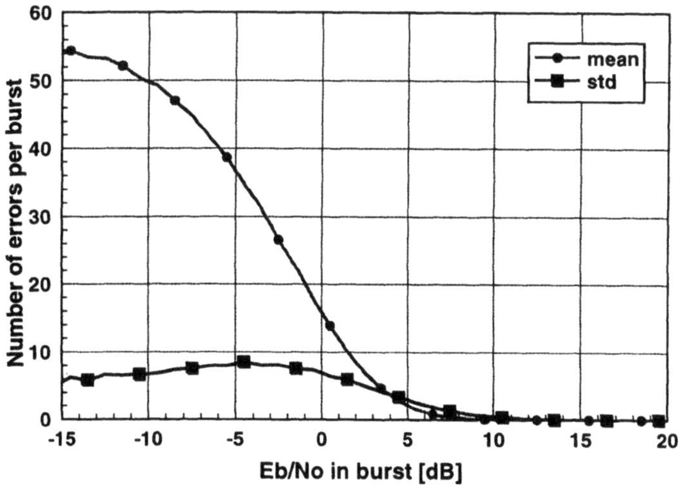

Figure 30. Mean and standard deviation of the number of errors per E_b/N_0 in one burst.

The mapping tables use the instantaneous *C/I* of a burst as input and give the number of errors in the burst as output. Each curve represents a different range of the fading depth. The fading depths are grouped in 4 as indicated in Figure 31 and below.

- The relative burst-RSSI[10] is greater than 0 dB. This means that the signal is not in a fade.

[10] The relative burst-RSSI is the received signal strength of the burst compared to the mean received signal level.

- The relative burst-RSSI is smaller than 0 dB, but greater than -6 dB. So the signal is in a fade, but it is not a deep one.
- The relative burst-RSSI is smaller than -6 dB, but greater than -9 dB. This means that the signal is in a medium deep fade.
- The relative burst-RSSI is smaller than -9 dB. This means that the signal is in a deep fade.

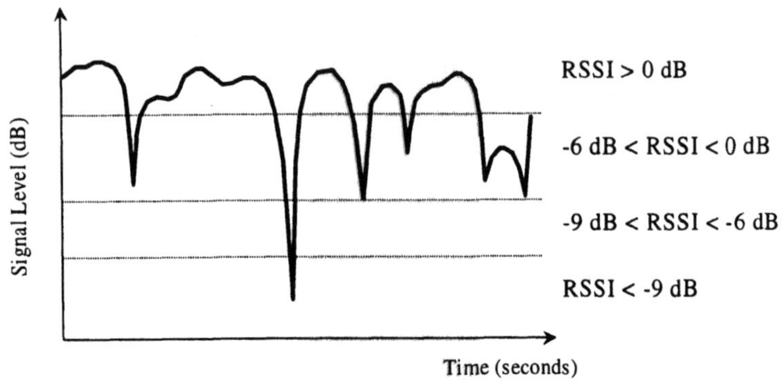

Figure 31. Graphical illustration of the 4 fading depths.

In Figure 32 the relation between the number of errors per burst and the instantaneous *C/I* in the burst is shown. These specific 4 classes are chosen because within these the number of errors as a function of the *C/I* does not change significantly. The curves in Figure 32 are used to produce the mapping values from each *C/I* to the corresponding number of errors in the same burst.

4.1.2 Lookup Table for the FEP

The mapping of the sum of the number of errors to the FEP, does, in contrast with the previous mapping to BER, depend on the speed of the mobile station, the network traffic load and on the type of hopping [149,235]. This dependency is caused by three factors, namely, interference diversity, frequency diversity and soft information. Both the gains from frequency and interference diversity have previous been described in Section 3.1. The soft information gain arises from using a maximum likelihood decoding type of data receiver. In a simplified form, using soft information is equivalent to exploiting an estimate of the probability of a correctly detected and decoded sequence of bits. For more information about such data receiver structures, see [26,112,150].

This mapping step has to be done per interleaving frame. Therefore the time resolution (unit) is one interleaving frame. This means that for traffic channels 8 half-bursts are mapped to one FEP, while in case of the SACCH 4 bursts are mapped to one FEP. The individual interleaving patterns used for respectively the SACCH and the TCH/F channel in GSM causes this difference.

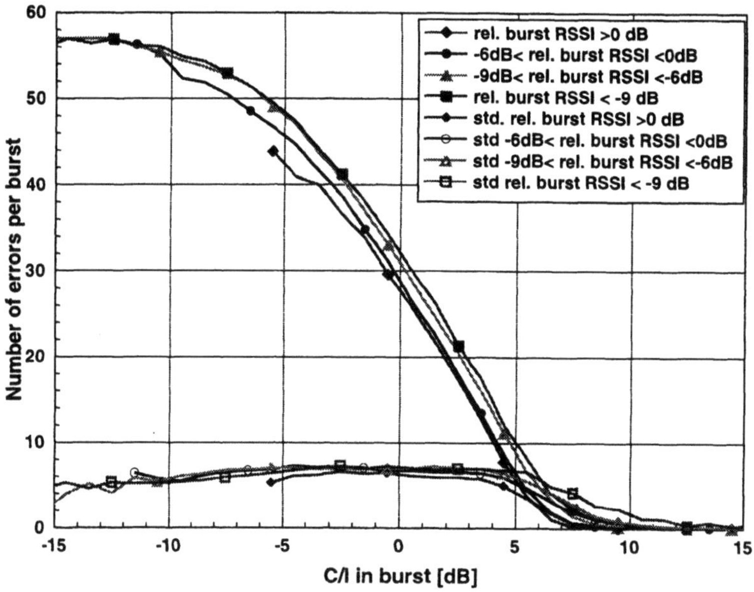

Figure 32. Mean and standard deviation of the number of errors as function of the *C/I* value for different fading depths of the desired signal.

By initially mapping the *C/I* value to a number of errors per burst, summing the number of bursts of eight half-bursts and map the result to a FEP, the gain from frequency and interference diversity have already been compensated for. This should be obvious if remembering that spreading the errors over 8 different half-bursts causes provides the gain from these two factors. However, the gain from soft information is difficult to predict. One way to do this, is by simulating different channel models and different hopping patterns to create statistics to make different mapping tables for each of the investigated situations. This has been done in [234]. Another approach is to generalise the soft information gain, and exploit the fact that the gain from soft information depends on the variation or difference in the number of errors per half-burst. Therefore not only the mean, but also the standard deviations of the number of errors in the 8 half-bursts (TCH) or the 4 whole bursts (SACCH) should be taken into account. This second method has been carried out and is described in the following [243].

For simplicity, it is assumed that the errors in a burst are uniformly distributed. In other words, it does not matter whether the entire burst or half a burst is used in the simulations. The number of errors in a burst therefore corresponds to twice the number of errors in a half-burst. This assumption is true for mobile speeds of less than 100 km/h. If the speed exceeds this value more errors starts to occur in the tail and head of bursts [143]. The mapping procedure is shown in Figure 33, using the assumption mentioned above.

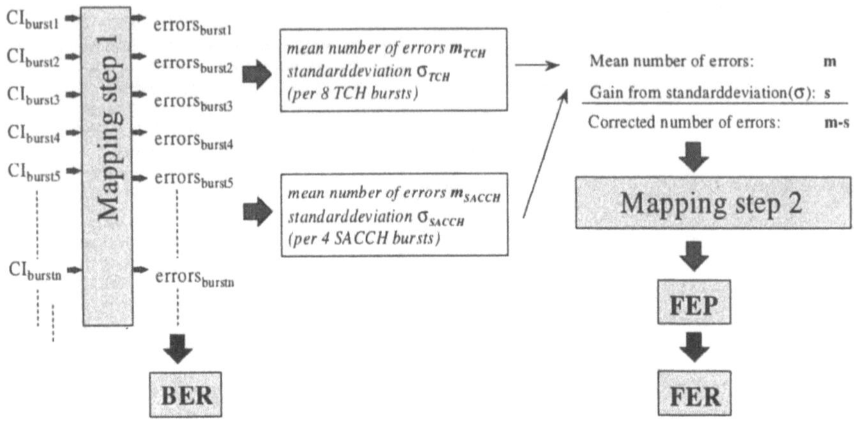

Figure 33. Procedure for mapping the number of errors in one frame to a FEP, for both the TCH/F and the SACCH.

Initially the mean number of errors of 8 bursts (TCH) or 4 bursts (SACCH) is calculated. Then the standard deviation of the number of errors in each burst is used to find the gain that depends on the variation in number of errors per burst. In Figure 33 this gain is denoted as 'gain (std)' and is an absolute number of errors.

Many (more than 100.000) link simulations using various hopping and non-hopping patterns under different channel conditions, have been run. The number of errors of every burst and the occurrence of a frame erasure of every frame was collected. By now sorting the data according to the mean and variance of the number of errors in the interleaving period, a statistical dependency was found. This dependency is seen in Figure 34, where the gain in terms of absolute number of errors is shown as a function of the standard deviation.

It can be seen that the gain from the standard deviation is different for the SACCH and TCH/F. This difference arises due to the difference in interleaving. In many situations the standard deviation of the 4 whole bursts is higher than the standard deviation of the 8 half-bursts.

The final step in the FEP estimation procedure is to correct the number of errors, i.e. the mean number of errors and subtract the standard deviation gain 'gain (std)' as shown in Figure 33. This corrected number of errors per burst is then used directly to find the FEP by mapping the corrected number of errors to a specific FEP using the curves in Figure 35.

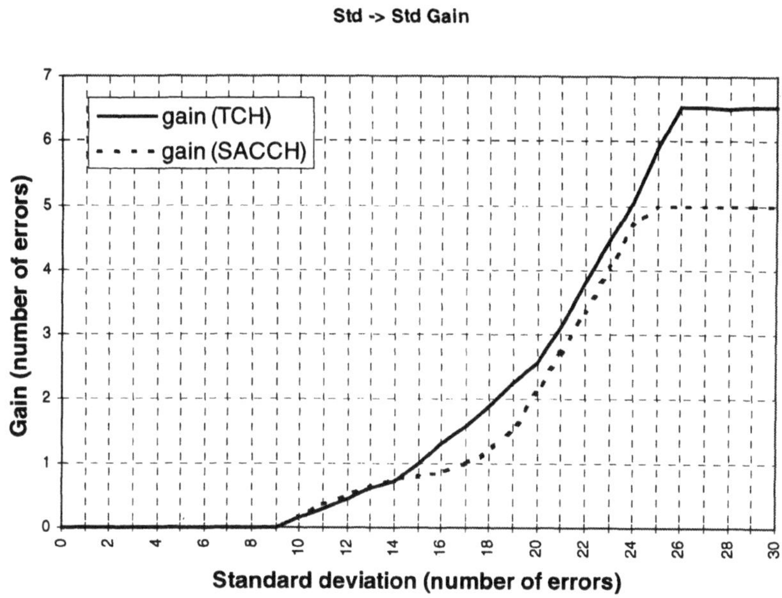

Figure 34. Gain (in absolute number of errors) from the standard deviation of the number of errors per burst.

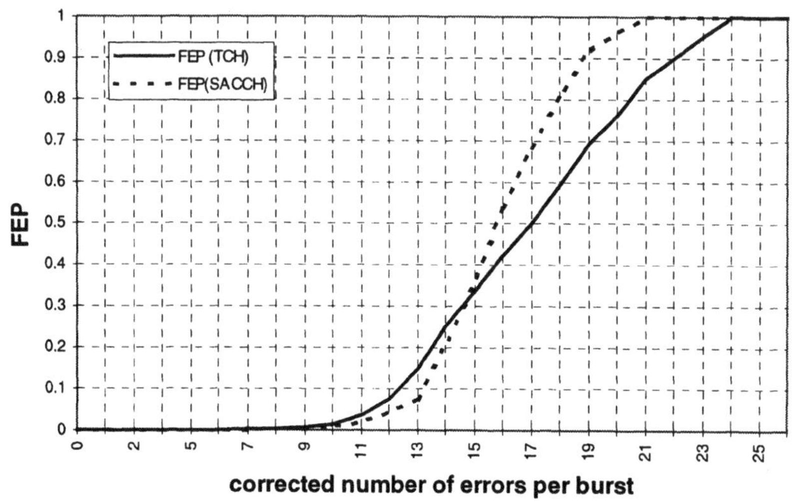

Figure 35. Curves mapping the corrected number of errors per burst to a FEP.

Again the curves are made by running many (more than 100.000) link simulations using various hopping and non-hopping patterns under different channels conditions to produce statistical data.

Finally the FEP of every frame is compared to a random number between zero and one to determine if there is a frame erasure. By running a large number of consecutive bursts (20.000) in each simulation the estimate/prediction of the FER of a connection could be made [243].

4.2 Accuracy of the BER/FER Prediction Method

Afterwards simulations have been carried out using specific *C/I* values to produce simulation output curves for both the BER and the FER. The mapping method has been evaluated by comparing the results with actual link simulation results. The comparisons are shown in the following, starting with the BER.

4.2.1 Prediction of the BER

An example of such a comparison between actually simulated data and the corresponding prediction values, using a TU3 type of channel, is shown in Figure 36. Here it is seen that the outcome of the comparison is very good. In this particular example using the prediction method it is possible to estimate the raw BER within 0.2 dB.

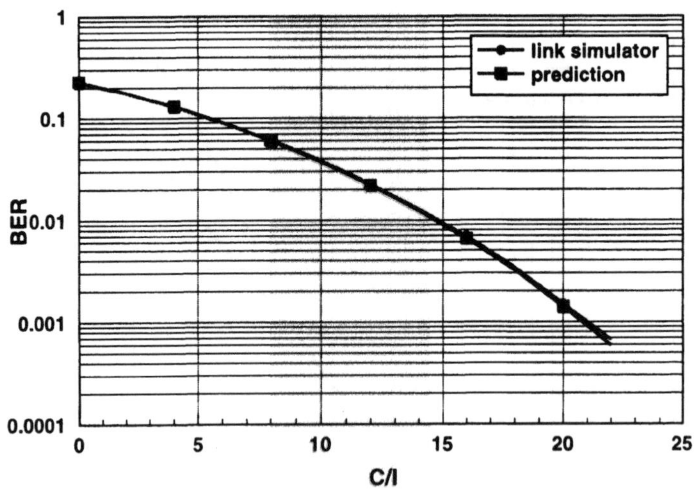

Figure 36. Example of the predicted and simulated raw BER of a TCH/F channel as a function of *C/I* for a TU3 type of channel.

The comparison was done for a TU3 with a hopping test profile. However since the raw BER does not depend on the hopping but only on the speed of the mobile, the evaluation looks the same for all TU profiles, i.e. the accuracy is always within 0.2 dB.

4.2.2 Prediction of the FER

The prediction of the FER has also proven quite accurate. The method is evaluated in the same way as with the BER by comparing output simulation results from the link simulator. The link simulator has been used to evaluate the TCH/F and the SACCH channels. A great variety of different frequency hopping configurations and *C/I* values have been simulated to produce the FER estimation for the comparison. In general, these comparisons show that the prediction is within 0.5 dB at the 2% FER threshold, when running simulating of 20.000 bursts. In a few situations the prediction error is a bit greater, but certainly within 1 dB. An example of a comparison between the predicted and the simulated FER for a TCH channel is shown in Figure 37 using random FH on 3 frequencies.

5. SUMMARY AND CONCLUSIONS

This chapter has concentrated on the GSM link level. The link level has been described in functional blocks. To study the behaviour of the GSM link in a frequency hopping environment a GSM link simulator was used, containing a transmitter, a channel and a receiver. The receiver is a soft decision Viterbi equalizer, which uses soft information, meaning that besides detecting every bit, also a probability of the right detection is given as output. The channel model has been a tapped delay-line with each of the taps experiencing Rayleigh fading. Interference or noise can be added to the channel.

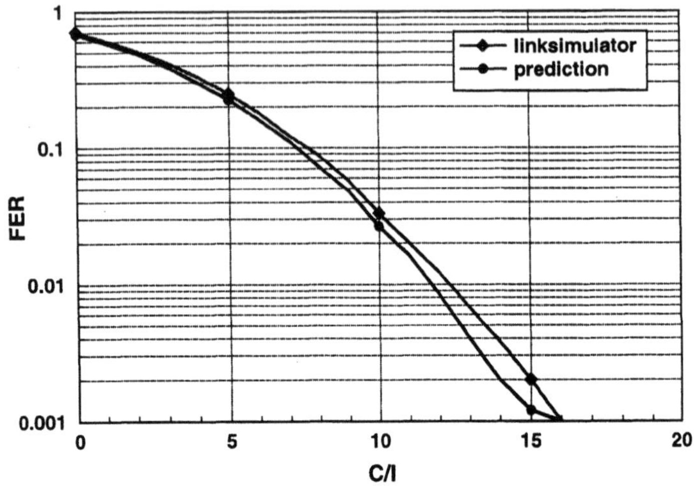

Figure 37. Comparison between the predicted and simulated FER for a TCH/F channel.

Frequency hopping provides a frequency and an interference diversity gain. It was found for the traffic channel that the frequency diversity gain is highest for slow moving mobiles, where it is marginal for fast moving mobiles. The reason for this is that the fading on the individual bursts of the fast moving mobiles already is uncorrelated. The frequency diversity gain on the SACCH channel has also been studied and found to be very small. This is due to the relatively large distance between the different bursts.

It was also found that not only the average *C/I*, but also the variance in *C/I*, influences on the FER. A higher variance in *C/I* at the same mean *C/I* results in a lower FER.

Mapping tables were presented, which can be used to map the *C/I* of a burst into an equivalent BER for that burst. The errors of 4 bursts in case of the SACCH or 8 bursts in case of the SACCH can then be used to find a frame erasure probability, which again can be used to find the FER. The produced mapping tables constitute the physical layer of the network system level simulator CAPACITY, as we will see in the following chapter.

Chapter 5

COMPUTER AIDED NETWORK DESIGN

One of the two main goals has been to develop a tool that can be used to design radio features for GSM networks. Since the developed tool in reality is a software program, the method has been called computer aided network design. In Section 1 an introduction is given followed by a description of the network simulator used for network aided design in Section 2. Section 3 describes the output parameters, which will be used frequently throughout the rest of this book. One of the very essential output parameters, the dropped call rate, is then treated in Section 4 by two different implementations. The accuracy of the simulator is the subject of Section 5, while the default network parameter settings are given in Section 6.

1. INTRODUCTION TO COMPUTER AIDED NETWORK DESIGN

The underlying idea of computer aided design is to use a computer model to design networks and new features for networks. Features as well as parameter settings can be simulated, before actually implementing them in the network. This way risky trials are avoided with the increased risk of degrading the network performance during the trial.

In order to be able to design new features of a GSM network, the modelling of the network functions and radio environment has to be accurate. For that reason the GSM network simulation tool, CAPACITY, has been developed at Center for PersonKommunikation at Aalborg University as described previously.[11] The tool is capable of modelling a frequency hopping GSM type network and can be used to create simulation results of a dynamically changing network. The program can

[11] Initially CAPACITY was developed as a tool to study the capacity of a frequency hopping GSM network and was programmed in C for the MS Windows-3.1 environment. Later CAPACITY has been expanded to cope with other network features. It is currently programmed in C++ and operating in the MS Windows-NT environment.

simulate several of the factors that affect the performance of a GSM system, such as frequency hopping and DTX, with mobiles moving randomly around in the network.

The radio propagation is modelled using a simple model, where path loss and fast and slow fading have been included. More advanced path loss models could have been used, but this would cause the simulation time to increase drastically, since the propagation has to be calculated for every burst for every mobile station.

In CAPACITY only the downlink is simulated, since this is believed to be the limiting link of the two [29,231].

In general, continuous development of CAPACITY takes place at Aalborg University by the Cellular System Group (CSG) at CPK. This has been the case during the entire time of the Ph.D. projects and has caused some of the simulation results to change slightly due to the more and more advanced and detailed modelling of the network.

Several documents have been produced in which the network simulator has been described. The initial implementation of CAPACITY is documented in detail in [105], while the latest documentation on CAPACITY is found in [140].

2. NETWORK MODELLING BY CAPACITY

The implementation of the GSM network simulator CAPACITY is described in this section. It is important to understand the choices that have been made and their limitations in order to analyse the results. Initially the general structure of the tool is described followed by an explanation of the content and design of each block.

2.1 The General Program Structure

A GSM BSS network consists of a number of base stations. The number of base stations as well as the cell radius can be varied. The simulations performed for this book are done with the base stations placed in a regular grid. The mobile stations are initialised at random places throughout the network and they move with a constant speed in a randomly chosen direction. Since the geographical area is limited and border effects are not desirable, a wrap around grid has been used. This means that mobiles leaving the simulation area at the east return in the west. The same holds for the south and north borders.

The frequencies, the kind of antennas to be used and the output power can be specified for every cell. In this book the output power and cell sizes are chosen in a way that the quality/capacity performance is limited by interference, not coverage.

The general structure of the tool is shown in Figure 38. Each block is simulated for all dedicated mobiles whereas non-active mobiles are not simulated.

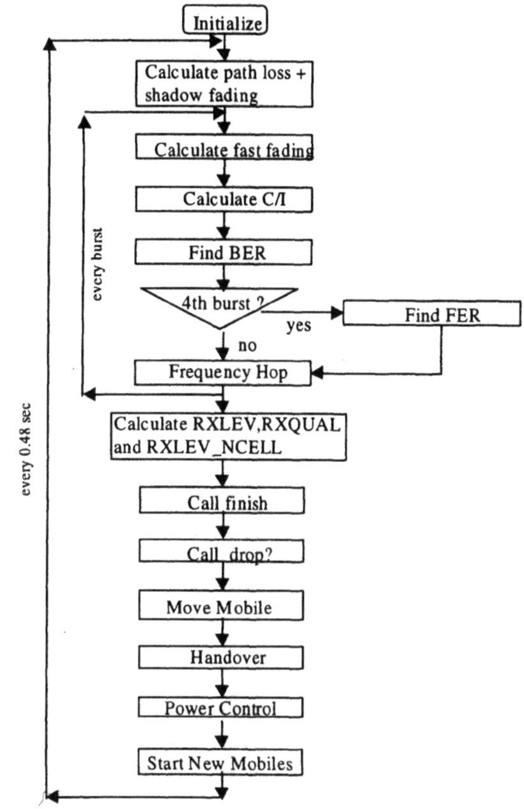

Figure 38. Flow chart indicating the basic structure of CAPCITY.

2.1.1 Path Loss Calculation

The path loss slope, describing how quickly the path loss decreases due to distance, depends on the distance between the transmitter and receiver and the obstacles in between. The path loss model used in CAPACITY is a simple Hata type of model [90], as seen in Equation (1). The path loss is simply a function of the distance between the base station and the mobile station and the path loss slope (γ).

$$L_{path\,loss\,CAPACITY}[\text{dB}] = 10 \cdot \log_{10}(d^{\gamma}) = 10 \cdot \gamma \cdot \log_{10}(d) \quad (1)$$

In Equation (1), d is the distance between the base station and the mobile and is measured in meters. The propagation over water can be viewed as free-space propagation with γ close to 2, where for land mobile communications the path loss slope is typically within the range of 3 to 5 [66]. γ can increase to higher values such as 7 or 8 in woods and areas with high buildings. In all the simulations the path loss slope has been set to 3.5, see [105, p. 71]. In CAPACITY the path loss is only updated once per 480 ms, i.e. the path loss is constant over a SACCH frame.

2.1.2 Slow Fading

Slow fading results from the shadowing of the radio signal by large obstacles. The received signal power at the mobile station varies on account of the alternative interruption and release of the line-of-sight between the receiver and the transmitter. The variation of the received signal power, due to shadow fading, is continuous and is correlated over a certain distance. This correlation decays with increasing distance between the locations [98].

Analysis of mobile radio propagation results from different surveys has shown that the local mean of the signal envelope, s, is well described by a log-normal distribution [99,182]. That is, the local mean, measured in dB, is a Gaussian random variable with probability density function given by [257]:

$$f(s) = \frac{1}{\sqrt{2\pi} \cdot \sigma_s} e^{\frac{-(s-\mu)^2}{2\sigma_s^2}} \tag{2}$$

In Equation (2) σ_s is the standard deviation of the local mean (in dB) and μ is the mean value of the random variable s, which is the local mean signal envelope of the received signal. The value of σ_s depends largely on the type of terrain. In suburban areas, containing macro cells, a typical value is 8 dB, while for urban micro cellular areas, the value is more like 4 dB [85]. σ_s has experimentally been found to be between 4 and 12 dB [76,188].

The log-normal fading is modelled as a Gaussian distribution, as can be seen in Equation (2). A standard deviation of 6 dB is used. Furthermore, to include the mobility of the mobiles it is necessary to include correlation between consecutive samples. The fading is correlated over a distance of 110 meter to simulate the slowly changing environment [29]. The de-correlation threshold criterion has been set to e^{-1}. This means that the correlation between the number of samples within the correlation distance is less than 0.37 for distances larger than 110 meters. The correlation between two consecutive samples has been implemented as a one pole filter, where the filter coefficient is a function of the speed of the mobile, the sampling period (480 ms) and the correlation distance [105]. The shadow fading is considered constant within each SACCH frame period, i.e. the shadow fading is only updated once every 480 ms.

2.1.3 Fast Fading

Often GSM network simulators are based on calculations in multi-frame resolution, i.e. one sample per period of 480 ms. To get more realistic results with respect to the user quality, the network simulator CAPACITY works on burst level. The primary difference lies in the fast fading, since slow fading is assumed stationary throughout the SACCH frame. The implementation of fast fading is described below.

The typical urban channel profile has been used [64]. As already described such a channel can be simulated using a tapped delay-line model with the taps corresponding to the impulse response. By taking the correlation in distance into account and assuming uncorrelated Rayleigh fading for the different taps (Rayleigh fading is completely uncorrelated over 0.4 wavelengths of 0.33 m. [220]), a file that describes the typical urban fading in time can be created. Such a file is denoted a TU-file and contains the signal envelope of typical urban fading. An example of such a file with 20 times oversampling is seen in Figure 39.

To each mobile (as well as each interferer) this type of fading is added. The speed of each mobile defines how large the jump in the file should be between two consecutive samples (two consecutive bursts). It is assumed that the fast fading does not change during the burst, which is valid for low and average speeds, but not for very high speeds like 100 km/h [143]. The fading of the individual mobiles is assumed uncorrelated.

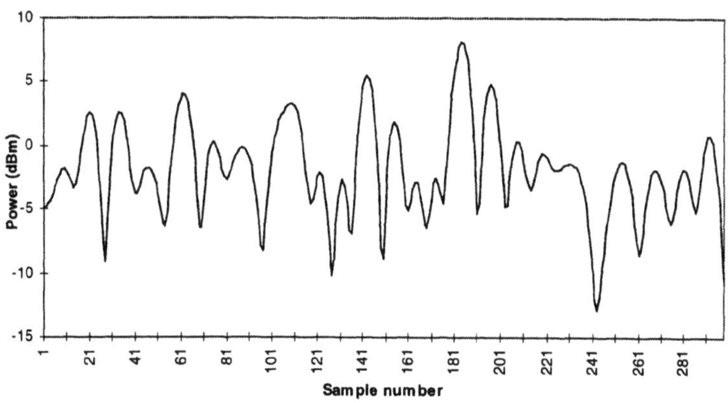

Figure 39. Example of a TU fading file with 20 times oversampling.

Since each contribution to the multi-path propagation depends on the carrier frequency, fast fading is said to be frequency selective. Depending on the channel, closely spaced (in frequency) channels will experience close to identical fading, whereas widely spread frequencies are more uncorrelated. This correlation between two (or more) signals is denoted the coherence bandwidth of the channel [135, p.77]. From reference to reference the precise definition of coherence bandwidth differs. In [110] it has been shown that for the GSM-900 system the fast fading on the different frequencies is almost uncorrelated in an urban macro cellular environment, whereas for micro cells a high correlation was seen. Since in CAPACITY a macro cellular environment is simulated, the fading on different frequencies is assumed uncorrelated.

2.1.4 *C/I* Calculation

The *C/I* is calculated for every mobile station in every burst. The received signal strength for every mobile is simply calculated by the path loss Equation (3) and then adding the shadow and fast fading as described in the previous sections. Now the *C/I* simply can be calculated by:

$$C/I\,[dB] = 10 \cdot \log\left(\frac{C}{\sum_{M} I_M}\right) \tag{3}$$

In Equation (3) M is the number of interferers.

2.1.5 BER and FER Calculations

The BER of every burst and the FER of every frame is calculated in CAPACITY. This is done for both the TCH/F and the SACCH channel. For this purpose the mapping tables, described in the previous chapter, have been implemented allowing an easy BER and FER estimation.

2.1.6 RXLEV, RXQUAL and RXLEV_NCELL Calculations

The 3 parameters RXLEV, RXQUAL and RXLEV_NCELL are parameters that are required by the power control and handover algorithms. RXLEV is a measure for the average signal level during 480 ms. It is simply calculated by averaging the individual signal level samples in dB and then mapping the result to an RXLEV class. Similarly the level of the neighbouring cells is measured, leading to a RXLEV_NCELL measurement for every neighbouring cell. The exact implementation of the neighbouring cell measurements is not described here, but in Chapter 8, as part of the handover algorithm study. RXQUAL is mapped directly from the average BER calculated once per 480 ms, mapping the linear BER to the logarithmic, discrete RXQUAL scale shown in Chapter 3.

2.1.7 Call Start, Finish and Move Functions

The call arrival process is modelled as a Poisson process [102], which can be described by the conventional Poisson distribution:

$$p(k) = \frac{(\lambda T)^k e^{-\lambda T}}{k!} \tag{4}$$

In Equation (4) *p(k)* is the probability of *k* arrivals in time *T* and λ is the arrival rate. In CAPACITY mobiles are started every 0.48 seconds, so *T* is equal to 0.48, where λ is depending on the load in the system.

The mobile stations start at a random place in the network and they move with a constant speed in a randomly chosen direction. Every 480 ms, their locations are updated according to their speed and direction. The call duration is assumed to be negative exponentially distributed. For each individual situation where this call duration expire a mobile station is removed from the simulations.

2.1.8 Handover Algorithm Implementation

The handover algorithm is able to perform handovers due to various reasons. The mobile can make a handover when the received signal level is too low or the quality is not good enough. Also when the signal level is good enough, but the quality is worse than a certain predefined threshold, a handover can be triggered. This last handover type is referred to as an interference type of handover. Furthermore, when the mobile station can receive the signal of a neighbouring cell better than its own, a handover can be made. The requirement in this latter case is that the received signal strength from the neighbouring cell is higher than what is received from the serving cell. This is the power budget type of handover. A description of the implemented handover algorithm, as well as handovers in general, is found in Chapter 8.

2.1.9 Power Control Algorithm Implementation

The idea of the power control algorithm is to minimise the output power of the base stations, when the mobile experiences surplus quality. It simply decreases the output transmission power, when the quality and received signal level are better than required. When on the other hand the quality or signal level is lower than a certain threshold, the power is increased (until the maximum is reached). In GSM the power control algorithm implementation is based on the received signal strength as well as on the received signal quality. Therefore the quickest power adjustments can be done using a SACCH frame time resolution. The issue of power control in GSM is treated in detail in Chapter 7.

2.1.10 Dropped Call Algorithms

Two different dropped call algorithms have been implemented. One based on calculations of the RXQUAL values experienced on the actual traffic channels (in the following denoted the 'RXQUAL dropped call algorithm') and one based on the decoding of the SACCH frames (in the following denoted the 'SACCH dropped call algorithm'). The first operates on the actual RXQUAL values experienced by the mobile user, while the latter corresponds to the `failures_on_radio_interface` counter specified in GSM and used in the typical dropped call algorithm previously mentioned in Chapter 3. The two algorithms are described in the following, starting with the RXQUAL dropped call algorithm whereas the performance of the two is treated in Section 4.

As described, in GSM the mobiles continuously report the experienced RXQUAL of the speech traffic time slot (when operating in the TCH/F mode). In CAPACITY the dropped call rate using the RXQUAL dropped call algorithm is updated each multiframe since a new RXQUAL value is calculated for each multiframe period. This algorithm uses two thresholds:

- **Call-drop-RXQUAL threshold.** This value indicates the highest RXQUAL value, which is still assumed to give acceptable speech quality. So a Call-drop-RXQUAL-threshold equal to 5 means that RXQUAL value 6 and 7 are not acceptable (i.e. bad quality).
- **Call-drop-threshold.** This is a counter which indicates how long a certain call can survive before being dropped.

The dropped calls are calculated in the following way. Each mobile station has a counter, which is initialised to the value specified by the call-drop-threshold. When the measured RXQUAL value of a mobile is worse than the call-drop-RXQUAL-threshold, the counter is decreased by 1. If the simulated RXQUAL is better or equal to the call-drop-RXQUAL-threshold, the counter is increased by 2. The counter never becomes higher than the initial value, the call-drop-threshold. When the counter reaches zero, the mobile is dropped. The principle of this algorithm for the TCH/F is shown in Figure 40. The call-drop-RXQUAL-threshold has been set to 5 and the call-drop-threshold to 19.

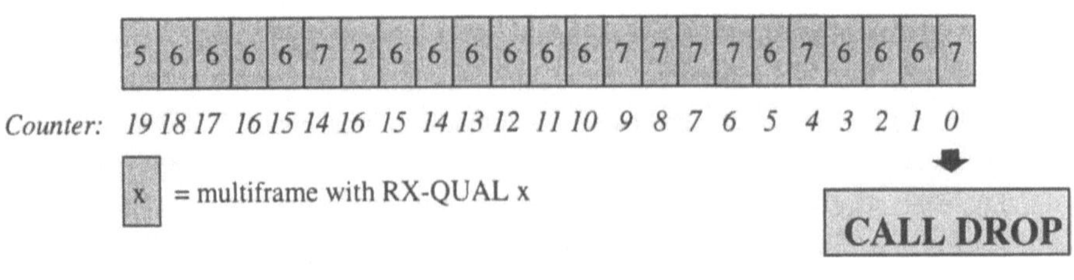

Figure 40. Example of the RXQUAL dropped call algorithm implemented in CAPACITY.

Due to the functionality of this algorithm the mobile never drops if it does not experience bad quality (RXQUAL 6 and 7) for at least 10 seconds.

In the live GSM network the dropped call algorithm is as stated in Chapter3 based on the decoding of the SACCH frames instead of the RXQUAL values. The second algorithm implemented in CAPACITY corresponds to this algorithm. The basic functionality is shown in Figure 41, and is equivalent to the first one except from the bad quality criteria. The countdown is here based on the decoding of the SACCH.

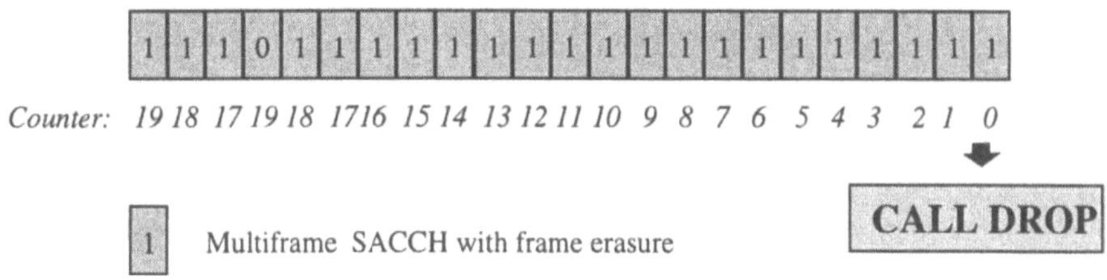

Figure 41. Example of the SACCH dropped call algorithm implemented in CAPACITY.

After having introduced the most essential functions of CAPACITY, the available output parameters are shown.

3. AVAILABLE OUTPUT PARAMETERS

The following parameters are available as output from CAPACITY:

- **C/I.** The signal to interference ratio distribution of the whole network and for each layer in case of IUO like in [166,242] is available. The statistics of the signal to interference ratios takes measurements for all mobiles, measured once per 480 ms, into account.
- **BER.** The bit error rate distribution of the whole network, as well as for the different layers individually, is calculated as with the *C/I.*
- **RXQUAL.** The distributions, measured throughout the whole network and for the separate layers, are again calculated. Statistics of actual RXQUAL values at the time of handover attempts are also available.
- **FER.** Again distribution throughout the entire network as well as for the different layers are calculated. All individual FER's measured per 480 ms per mobile station are taken into account as with the RXQUAL and the *C/I.*
- **Dropped call rates**, calculated from the algorithms described in the previous section.

- **Blocked calls**, the number of blocked new calls is denoted as blocked calls and is an important parameter for the trunking efficiency.
- **Blocked Handovers,** just like blocked calls, this is an important parameter when concerned with trunking efficiency.
- **Number of Handovers per mobile station**, since each handover comes with the risk of a handover failure, the average number of handovers per mobile station is an important parameter.

4. DROPPED CALL ALGORITHM COMPARISON

In this section the two dropped call algorithms implemented are compared with each other and with the dropped call algorithm in the live network. Different simulations using the two dropped call algorithms have been carried out. From these results some conclusions can be drawn when comparing the RXQUAL dropped call algorithm and the SACCH dropped call algorithm.

The two algorithms generate different results, when using the same call-drop-threshold. The RXQUAL dropped call algorithm produces simulation results where the amount of dropped calls comparable to the live network. The SACCH dropped call algorithm gives less dropped calls. The reason for this will be treated in detail further.

A way of making the SACCH dropped call algorithm produce higher (more usable) dropped call rates could be by adjusting the values of the call drop threshold. However, it has been found that, while using a call drop threshold equal to 19 for the RXQUAL dropped call algorithm, the SACCH dropped call algorithm requires a call drop threshold of as little as 3 to get the equivalent amount of dropped calls. That is, while the SACCH dropped call algorithm implementation itself is realistic, the value of the call drop threshold is sure not! A call should not be allowed to drop after only 1.5 sec.

Table 7. Comparison of the bad criteria in the two dropped call algorithms.

	Using the RXQUALDropped Call Algorithm	**Using the SACCH Dropped Call Algorithm**
The bad criteria :	RXQUAL value 6 and 7	Frame erasure on SACCH
Measured using :	104 TCH bursts	4 SACCH bursts
The minimum average number of bit errors per burst, when non acceptable:	7.3	≈ 16

Using the mapping tables in [243], the two dropped call algorithms can be compared in a different way. The criteria chosen to determine that a multiframe is not acceptable can be found. These criteria's are shown in Table 7 for each of the two algorithms. The first line shows the bad criteria (determine whether the quality is bad or not), while the second line shows on what measurement this is based. The last line is the result, a direct comparable value. It is the minimum average number of bit errors per burst needed to fulfil the bad criteria. These numbers are raw bit errors, so the difference in coding/decoding and interleaving/deinterleaving has been compensated for.

It can be seen that the minimum average number of raw bit errors per burst required to denote the criteria bad, is a lot higher in the case of the SACCH dropped call algorithm.

If looking at the problem by tracing a call of a mobile and looking at the BER or FER during the last 10 seconds before the call is dropped, a great difference between the two algorithms is seen as well. In Figure 42 the average BER and FER distribution on the TCH/F of the last 10 seconds (before a dropped call) is seen for both algorithms. For the RXQUAL dropped call algorithm the call drop RXQUAL threshold is set to 5 and the call drop threshold to 19. The SACCH dropped call algorithm uses a call drop threshold of 8, i.e. after approximately 4 seconds the call can be dropped.

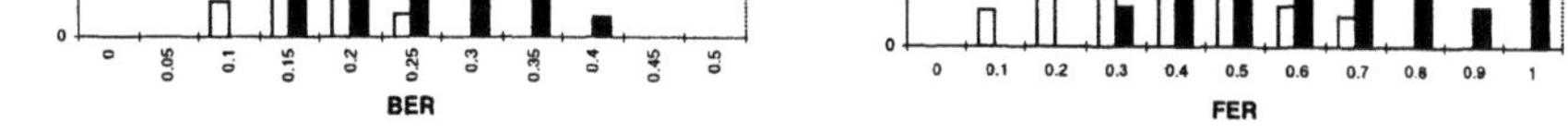

Figure 42. BER and FER distribution for the last 10 seconds before a dropped call in case of the RXQUAL and the SACCH dropped call algorithm. The call drop thresholds are 19 and 8 respectively.

A great difference between the two distributions can be seen. With a bad criteria using a call drop threshold of as little as 8 (in the SACCH case), the quality right before the call is dropped, is still a lot worse than in the RXQUAL case (where 19 has been used as call drop threshold).

The performance of the SACCH dropped call algorithm, should actually rather than being compared to the complete dropped call calculation used in the live network, be compared to the radio link failure measurement counter (based on the radio link time-out) itself. Remember that the radio link failure measurement counter (denoted `failures_on_radio_interface`) is part of the calculation. Therefore measurements have been carried out to see how large a percentage of the total amount of dropped calls arise from the `failures_on_radio_interface` counter. In both a baseband frequency hopping and a non-hopping area containing 30 cells measured in busy-hour over a period of 9 days, 67 % of the reported measurements resulting in dropped calls, arise from this counter. If this value represents the general trend, it means that the SACCH based dropped call algorithm in CAPACITY (of compensated by this factor) still produces less dropped calls than the RXQUAL dropped call algorithm. The study therefore continued. Since the radio link failure in the live network counts all failures on the radio interface, failures cause by lack of coverage, dropping the battery etc., they are all included in the measured 67 %. However in CAPACITY the SACCH dropped calls can only be caused by interference. Therefore, using A-bis measurements it was investigated how many of the 67 % of dropped calls were dropped due to interference. It was found that only approximately 20 % of these dropped calls were dropped due to interference. In other words, if using the SACCH based dropped call algorithm and comparing to the live network, the comparison should be done to only 13 % of the amount of measured dropped calls !

That means, the SACCH dropped call algorithm fits quite well with the dropped calls arising only due to interference in a real live network. However, since the RXQUAL dropped call algorithm generates dropped call rates, which are more easily comparable to the dropped call rates in a real network, it has been chosen to use this algorithm throughout this book.

5. ACCURACY OF SIMULATION RESULTS

Of course the accuracy of the simulations carried out in CAPACITY is of importance. The simulated real time has to be of a certain length to achieve reliable estimations in terms of blocking and dropped calls. There are two basic requirements:

- **A minimum simulated real time.** The mean call holding time for a call is set to 80 sec. The holding time is specified as negatively exponentially distributed. This means that 1% of the mobile stations have a call length of more than 370 sec. Therefore, the simulation time has to be of a certain length in simulated real time to get reliable results, where also the mobile stations with the longest holding times are taken into account. This is especially important when considering the hard blocking probability, since it is the long calls that produces a large part of the hard blocking.
- **A minimum simulated number of mobile stations.** This number is of importance to be able to rely on the dropped call percentage. The necessary number of mobile stations, *N*, can be found using Equation (5) [105].

$$N = \frac{1 - P(drop)}{P(drop) \cdot (E_R)^2} \tag{5}$$

where *P(drop)* is the dropped call percentage and E_R is the relative error. That is, if an accuracy of 10 % is desired for a dropped call rate of 2 %, the minimum number of mobiles to be simulated is 4900.

If lowering the dropped call rate while still requiring the same accuracy an increased minimum number of mobiles correspondingly has to be simulated. In each simulation a minimum number of mobiles of 6000 and a simulated real time of at least 400 seconds have been used. Typically a lot more mobiles are simulated.

6. DEFAULT SIMULATION PARAMETERS

Running extensive simulations realistic power control and handover algorithm configurations have been found as will be described in Chapter 7 and 8. In these chapters also the default parameters for these algorithms are given. However, many other remaining network parameters have to be specified also to be able to operate CAPACITY properly. The most important ones are referred to as the default network configuration parameters and are given in Table 8.

Table 8. Default network configuration parameters in CAPACITY.

Path loss	$L_p = 35 \log d$
Shadow fading standard deviation	6 dB
Shadow fading correlation distance	110 m
De-correlation criteria at corr. distance	Exp(-1) $\cong$ 0.37
Call mean hold time	80 sec (neg. exp. dist.)
Timeslots simulated per frequency	8
Cell radius	3000 m
Max. effective BS output power	34 dBm
Min. effective BS output power	4 dBm
Antennas	90° sectorized
Frequency hopping mode	Random
DTX factor	0.5

Chapter 6

INFLUENCE OF FH ON A GSM SYSTEM

The focus of this chapter is on the performance of frequency hopping in GSM networks. First the capacity limits of such a system is described in Section 1 using analytical considerations. In Section 2 the reference frequency hopping results from network simulations are found, while Section 3 deals with the interaction of some of the quality parameters in GSM. Both comparisons from simulations as well as from live measurements are included. In Section 4 a way of implementing frequency hopping in systems with a small frequency band is presented. In Section 5 the experiences achieved from a live introduction of frequency hopping in a non-hopping network is described. Finally a short summary can be found in Section 6.

1. CAPACITY LIMITS OF A FH GSM NETWORK

The question on how to determine the maximum capacity of a GSM network with a given number of available resources seems a simple task. However, in practice, this can be quite complex. First of all the coverage has to be determined, since a coverage limitation is a capacity limitation. Mature GSM networks, as the GSM-900 operators in Denmark, typically provide a on-street coverage at least 95%.

Besides by coverage, capacity of a system is also limited by the available number of channels or the level of interference. The first limitation is referred to as hard blocking and describes the blocking that occurs due to congestion, i.e. lack of resources on the available traffic channels. The other type of limitation, is denoted soft blocking. Soft blocking is a measure for the amount of bad quality in the network. Several types of measurements can be used to describe soft blocking, as will be described in Section 1.3.

Some kind of compromise between the two types of blocking has to be made. Figure 43 illustrates this in terms of soft/hard blocking as a function of the offered traffic per cell, assuming a fixed percentage of blocking for both blocking types and a fixed number of available channels.

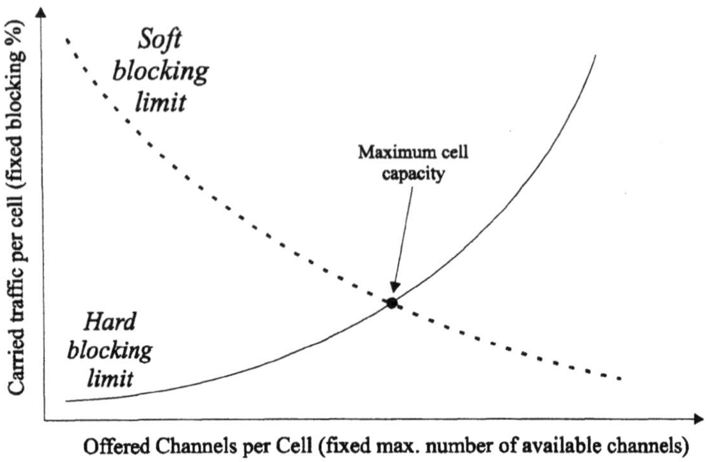

Figure 43. The two different types of blocking. The dotted line is the soft blocking limitation, while the other show the hard blocking traffic limitation.

The hard blocking curve describes the maximum amount of carried traffic given a certain number of physical channels. This limitation is typically found from classical tele-traffic theory.

The soft blocking limit on the other hand is entirely based on the radio communication limitations. From Figure 43 it can be seen that for a small amount of offered channels per cell, the capacity limitation from soft blocking is very little. For a higher numbers of available channels per cell, on the other hand, the amount of traffic that can be carried is quickly limited by soft blocking. This is caused by the fact that the soft blocking capacity limit is directly related to the level of interference or frequency reuse used in the cellular radio communication system. Note that the blocking curves are not fixed curves. Certain network features can shift both curves.

The overall maximum cell capacity is found at the highest point where both blocking criteria's are satisfied. Since both blocking criteria's always have to be satisfied it would be possible to draw one curve describing the maximum capacity as a function of the number of channels per cell. An example of such an estimation is found in [105, p. 97].

The teletraffic term often used to describe the quality of a network is Quality of service (QoS), which combines both the hard blocking (Blocking_Percentage and the soft blocking in the network. In [66] soft blocking is determined by the number of dropped calls, why the QoS has been defined as:

$$\text{QoS} = \text{Blocking\_Percentage} + \beta \cdot \text{Dropped\_Call\_Percentage} \quad (1)$$

In the equation, β is a weighting factor in the range between 1 and 2, i.e. the dropped call rate can be weighted higher than the hard blocking if desired.

The rest of this section is split into three parts. Coverage is the subject of Subsection 1.1, while hard blocking is described in Subsection 1.2 and soft blocking in Subsection 1.3.

1.1 Defining Coverage

Traditionally the probability of coverage has been defined as a cell area probability or a cell edge probability. The cell area should be understood as the part of the locations within the cell where the receiver experiences a signal strength above a (pre-defined) threshold. The cell edge probability would then refer to the probability that a receiver would experience a signal strength above a (the pre-defined) threshold at the cell edge. In order for an operator it is essential to know the relationship between the cell area and cell edge coverage probability, since typically only the cell edge coverage probability can be drectly dertermined from the network statistics. W.C. Jakes has solved this problem and derived the solutions in [104].

1.2 Determining the Hard Blocking

Looking at a typical daily traffic profile for a GSM cell, as shown in Figure 44, it is seen how the cell traffic varies over the day. During the night almost no traffic exists while later during the day the traffic increases. Typically measurements are reported once per hour. The hour that carries the most traffic is usually referred to as the busy hour. In the example the busy hour is between 7 and 8 p.m. Usually it is the amount of busy hour traffic which is used to configure the number of channels in the cells. The term traffic load is equivalent to the amount of occupied channels compared to the available number of channels.

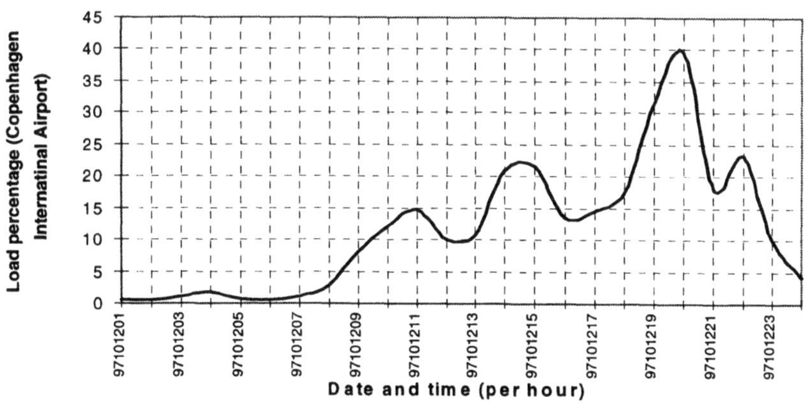

Figure 44. Example of a daily traffic profile for a GSM cell using hourly measurements.

The term traffic intensity indicates the amount of traffic in the network. The unit of traffic intensity is Erlang. Different traffic parameters exist. The term carried traffic is equivalent to the traffic handled by the networks. The term offered traffic is equivalent to the traffic that would be carried if no traffic were rejected due to lack of resources (channels).

A limited number of channels are available in each cell, so traffic can be rejected. The amount of rejected or blocked trafficis equivalent to the offered traffic subtracted the carried traffic This relation is shown in Figure 45, where N is the number of available channels.

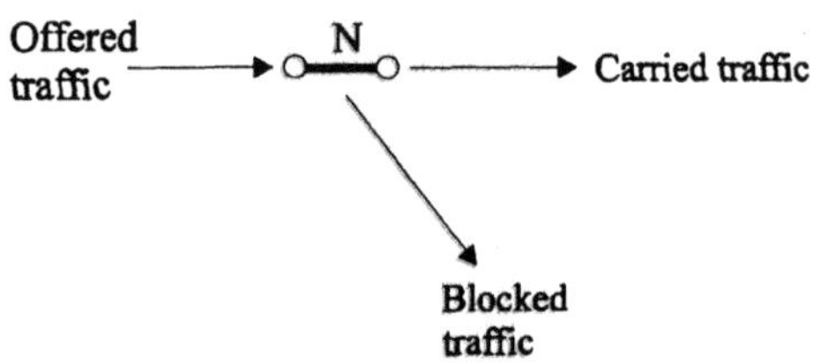

Figure 45. Relationship between offered and carried traffic calculated from the amount of blocked traffic.

To calculate blocking in networks as a function of the number of channels, the Erlang B formula can be used when the following assumptions are satisfied:

- All call attempts are Poisson distributed.
- Blocked calls are cleared (BCC) in the system.

The Erlang-B formula is shown in Equation (2) [103,134]. The equation is based on analytical probability theory.

$$P_B = \frac{A^N / N!}{\sum_{i=0}^{N} (A^i / i!)} \tag{2}$$

In (2) P_B = probability of blocking,
A = offered traffic in Erlang and
N = number of channels in the cell.

In systems where the blocked calls are put on hold and waits in a queue, the Erlang-C formula, can be used [103,134].

Another very important traffic term is the trunking efficiency which is linked very tightly to the Erlang formulas described above. From the Erlang-B formula it is seen that the blocking is a function of the number of available channels and the offered load. In Table 9 the amount of carried traffic is calculated along with the carried load (shown in the brackets). The load corresponds to the trunking efficiency.

Table 9. Trunking efficiency of the Erlang-B formula [66].

Channels per cell	**Carried Traffic (Erl. B - 0.2 %)**	**Carried Traffic (Erl. B - 2.0 %)**	**Carried Traffic (Erl. B - 20 %)**
8 (100 %)	2.3 (28 %)	3.6 (45 %)	5.9 (74 %)
16 (100 %)	7.2 (45 %)	9.6 (60 %)	13.4 (84 %)
24 (100 %)	13.0 (54 %)	16.3 (68 %)	21.2 (88 %)
32 (100 %)	19.2 (60 %)	23.2 (73 %)	29.0 (91 %)
64 (100 %)	45.7 (71 %)	53.3 (82 %)	60.7 (95 %)

The trunking efficiency depends heavily on the number of channels. When only few channels are available the trunking efficiency is low, while the availability of many channels leads to high trunking efficiency.

In traditional flat cellular networks,[12] the Erlang-B formula is generally used to describe the hard blocking limit, i.e. the limitation in physical resources independent of the quality experienced on the radio interface. Recently operators have started doubting this use of the Erlang-B formula, since some has found it to overestimate the capacity [16]. The reason for this is that with the increasing complexity of GSM networks the required assumptions are no more valid.

1.3 Determining the Soft Blocking

Soft blocking is related to the amount of interference in a network. In general two types of interference are considered: co-channel and adjacent channel interference. Co-channel interference describes the effect arising from using the same frequency repeated in other cells. In the same way adjacent channel interference is the effect from adjacent frequencies being reused.

Especially the carrier to interference (*C/I*) ratio is of interest. For a fixed frequency allocation strategy employing homogeneous frequency reuse schemes (assumed for simplicity), the distance to all nearest neighbouring co-channel cells is the same. An ideal cluster size of e.g. 7, corresponds to having exactly six

[12] Throughout this book the term 'flat network' means a network *not* using multiple layers.

interfering cells located in the first tier (omni-directional antennas). The frequency reuse distance *(D)*, can in such a situation be approximated by [122]:

$$D/R = \sqrt{3K} = q \tag{3}$$

The level of co-channel interference is proportional to q, which is referred to as the co-channel reduction interference factor or normalised reuse distance. When only looking at the distance path loss, not considering fading, the *C/I*-ratio at the cell border can be approximated to [122]: [13]

$$\frac{C}{I}\ [dB] = 10 \cdot \log_{10}\left(\frac{1}{6 \cdot q^{-\gamma}}\right) \tag{4}$$

In Equation (4) the parameter γ is the propagation path loss slope. It is observed that the *C/I* is a function of the cluster size and the path loss slope. The *C/I* values for a path loss slope (γ) of 3.5 and different cluster size *(K)* are shown in the second column of Table 10.

Table 10. *C/I* at the cell edge for an omni directional base station configuration.

Cluster size (K)	*C/I* without shadow fading	*C/I* with log-normal fading (50% outage)	*C/I* with log-normal fading (25% outage)
1	0.5 dB	-2.2 dB	-6.1 dB
3	8.9 dB	6.2 Db	2.3 dB
4	11.1 dB	8.4 dB	4.5 dB
7	15.3 dB	12.6 dB	8.7 dB
9	17.3 dB	14.6 dB	10.7 dB
12	19.4 dB	16.7 dB	12.8 dB

The *C/I* calculated by Equation (6) does not take log-normal fading into account. When including log-normal fading, the mean power m_u of the summed interfering signal will be different. This is because the log-normal fading in the linear power exhibit an asymmetrical PDF.

In [154], [44] and [192] it is described how an approximate probability density function can be obtained using a method described by L.F. Fenton [67]. This method is, in the literature, referred to as the classic method. Fenton shows that the sum of n stochastic independent log-normal distributions can be approximated by another log-normal distribution [192]. However, in [192] it is described how Fenton's method is best suitable for probability density functions of low standard deviations. With high

[13] This assumption is only approximate and not valid for very low cluster sizes since it is assumed that D >> *R*.

standard deviations (above 4 dB) the method will make to positive results for the joint co-channel interference [192]. In [211] exact expressions for the mean and variance for the sum of two log-normal variables has been derived, by Schwartz and Yeh. In the same paper a recursive approach to approximate the mean and variance of *n* variables has been taken. The Schwartz and Yeh method is valid for standard deviations (σ_s) in the range: 6 dB < σ_s < 12 dB. Since in this study a standard deviation of 6 dB is used, this latter approach becomes relevant.

The method of Schwartz and Yeh is somewhat complex and is based on the idea that if looking at only two interfering signals, the exact value of the mean and standard deviation of the summed signal, which by approximation is log-normally distributed, can be found. The summed output signal is then used as input to the next summation using the third interferer, which again gives the mean and standard deviation of a new resulting signal (also log-normally distributed). This process, shown in Figure 46, continues as long as interferers exist and the total mean and standard deviation from all interferers are found and the complete interference can be described.

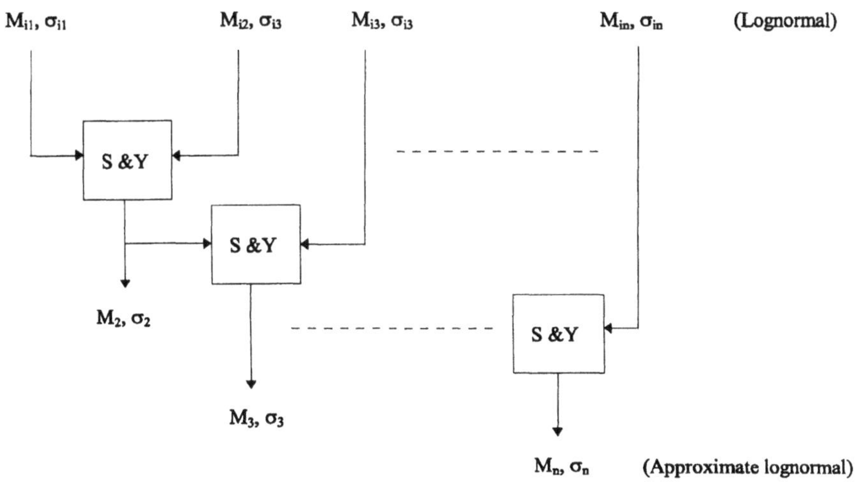

Figure 46. The summation principle of the method by Schwartz and Yeh used to sum multiple numbers of log-normally distributed random variables (interferers).

In [191] the mean and variance of the sum of interfering signals have been calculated for signals with a standard deviation (σ_i) of 6 dB and a mean (m_i) of 0 dB. The summed results (σ_u and m_u) are given in Table 11.

Table 11. Standard deviation (σ_u) and local mean power (m_u) of 1 to 6 interferers with log-normal fading [191].

$\sigma_i = 6$ dB, $m_i = 0\ dB$		
n	σ_u	m_u
1	6.0 dB	0.0 dB
2	4.6 dB	4.6 dB
3	3.9 dB	6.9 dB
4	3.5 dB	8.4 dB
5	3.3 dB	9.6 dB
6	3.0 dB	10.5 dB

The probability of the local *C/I* being below some threshold value, x_o, at the cell edge is [66]:

$$p_{x_o} = \frac{1}{2} \cdot erfc\left(\frac{x_o - CIR_{50}}{\sigma_{CIR} \cdot \sqrt{2}} \right) \tag{5}$$

where CIR_{50} is the median (50% outage) and

$$\sigma_{CIR}^2 = \sigma_d^2 + \sigma_u^2 \tag{6}$$

with σ_d and σ_u equal to the standard deviation of log-normal fading for the desired, and sum of interferers respectively.

A coverage level of 90 % has in the past typically been used as a goal for network planning in mobile communications. The 90 % coverage probability is an average for the entire cell, which can be found to correspond to a location probability of approximately 70% at the edge of the cell, when $\sigma = 6$ dB and $\gamma = 3.5$. Results are given in Table 10 for both 50 % and 25 % outage probability at the edge of the serving cell. In a cellular network, a mobile station located at the cell edge may often have more than one base station as serving candidate. Provided the radio signal from neighbouring cells are exposed to decorrelated log-normal fading (often the case) an improvement in outage probability will be achieved from the handover functionality by macro diversity.

Besides the *C/I* some other parameters can be used to describe the soft blocking. Along with the parameters based on physical link and network performance such as RXQUAL and dropped calls, other more subjective measures such as e.g. the number of unsatisfied customers could be used [57]. However, in this book the definition of soft blocking have been limited to be determined exclusively by the following 4 engineering parameters, which are all non-subjective parameters.

1. The dropped call ratio (DC).
2. The raw bit-error-rate (BER), the RXQUAL value in GSM.
3. The carrier-to-interference ratio (*C/I*).
4. The frame erasure rate (FER).

2. NETWORK SIMULATION RESULTS

In this paragraph the most important system level simulations, used to describe the performance of a GSM network, are shown. These results are the primary reference in determining the capacity enhancement of the other studied network features like advanced handover algorithms, IUO, power control and DTX. They present the benchmark capacity of a FH network.

Two different sets of reference simulations are carried out. Networks with a frequency reuse of respectively 1/3 and 3/9. Both networks are simulated for a mobile speed of 3 and 50 km/h. The performance, in terms if soft blocking as well as hard blocking, is found for various network load percentages. The section ends with a comparison of the two different network topologies that have been tried out in CAPACITY.

2.1 Introduction to the Network Simulations

The default network parameters, introduced in Chapter 5, have been used. Furthermore, the default power control and handover parameters, found in Chapter 7 and 8, are used. The network has been simulated with random frequency hopping in combination with power control and DTX. Concerning these reference simulations they should, in terms of network capacity, *not* be compared directly to live network capacity measurements. It is unrealistic to believe a simulator can model the real live network exactly, i.e. absolute comparisons to live measurements should not be done. Instead the idea is, throughout the book, to find the relative capacity gain by finding the relative gain from direct comparisons to reference simulations.

2.2 The CAPACITY Network Simulation Results

In the following the reference network simulation results are presented for 4 simulations. An available bandwidth of 7.8 MHz (39 frequencies) is considered as example. In the simulations the BCCH (with a reuse of 4/12) is not simulated, resulting in 27 simulated frequencies. For the 3/9 reused case, each cell has 3 allocated frequencies (baseband FH). In case of the 1/3 reuse, each cell should have 9 frequencies allocated per cell. However, to simplify the simulations to be carried out, only 4 frequencies are allocated. This has no impact on the results since most of the gain from FH is already achieved when hopping across 4 frequencies.

2.2.1 Simulating a Tight (1/3) Frequency Reuse

Using the simulation parameters described in the previous section, CAPACITY simulations using a network with a 1/3 frequency reuse and a mobile speed of 3 km/h have been carried out. The statistics are presented in Table 12 and Figure 47. The table contains selected data on all blocking parameters except the FER, which is presented in Figure 47.

Table 12. Selected statistics of the hard blocking, dropped calls, *C/I* and BER for the simulation results of a 1/3 reused network at a mobile speed of 3 km/h.

Name of simulation	Blocked calls	Dropped calls	BER> 0.064 (RXQUAL 6&7)	CIR (10% outage)
1/3 reuse, 20% load	0.0%	0.3%	1.0%	11.6 dB
1/3 reuse, 25% load	0.0%	0.4%	1.2%	11.2 dB
1/3 reuse, 30% load	0.0%	1.0%	2.0%	10.5 dB
1/3 reuse, 35% load	0.0%	1.2%	2.5%	10.1 dB
1/3 reuse, 40% load	0.0%	2.8%	3.1%	9.6 dB

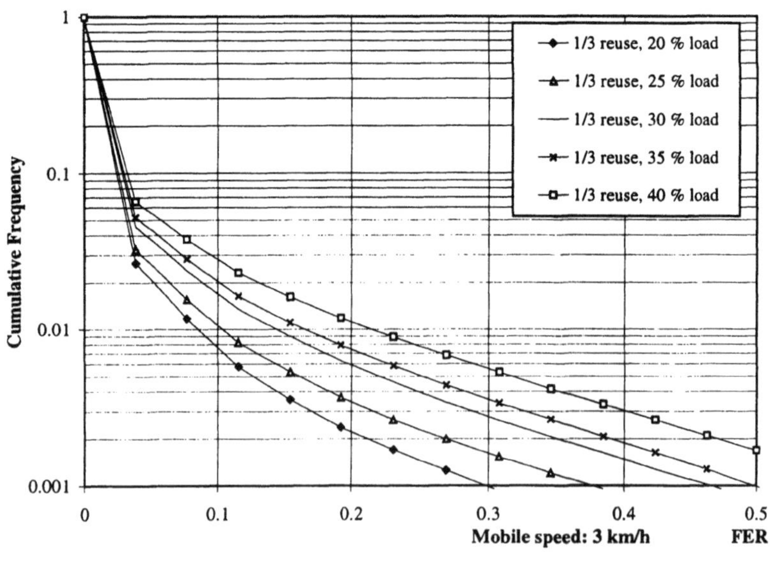

Figure 47. Cumulative frequency of the FER for the 1/3 reused network at 3 km/h.

Throughout the entire book the FER statistics are presented using the same type of accumulated figure as shown in Figure 47. To clarify how to real the figure an example is given using Figure 47. Looking at the FER on the abscissa the corresponding ordinate value represents the outage percentage of FER. If e.g. looking at the 10 % FER for the simulation of 20 % load, the resulting outage is approx. 0.8 %. In other words 0.8 % of the simulated FER values are worse than 10 %. The simulations using the same network, but with a mobile speed of 50 km/h have furthermore been carried out. These statistics are presented in Table 13 and Figure 48.

Table 13. Selected statistics of the hard blocking, dropped calls, *C/I* and BER for the simulation results of a 1/3 reused network at a mobile speed of 50 km/h.

Name of simulation	**Blocked calls**	**Dropped calls**	**BER> 0.064 (RXQUAL 6&7)**	**CIR (10% outage)**
1/3 reuse, 20% load	0.0%	0.4%	5.0%	10.0 dB
1/3 reuse, 25% load	0.0%	0.4%	6.6%	9.1 dB
1/3 reuse, 30% load	0.0%	1.1%	8.0%	8.5 dB
1/3 reuse, 35% load	0.0%	1.8%	10.2%	8.0 dB
1/3 reuse, 40% load	0.0%	3.0%	10.2%	7.5 dB

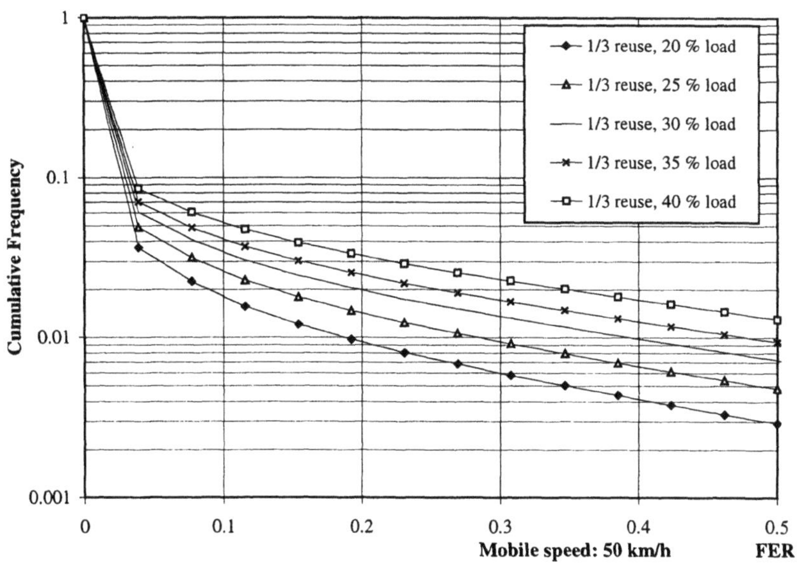

Figure 48. Cumulative frequency of the FER for the 1/3 reused network at a speed of 50 km/h.

2.2.2 Simulating a Loose (3/9) Frequency Reuse

Again using the default CAPACITY parameters, simulations using a network with a 3/9 frequency reuse and a mobile speed of 3 km/h have been carried out. The statistics are presented in Table 14 and Figure 49.

Table 14. Selected statistics of the hard blocking, dropped calls, *C/I* and BER for the simulation results of a 3/9 reused network at a mobile speed of 3 km/h.

Name of simulation	Blocked calls	Dropped calls	BER> 0.064 (RXQUAL 6&7)	CIR (10% outage)
3/9 reuse, 50% load	0.1%	0.1%	0.3%	12.7 dB
3/9 reuse, 60% load	0.9%	0.1%	0.3%	12.6 dB
3/9 reuse, 70% load	2.3%	0.1%	0.4%	12.5 dB
3/9 reuse, 75% load	4.4%	0.1%	0.4%	12.4 dB
3/9 reuse, 80% load	6.1%	0.1%	0.4%	12.3 dB

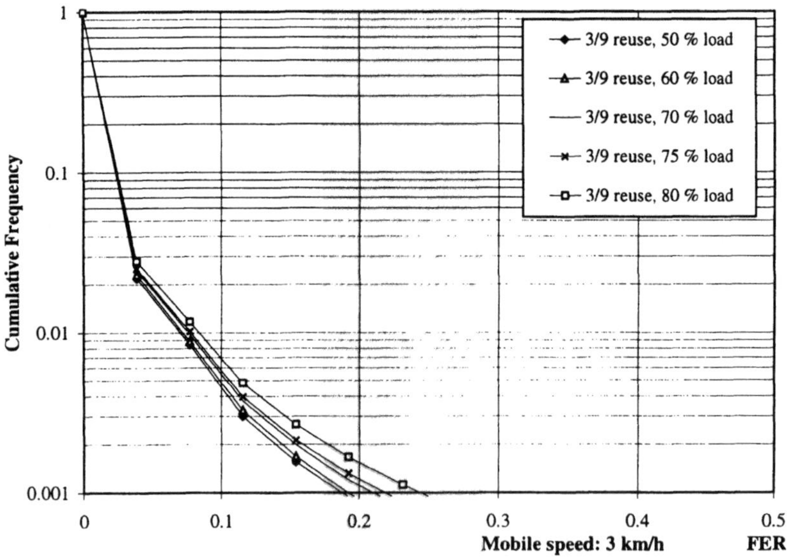

Figure 49. Cumulative frequency of the FER for the 3/9 reused network at 3 km/h.

The simulations using the same network, but with a mobile speed of 50 km/h have been carried out. These statistics are presented in Table 15 and Figure 50.

Table 15. Selected statistics of the hard blocking, dropped calls, *C/I* and BER for the simulation results of a 3/9 reused network at a mobile speed of 50 km/h.

Name of simulation	Blocked calls	Dropped calls	BER> 0.064 (RXQUAL 6&7)	CIR (10% outage)
3/9 reuse, 50% load	0.0%	0.1%	1.2%	10.8 dB
3/9 reuse, 60% load	0.6%	0.1%	1.4%	10.6 dB
3/9 reuse, 70% load	2.2%	0.1%	1.5%	10.5 dB
3/9 reuse, 75% load	3.4%	0.2%	1.7%	10.4 dB
3/9 reuse, 80% load	6.5%	0.3%	1.9%	10.3 dB

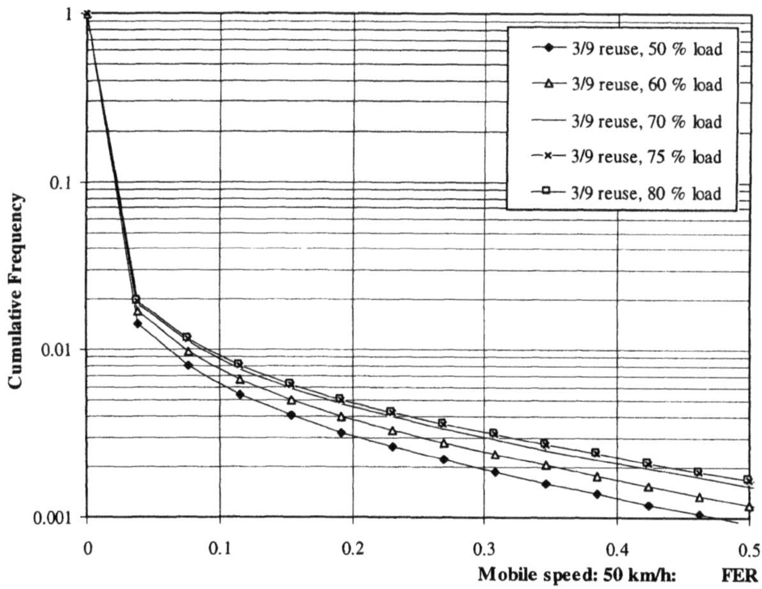

Figure 50. Cumulative frequency of the FER for the 3/9 reused network at 50 km/h.

2.2.3 Summary on the Network Simulations

Some interesting results can be drawn from the simulations above, of which the most important ones are:

- ***The 1/3, 3 km/h case:*** This network seems to be soft blocked. The dropped call rate is very sensitive to the load percentage. It is low while the load is below 35% and then is quickly accelerates. The BER (RXQUAL 6 and 7) outage degrades linearly (measured in percentage). The *C/I* statistics also degrade linearly (on a dB scale) as a function of the network load. The cumulative frequency of the FER is highly dependent on the network load. Step by step, ranging from 20% load to 40 % load the network performance continuously degrades.

- ***The 1/3, 50 km/h case:*** This network also seems to be soft blocked. The dropped call rate is very sensitive to the load percentage above 35 %. The BER (RXQUAL 6 and 7) outage degrades up to 35% load. From 35 to 40 % load the BER is unchanged. It is believed that this is caused by the high dropped call rate of the 40% loaded network, which removes some of the mobiles with high BER. The *C/I* statistics degrade linearly (on a dB scale) as a function of the network load. The cumulative frequency of the FER is highly dependent on the network load. Step by step, ranging from 20% load to 40 % load the network performance continuously degrades.

- ***The 3/9, 3 km/h case:*** This network is hard blocked. Already at a load of 70 % the hard blocking is 2.3 %, which corresponds to what can be found with the Erlang B tables. The dropped call rate is completely unaffected by the load. Furthermore, both the BER, *C/I* and FER statistics are not very sensitive to the load percentages used. Changing the load from 50 % to 80 % e.g. only degrades the 10 % *C/I* outage 0.4 dB.

- ***The 3/9, 50 km/h case:*** Again the network is hard blocked. At a load of 70 % the hard blocking is 2.2 %. The dropped call rate affected a little bit by the load, however even with 80 % load it is very low. Again the BER, *C/I* and FER statistics are not very sensitive to the load percentages used. Changing the load from 50 % to 80 % e.g. only degrades the 10 % *C/I* outage 0.5 dB.

- The network with faster moving mobiles performs significantly worse than the one with the slowly moving mobiles. This is believed to be due to the fact that some macro diversity gain is achieved in the 3 km/h cases. The slowly moving mobiles close to the cell border are often able to make a handover to a neighbouring cell, when they are in a fade. They make a handover back to the original cell, when another fade occurs.

- If comparing the 1/3 reused network with 25 % load to the 3/9 reused network with 75 % load, the 3/9 reused network is hard blocked, while the 1/3 reused network has almost no dropped calls and no hard blocking at all. If, on the other hand, comparing the *C/I*, BER and FER statistics the 3/9 performs better.

2.3 Alternative Network Topologies

Traditionally the site acquisition shown in Figure 51 has been used in CAPACITY. For each of the three sectors of the sites, a 90° antenna, with the antenna radiation pattern shown in [105], has been used. From here on this topology is denoted the 90° antenna network.

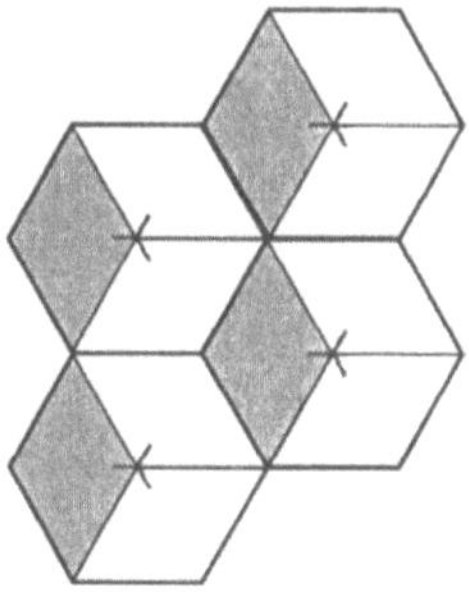

Figure 51. The network topology referred to as the 90° antenna network.

However, it has been discovered that, if rotating each antenna 30° clock-wise, while changing the radiation pattern of all antennas to 65° antennas [105], the overall network performance can be improved. The new network topology is shown in Figure 52. This topology is denoted the 65° antenna network.

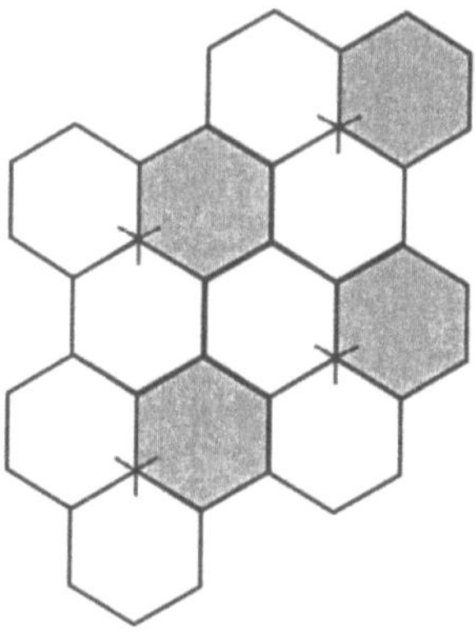

Figure 52. The network topology referred to as the 65° antenna network.

To illustrate the network enhancement that can be achieved, one example is given here. A frequency reuse of 1/3 with a network load of 40 % has been used. The mobiles are slowly moving.

The two topologies are described by the cumulative frequency of the *C/I* shown in Figure 53.

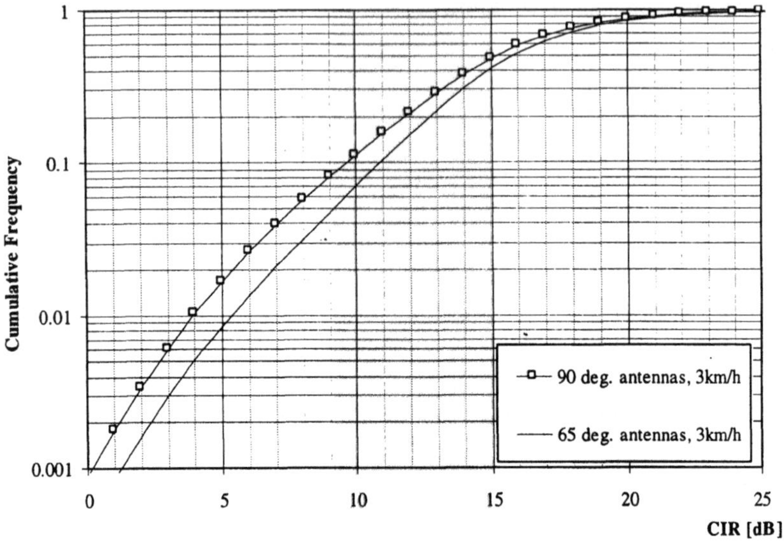

Figure 53 . Cumulative frequency of the *C/I* for respectively the two network topologies.

At the 10 % outage a difference around 1.3 dB is seen. The cumulative frequency of the BER is seen in Figure 54 and for the FER in Figure 55.

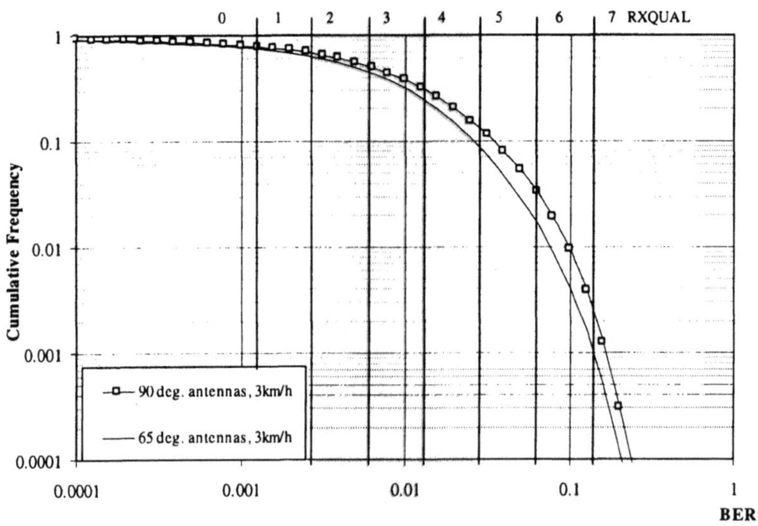

Figure 54. Cumulative frequency of the BER for the two network topologies.

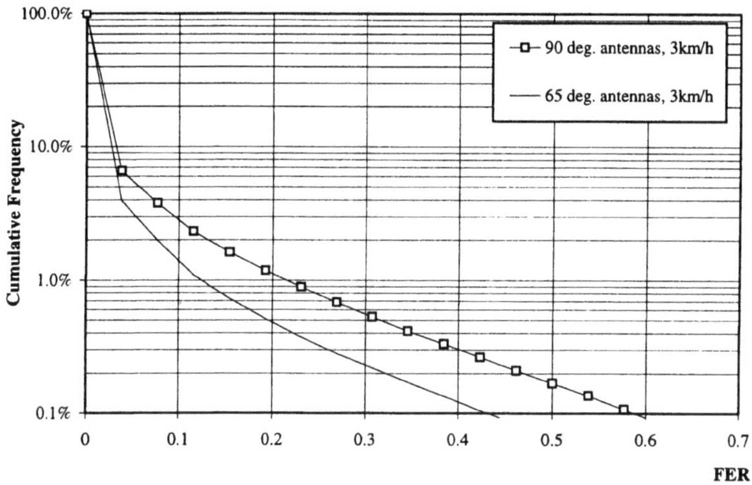

Figure 55. Cumulative frequency of the FER for the two network topologies.

In general quite a significant difference between the two networks is seen. The difference is most significant when using the FER statistics.

For the scenario used above, with a frequency reuse of 1/3, the hard blocking limit is of course not reached. The dropped call rate, on the other hand, changes significantly from the 90° topology to the 65° topology. This is shown in Table 16, where the dropped call rate, for the 65° topology, is reduced to less than one third of the one from the 90° topology.

Table 16. Statistics of the hard blocking and dropped calls for the two networks.

NAME OF SIMULATION	BLOCKED CALLS	DROPPED CALLS
90° antenna, a MS speed of 3 km/h	0.0 %	2.8 %
65° antenna, a MS speed of 3 km/h	0.0 %	0.9 %

Throughout all the CAPACITY simulations carried out in relation to the book, it has been chosen to use the 90° antenna topology. As stated several times the absolute capacity estimations found using CAPACITY should *not* be used in direct comparison to live network capacity measurements. Therefore whether using the 90° or the 65° antenna topology is not essential. To avoid confusion, it has been decided that all simulations using CAPACITY should be carried out using the same network topology. Since many of the simulations where carried out using the traditional 90° antenna topology, it has been decided to use this in general throughout the book.

3. INTERACTION BETWEEN NETWORK QUALITY PARAMETERS

In this section the interaction between the different quality measures has been looked at. First the dependency between the dropped call rate and RXQUAL (bad quality percentage) is studied in a non hopping as well as a frequency hopping network. This is done using simulations in Section 3.1 and with data from a real GSM network in Section 3.2. Then in Section 3.3 the relationship between the FER on the SACCH and on the TCH is treated by simulations.

3.1 Simulations on Dropped Calls versus RXQUAL

Network operators often use the reported RXQUAL values to evaluate their network. Usually RXQUAL 5, 6 and 7 are referred to as the 'RXQUAL-bad' percentage in a non-hopping network, while RXQUAL 6 and 7 are used as the RXQUAL-bad percentage in a frequency hopping network. Below the relation between this bad quality parameter and the dropped call rate has been studied.

A 1/3 reused non-hopping network has been used, with a mobile speed of 3 km/h. The offered network load was equal to 15%. The relationship between the dropped call rate and the bad quality can be seen in Figure 56, where a linear regression of all the samples is also included as a straight line. It can be seen that when the RXQUAL-bad percentage increases, the dropped call rate also increases.

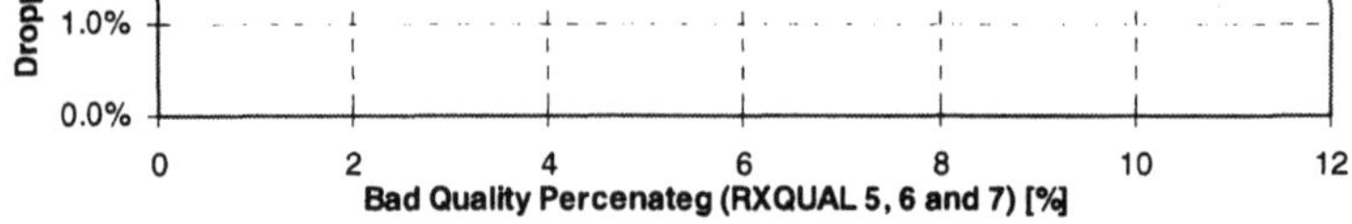

Figure 56. Relationship between bad quality (using RXQUAL 5, 6 and 7) and dropped calls when simulating a non-hopping network in CAPACITY.

The same has been done for a frequency hopping network. Again a 1/3 reused and a mobile speed of 3 km/h has been used, while the offered load was equal to 30 %. It is seen in Figure 57 that the amount bad quality (using RXQUAL 5, 6 and 7) is not such a good indicator for the dropped call rate as in the non-hopping situation.

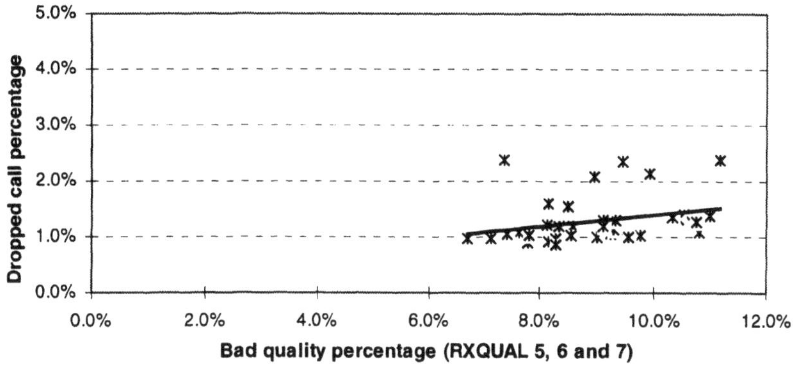

Figure 57. Relationship between bad quality (using RXQUAL 5, 6 and 7) and dropped calls when simulating a frequency hopping network in CAPACITY.

3.2 Live Network Measurements on Dropped Calls versus RXQUAL

The dependency between the dropped call rate and the RXQUAL-bad percentage has also been studied with real network data. Both a hopping and a non-hopping network have been used.

The measurements have been carried out using 30 cells in respectively Copenhagen and Århus, in October 1998. The measurements where carried out on a daily basis using busy-hour measurements over a period of two weeks.

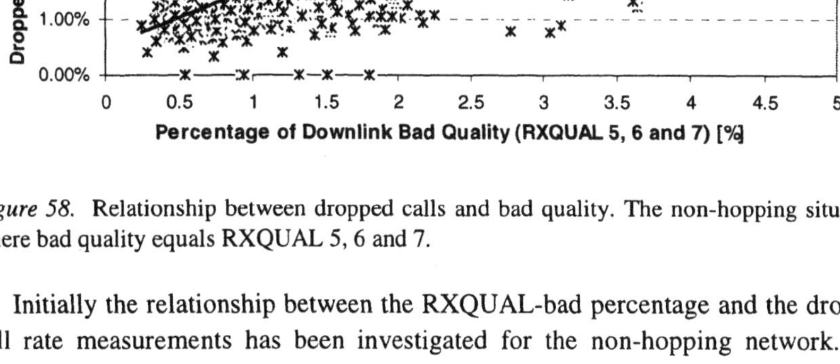

Figure 58. Relationship between dropped calls and bad quality. The non-hopping situation, where bad quality equals RXQUAL 5, 6 and 7.

Initially the relationship between the RXQUAL-bad percentage and the dropped call rate measurements has been investigated for the non-hopping network. The result is shown in Figure 58. As was the case in the simulation results the dependency is quite good. In other words the RXQUAL-bad percentage can be used as indicator of the amount of dropped calls.

The same situation has been studied in a frequency hopping part of the network. As stated above, with frequency hopping usually only RXQUAL values 6 and 7 are used as bad quality, however here the bad quality has been calculated in three different ways. Initially bad quality is calculated based on RXQUAL 7 only. Then bad quality is found using both RXQUAL 6 and 7, which is shown in Figure 59 b, and finally bad quality is calculated based on RXQUAL 5, 6 and 7, as in the non-hopping case. This result is shown in Figure 59 c.

In the later two cases it is seen how the linear regression has a slope, which is quite a lot smaller than for the non-hopping network. That is, the RXQUAL-bad percentage in a frequency hopping network as indicator of the amount of dropped calls is not as good as in a non-hopping network. When only using RXQUAL 7 the slope is steeper, see Figure 59 a. Concerning Figure 59 a great care should be taken in drawing any conclusions due to the limited amount of samples.

One other conclusion can be drawn from these results: RXQUAL 5 actually is acceptable quality, when using frequency hopping. Even part of the RXQUAL 6 measurements can still be considered as good quality.

These measurements supports the fact that one should be very careful in using the RXQUAL measurements in a frequency hopping network.

3.3 FER on the SACCH versus FER on the TCH

The FER on the SACCH is, as described in Chapter 3, used as the `failures_on_radio_interface` counter in the live network and is therefore related closely to the dropped call ratio in the live network. In the same way the FER on the traffic channels describes the subjective speech quality experienced by the mobile user. The relationship between the two has been studied in CAPACITY. A 1/3 reused network and a mobile speed of 3 km/h has been used, while the offered load was equal to 30 %. The FER values are averaged over 9.6 seconds.

The result is shown in Figure 60 where a high dependency between the two types of FER is seen. Using the linear regression also included, it is seen that the FER on the SACCH is in average higher than the FER on the TCH. This reflects the difference in coding and interleaving of the two types of channels. The correlation coefficient of the samples have also been found to a value of 0.80, i.e. a very high correlation between the two was found.

This means that the failure rate of the control channel is at least as high as the failure rate on the SACCH. This can lead necessary repetition of control information send on the SACCH, such as neighbour measurements and power control commando's. It should be noted though that the SACCH is used for non urgent control information.

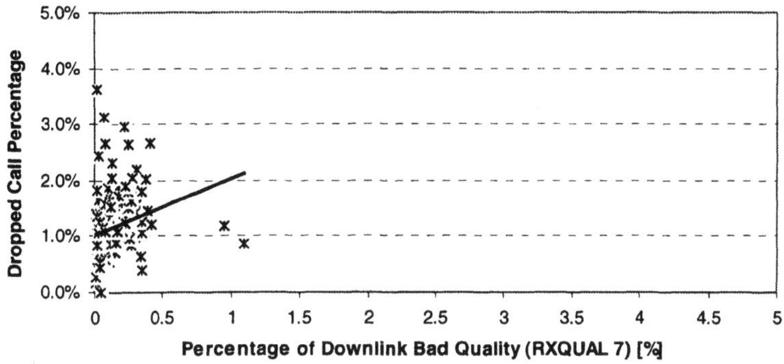

a).

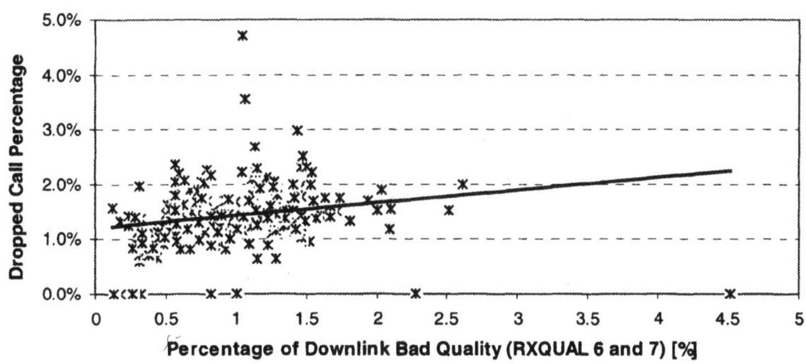

b).

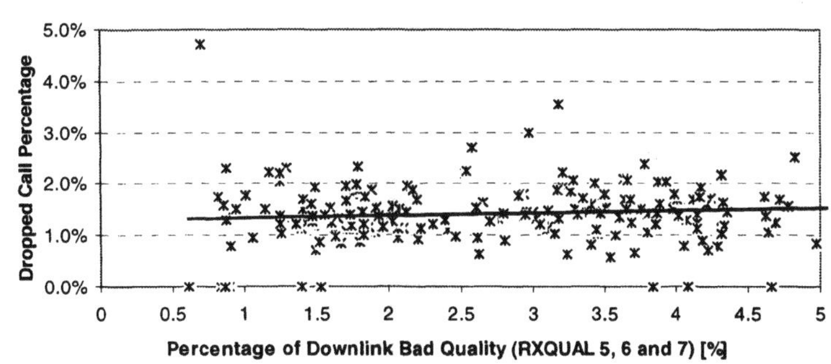

c).

Figure 59. Relationship between dropped calls and bad quality. The hopping situation only using RXQUAL 7 as bad quality (a), using RXQUAL 6 and 7 as bad quality (b) and using RXQUAL 5, 6 and 7 as bad quality (c).

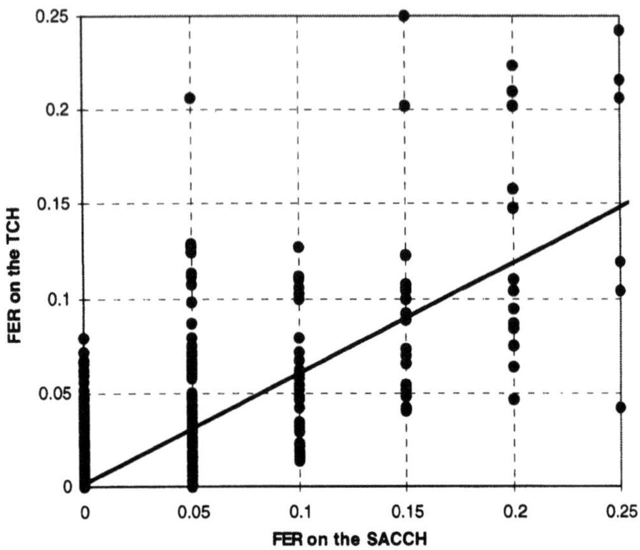

Figure 60. Relationship between the FER on the SACCH and the FER on the TCH when simulating a frequency hopping network in CAPACITY.

4. USING FREQUENCY HOPPING IN BAND LIMITED ONE LAYER NETWORKS

As indicated in the previous parts of this chapter, it is by hopping over as little as 3 to 4 frequencies, possible to achieve a large part of the frequency diversity gain. On the other hand, if hopping over only 2 frequencies, it can result in degradation, when compared to the non-hopping case, where the intra-cell type of handover can be used.

In some situations it can be difficult to apply frequency hopping successfully, since 3 to 4 frequencies per cell may not always be available. A method for overcoming this problem, with quite small bandwidth and sectorized sites, is presented in this section. The following description is based on the article [176].

4.1 The Basic Problem

To describe how relevant (to a typical European GSM operator) the problem of only having narrow bandwidths available, 2 network situations where the problem exists are initially given. They are:

CASE 1: At country cell borders each network operator has to co-ordinate the frequency allocation with every influencing operator in the region, in order to avoid network quality degradation due to adjacent/co-channel interference. This includes national as well as international operators. Therefore the spectrum available to each operator is divided into subgroups. One group of frequencies not to be used in the border areas and another with frequencies that can be used. These latter frequencies are denoted preferred frequencies. Depending on the amount of country neighbours, the number of preferred frequencies is limited to a fraction of the total spectrum of the operator. In several Nordic countries the number of preferred frequencies is limited to 18 or even less. With only 18 frequencies it can be difficult to apply FH successfully.

CASE 2: Another situation, in which it can difficult to exploit the gain from FH, is in traffic dense urban areas, where it is necessary to exploit both micro and macro cells. In order to accommodate the required traffic density (Erlang/km^2) micro cells are used and a splitting of the available frequencies into two groups is often used. Such a splitting of the spectrum is shown in Figure 61, using a spectrum bandwidth corresponding to a typical European GSM 900 operator of 7.5 MHz (equivalent to 37 frequencies).

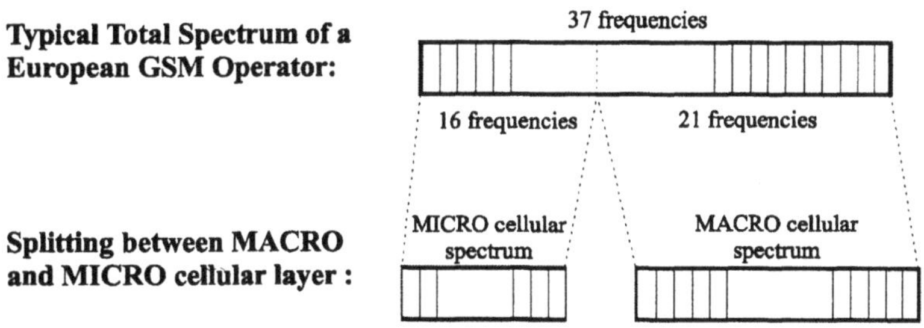

Figure 61. The available spectrum after having divided the total spectrum in a micro and a macro cellular layer.

In the case of a two layer network, typically a minimum of 16 channels is used for the micro cellular layer (corresponds to having 2 TRX's per micro cell base station), leaving 21 frequencies for the macro cellular layer. With as little as 21 frequencies it is hard to achieve a satisfactorily gain from frequency hopping, i.e. to hop over a minimum of 3 frequencies. Two fundamentally different methods, MAIO-management and using a lower frequency reuse (as e.g. 1/1 or 1/3) with load admission control, which both can exploit the gain from frequency hopping in situations like the two described above, are compared.

4.2 The MAIO-Management Concept

In the following, CASE 2 with a network of 21 frequencies is used for illustration. That is, only the macro cellular layer is considered. With a total of 21 frequencies, if considering baseband frequency hopping, no more than 2 TRXs can be allocated per cell in the macro cellular layer,[14] see Figure 62. Hence, only 2 frequencies can be used in the FH hopping sequence.

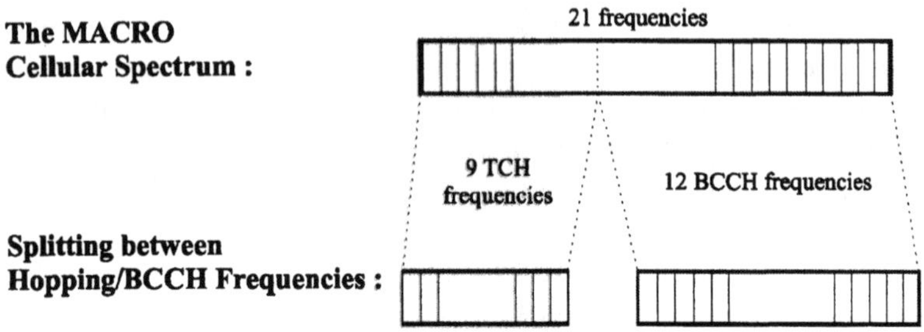

Figure 62. Splitting of the macro cellular spectrum in CASE 2.

Some GSM network field trials have indicated that baseband frequency hopping using only 2 frequencies may not be a lucrative network feature. It has to be ensured that both of the two frequencies always have a good quality, since the frequency and interference diversity gain from frequency hopping is relatively small. Therefore, instead of using baseband frequency hopping over two frequencies, two alternative approaches both applying synthesised frequency hopping, can be taken. They are:

1. Besides the required BCCH frequency (with a reuse of 4/12) allocate the remaining hopping frequencies using a tight frequency reuse pattern, as e.g. 1/3 and limit the co-channel interference in the network using a load limiting technique. This technique will here be denoted soft capacity.
2. Along with the required BCCH frequency, use the one TCH frequency channel (using a 3/9 reuse pattern) allocated to each of the three sectors orthogonal in all three cells. This could quite easily be implemented by giving channels in the different sectors of a site the same HSN, but a different MAIO [62]. This technique is denoted MAIO-management.[15]

In solution 1) a frequency reuse pattern of 1/3 is applied to allow 3 frequencies to be included in each hopping sequence in each cell (the BCCH is not included in the hopping sequence). Network simulations have shown that for a 1/3 reused network configuration, the average load on each freqe uency channel may not exceed 30-35% [238]. This load limitation can be achieved using two different approaches:

- Allocate 1 TRX hopping over 3 frequencies, i.e. a *hard limitation* of 33 %.

[14] One frequency has to be used as BCCH (reuse 4/12) and since nine channels remains and the minimum frequency reuse of the TCH channels is 3/9 [69], only one more frequency can be allocated per cell.

[15] Nokia Telecommunications has originally proposed the MAIO-Management feature.

- Allocate 2 or 3 TRXs in every cell, and exploit a software algorithm to decide whether a free channel can be used or not. Such an algorithm is based on an estimation of the quality and is often denoted a *soft capacity* load limiting technique [20].[16]

Simulations have shown that the use of a soft capacity limitation technique may improve trunking efficiency as it better serves peak traffic in a cell. However in practice this should be compared to implementing alternative existing GSM features such as directed retry and traffic reason handover [66], which even requires less TRX's in the network.

Solution 2) has not previously been investigated.

If using synthesised frequency hopping and a common controller for all sectors of a base station, there are no GSM system limitations on allocating the same frequency group to TRX's in different sectors. The MAIO-management will prevent a frequency from being used in more than one of the sectors at the same time. By using MAIO-management, it is thus possible to hop over 3 times as many frequencies (for 3 sector sites). In case of only one available frequency channel per cell (apart from the BCCH frequency) the situation changes from not being able to apply FH at all, to being able to hop over 3 frequencies in the cell, with each frequency channel loaded no more than 33% in each sector. From the basic functionality it is obvious that MAIO-management can only be applied to sectorized sites.

To clarify the functionality, an example is shown in Figure 63, where 6 frequencies are used on each site, i.e. each sector having 2 TRX's of which one is the BCCH frequency (b_4, b_5, b_6). For each TDMA frame the frequency of the TRX not carrying the BCCH channel is shifted in a cyclic (for simplicity) manner. In practice random frequency hopping should be used. In time instant t = 0, sector 1 uses the frequencies b_1 and f_4. In the next TDMA frame (t = 1) the frequencies have switched to b_1 and f_5, and to b_1 and f_6 at t = 2.

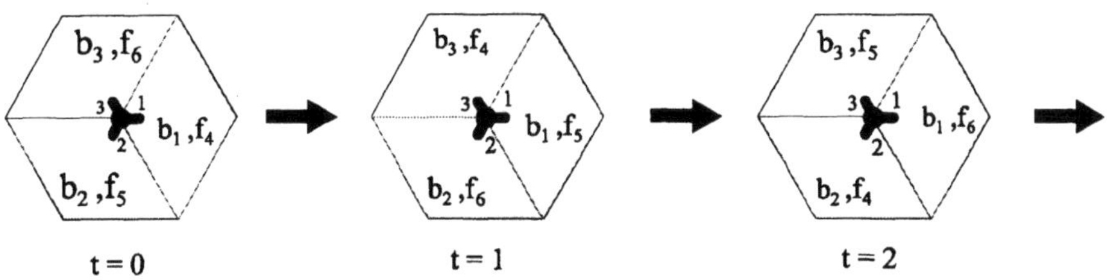

Figure 63. Example of the frequency allocation when using sequential MAIO-Management.

[16] The soft capacity technique is sometimes also referred to as admission load control.

Despite now having 3 frequencies to hop across (f_4,f_5, f_6), and thereby getting the desired frequency and interference diversity gain, the frequency planning has hardly changed.

The penalty of applying MAIO-management shall be found in the co-channel interference introduced, since the reuse has become worse. By turning the frequencies over the 3 sectors of a site, the reuse corresponds to the reuse of a omni-directional site. However since only one frequency can be used at a time in one of the three sectors, the maximum load is 33%. Therefore, from an interference point of view, the 3/9 reuse becomes a 3/3 reuse loaded up to 33% by using MAIO-management. When using a 90% area outage, the *C/I* of a 3/9 reused network is 7.0 dB, while it is 2.3 dB for a 3/3 network [66]. Limiting the load to a maximum of 33% adds a 4.8 dB gain and compensates completely for the loss of having a worse reuse.

4.3 Soft Capacity versus MAIO-Management

In the following only the hopping frequencies are considered, i.e. the BCCH frequencies are left out. If including log-normal fading and using a 90% area outage, the *C/I* for the 1/3 reuse is -1.3 dB while 7.0 dB for the 3/9 case [66]. That is, if assuming 100 % load in both situations a difference in *C/I* (90% area outage) of 8.3 dB between the two. This means that the 1/3 reused network can not be loaded as much as the 3/9 network, if the quality is to be the same. So 8.3 dB has to be won by a lower load and by the gain from being able to hop over 3 frequencies.

As described in the previous section, in the MAIO-Management case, going from 3/9 to a 3/3 reuse gives a loss of 4.8 dB, which was fully compensated by just having a lower load (33%). On top of that a gain from frequency hopping is achieved.

However, such an evaluation of MAIO-management versus the soft capacity concept, based on simple *C/I* distributions assuming ideal clusters may not be accurate. Using MAIO-management or soft capacity in combination with DTX, the maximum load on each frequency in a sector will be approximately 16%. Such a low traffic load causes additional interference diversity from frequency hopping due to the coding/interleaving and soft detection in GSM. This gain may be different for the two solutions because of the difference in *C/I* distribution. Therefore network simulations, using CAPACITY, have been carried out to a better performance estimation.

From a practical point of view, a large advantage lies in the frequency planning procedure when using soft capacity. It is quite simple, since the frequencies are simply divided into 3 groups and then (all of them) allocated to each site.

4.4 Network Simulation Results using CAPACITY

The two cases described in the introduction have been simulated for a mobile speed of 3 km/h and 50 km/h. Therefore, the potential capacity gain using an available spectrum of 4.2 MHz is considered for both of the two scenarios:

1. Synthesised frequency hopping with a frequency reuse pattern of 1/3 and 33 % load limitation using a soft capacity load limitation technique, while hopping over 3 frequencies. The frequency reuse on the BCCH frequency is kept at 4/12.
2. Synthesised frequency hopping using MAIO-Management with a reuse pattern of 3/9 on the hopping frequencies and 4/12 on the BCCH frequency.

Furthermore, for comparison, the quality of a non-hopping network is simulated in all cases as well. Discontinuous transmission is excluded in the simulations. The traffic load used in all simulations is equivalent to 30% of the available traffic channels. For simplicity, the BCCH frequencies are not simulated. The results are presented using the cumulative frequency of the FER of each mobile station reported once per SACCH frame.

Two sets of simulations have been carried out. At first, simulations with a mobile speed of 3 km/h, where the frequency diversity gain from frequency hopping is expected to be relatively large (compared to the non-hopping case) are carried out. These results are shown in Figure 64. With MAIO-management the 1% outage is reached at a FER of about 0.04, i.e. 1% of the reported FER values are worse than 0.04. The performance of both soft capacity and the non-hopping network is worse. The 1 % outages are reached at a FER of respectively 0.10 and 0.20.

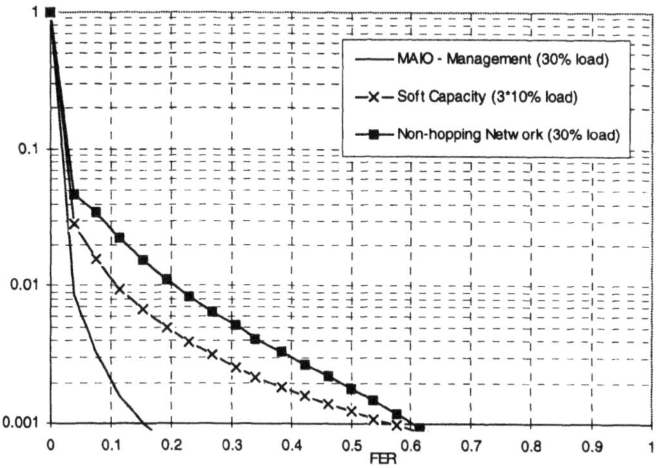

Figure 64. Cumulative frequency of the FER on traffic channels at 3 km/h.

Secondly simulations with a mobile speed of 50 km/h are carried out. The results are shown in Figure 65. At the 1% FER outage soft capacity is performing the worst. With MAIO-management, 1% of the FER measurements are worse than 0.10, while for the soft capacity network 1% of the measurements show a FER of 0.25 or worse. In the non-hopping case the FER is 0.20 or worse in 1% of the cases.

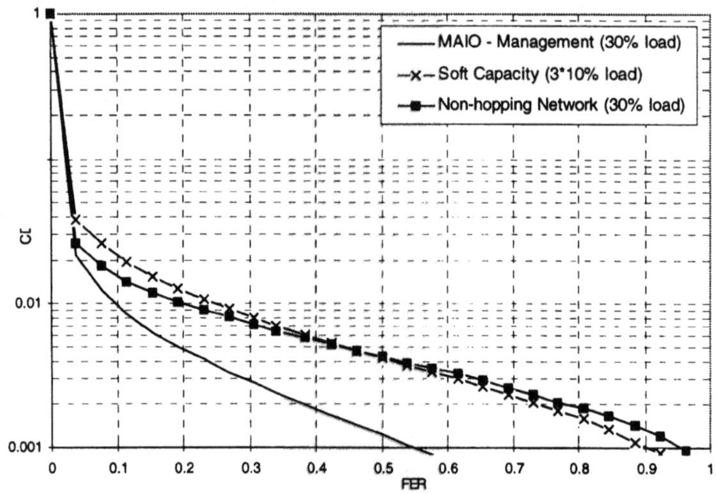

Figure 65. Cumulative frequency of the FER on the traffic channels at 50 km/h.

Therefore, also with the mobile speed moving at a speed of 50 km/h, the MAIO-management seems to give a better quality than soft capacity or a non-hopping network. When comparing the simulations of slow and fast moving mobiles, the quality is degraded with increased speed, as was the case previously.

Soft capacity seems to perform better than non-hopping in the case of slow moving mobiles, while it seems to be worse than non-hopping, when the mobile speed is relatively high. Furthermore, soft capacity has the advantage that it may be capable, as mentioned before, of solving local trunking problems.

4.5 Concluding Remarks on MAIO-Management

The network quality, in terms of FER, has been evaluated for two frequency reuse concepts, which allow utilisation of slow frequency hopping at very limited bandwidths. In the evaluation the case of a three sectorized base station configuration with 2 TRX's per sector, where one TRX carries the BCCH and only the second TRX is hopping, is considered.

The concept of soft capacity by using a dense frequency reuse of 1/3 is compared to the concept of MAIO-management using a nominal frequency reuse of 3/9. Both concepts allow 3 frequencies to hop over.

The penalty of a 1/3 soft capacity concept is the very dense frequency reuse, which potentially worsens the mean co-channel interference level. A mean load factor of 33% has been used to compensate for this problem.

The potential advantage is the gain from both frequency and interference diversity by applying frequency hopping.

From interference and frequency planning point of view the MAIO-management changes the nominal frequency reuse of 3/9 for a sectorized cell configuration into an omni-directional cell configuration of 3/3 with 33% relative load.

The simulation results show that the concept of 1/3 soft-capacity with 33% relative load gives worse network quality than a non-hopping network configuration using a reuse of 3/9. However, the soft-capacity feature may have the potential of improved trunking efficiency if installing more TRX's.

In contrast to the 1/3 reuse, the 3/9 MAIO-management concept gives much improved performance over the non-hopping 3/9 network configurations.

The MAIO-management feature seems to be an attractive solution for achieving the benefits of frequency hopping at low bandwidths, without the drawbacks of increased mean interference level which is the case for soft capacity with a 1/3 reuse.

Continued studies using live measurements have also shown that with larger bandwidths the MAIO-Management feature will also improve the network quality. The feature enables a possibility of completely removing adjacent channel interference between neighbouring cells on the same site, and this has proven to improve the performance.

5. EXPLOITING FREQUENCY HOPPING IN A LIVE NETWORK

To enhance the network performance of the SONOFON network, it was in the spring of 1997 proposed to use frequency hopping as a capacity/quality enhancement feature. Up to that time the intelligent underlay overlay (IUO) algorithm[17] had been used with great success, to gain capacity. This section gives a brief summary and describes the essential results achieved from the trial. All documented details are found in [169].

[17] The intelligent underlay overlay algorithm is treated in detail in Chapter 9.

5.1 Introduction

When doing live tests the one important thing to remember is that customers always have to be able to use the network during the test. Therefore, extreme network degradations are not acceptable. This was especially relevant for this trial, since FH was tried out in the most dense tele-traffic region of the network.

5.1.1 Aim of the Trial

The main purposes of the frequency hopping trial were:

1. Learn about the practical side of frequency hopping live network. How to optimise relevant parameters and features, while finding the general pros and cons.
2. Successfully substitute IUO with frequency hopping, while attempting an increase in network quality and no loss in network capacity.
3. Prepare for further network performance enhancements where other features, such as downlink power control and discontinuous transmission (DTX), are combined with frequency hopping.

The third item it not treated in this chapter, since the issue of power control and DTX in frequency hopping GSM networks is studied in Chapter 7.

5.1.2 Procedure of the Trial

The introduction of frequency hopping by itself was carried out over several steps, involving continuous network optimisation over a period of around half a year. Initially a functionality test using a limited, non-dense tele-traffic, region (the centre of the Danish Island Fyn) was carried out. Here the functionality of random as well as sequential baseband frequency hopping was carried out, trying different frequency planning strategies using 9 3-sector base stations. Having confirmed the proper functionality, the trial could continue with the primary test using a larger region of high tele-traffic density. This primary part of the test was carried out in the centre of Copenhagen. The complete test region, in the future denoted the large area, consisted of 179 cells (601 TRX's). Within this large area, a smaller cluster of 61 cells (212 TRX's) was chosen for the detailed frequency hopping study. This smaller area will from here on be denoted the small area. The reason for dividing the region in two regions was to be able to control the border region of the small area where FH were studied. A draft of the two regions is shown in Figure 66.

Baseband FH was used. In the small area all cells have at least 3 TRX's, avoiding the potential problem that could arise by hopping across only 2 frequencies. A frequency plan, with an average frequency reuse of approximately 13.3 for the large area and 12.9 for the small area, were used.

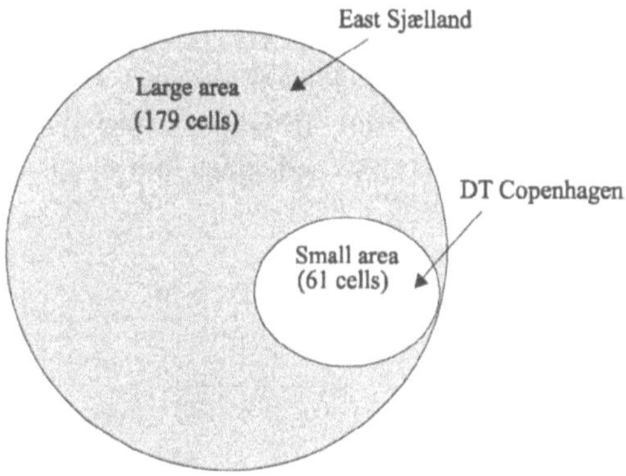

Figure 66. Illustration of the two frequency hopping regions of the field trial in Copenhagen.

5.2 Frequency Hopping Trial Results

The trial measurements described in the following where carried out over a period of 6 weeks. Throughout this period the network load was almost constant when looking at the daily busy hour measurements.

The 1st of August a completely new frequency plan was introduced in the entire large area. Initially this frequency plan was used without frequency hopping to gather statistics of a non-hopping network. Random frequency hopping were then activated on the 7th with no other changes to the network. Two effects were seen immediately: the number of handovers increased with about 25%, while the RXQUAL distribution changed. In Figure 67 it is seen how the percentage of all the RXQUAL values ranging from 1 to 7 increase for the frequency hopping case. This also includes the parameters used to describe bad quality (RXQUAL 6 and 7) in the network.

Note! Since the network has significantly higher probability of RXQUAL 0 than all the remaining RXQUAL values, a separate axis in Figure 67 has been devoted to RXQUAL 0 (primary axis). The secondary axis is the reference for RXQUAL 1-7. There were 2 reasons for these changes:

1. The quality thresholds for the handover algorithm were kept the same. An RXQUAL value equal to 5 corresponds to good quality in a FH network, while it is just acceptable in a non-FH network. Therefore the related handover thresholds should be shifted something like 1 RXQUAL value. Since this has initially not been done, more handovers are created. The RXQUAL distribution changes to the middle by frequency hopping, since the quality of the hopping frequencies are averaged out. For example a cell with two frequencies with

RXQUAL values equal to 6 and 0, give an averaged RXQUAL value of 5 in the FH case.

2. The intra-cell handover functionality was still being used, while there is little point in using it for the FH network. When an intra-cell handover is made, the frequency is changed, but since hopping is used, the frequency already changes for every burst. On the other hand, an intra-cell handover to another timeslot might improve the network in some situations, but this issue is not considered further.

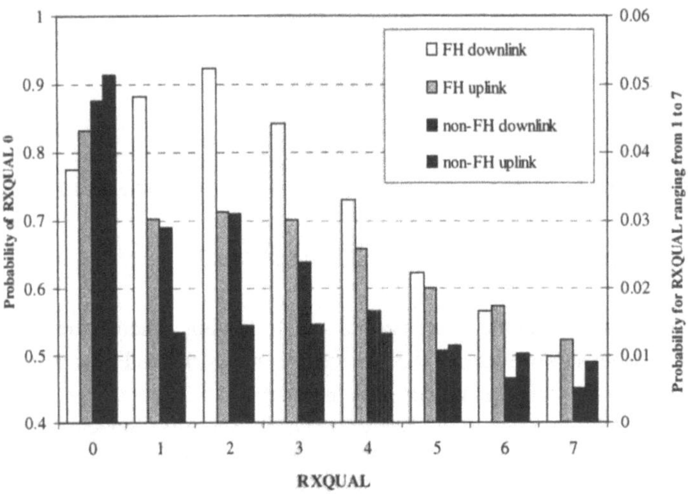

Figure 67. RXQUAL distribution of the small area with and without FH. No other changes have taken place.

Three things were changed:

- The quality thresholds for the handover algorithm were shifted (from 4 to 5) to overcome the problem described above.[18]
- The intra-cell handover functionality was turned off.
- A 2 dB handover hysteresis was introduced, since it was discovered that no hysteresis was being used. This was not done to cope with any effect from FH, but as a general network improvement to avoid handover ping-pong.

The first effect from the changes is seen in Figure 68. The time of the changes is denoted with 'RXQUAL & LEVEL margin'. In the beginning, when going from a

[18] More about the handover algorithm and handover thresholds can be found in Chapter 8.

network with IUO to a non-FH network the average number of handovers per call were reduced from around 2.4 to around 1.3.

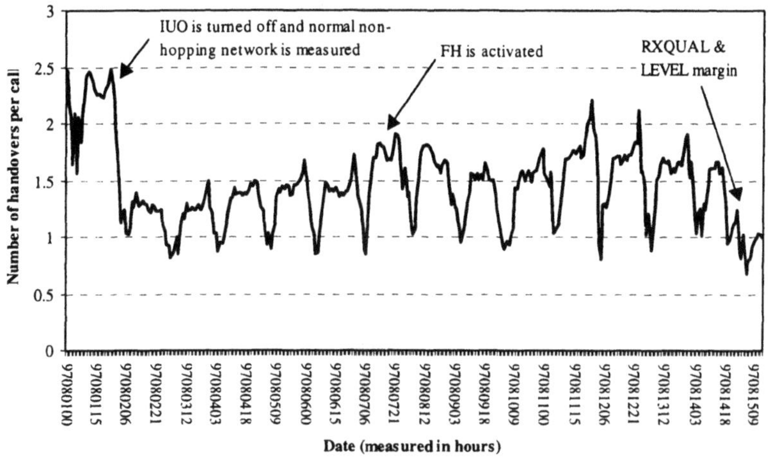

Figure 68. The number of handovers per call during the field trial.

When introducing FH (with no other changes) the number of handovers per call increased a bit around 1.6. After the changes and optimisation the number of handovers per call were reduced again to around 1.1. Just by introducing FH the dropped call rate was also decreased. After the changes, the dropped call rate decreased again. In Figure 69 the change is marked with '2 dB margin on quality'.

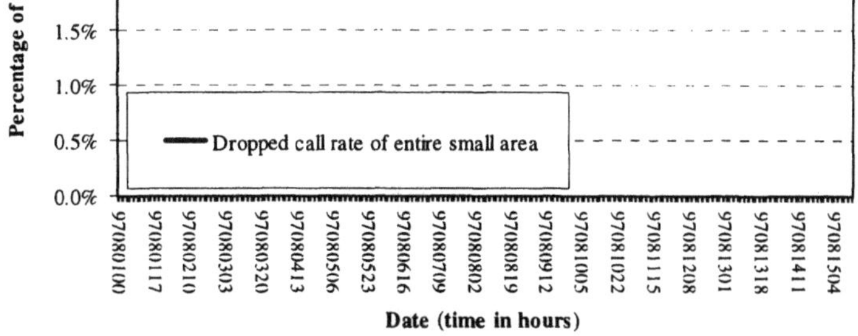

Figure 69. Dropped rate during the field trial. Measurements are from the entire small area.

The RXQUAL distribution of the optimised FH network is shown in Figure 70.

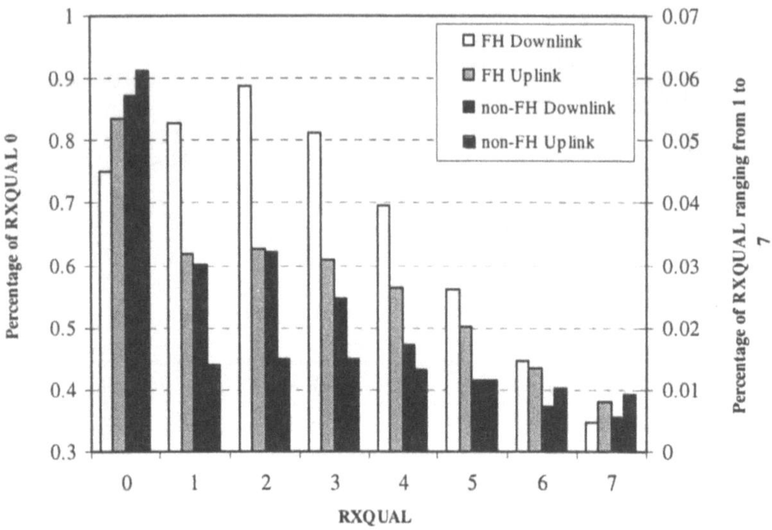

Figure 70. RXQUAL distribution of the small area with/without FH after the optimisation.

It is seen how the FH RXQUAL distribution has changed when compared to the first one, shown in Figure 67. For RXQUAL 7 the non-FH and FH networks are almost identical, while for RXQUAL 1 to 6 the FH network has still got higher values, as in Figure 67.

The following formula has been used to estimate an overall average BER for the network before and after the introduction of FH:

$$\text{Average\_BER} = \sum_{i=0}^{7} \text{BER}_{\text{RXQUAL}_i} \cdot \text{Probability\_of\_RXQUAL}_i$$

This average BER is calculated for each of the four conditions in Figure 70. The parameter $\text{BER}_{\text{RXQUAL}_i}$ refers to the nominal value specified in the GSM standard for each RXQUAL interval [65]. From the statistics used for Figure 70 the following four average network raw BER were found:

FH network, downlink	:	**0,6 % BER**
Non-FH network, downlink	:	**0.4 % BER**
FH network, uplink	:	**0.6 % BER**
Non-FH network, uplink	:	**0.5 % BER**

Frequency hopping does not effect the raw BER, so if only frequency hopping was introduced, these number should be the same. However in the non hopping network, the intra cell handover was used, plus the network was re-tuned by changing some of the handover parameters, so the difference in BER reflects these changes. However the BER does not reflect the actual speech quality.

Test drives have been made to compare the FER in the non-hopping and hopping network. These test drives showed that the FER in a frequency hopping network is lower than in a non-hopping network. An absolute gain was not found, since this would require more test drives.

Continuous optimisation, better frequency planning, antenna tilting etc. were carried out for the next couple of months, along with ongoing site acquisition. The dropped call rate where reduced to around 1.3 % for the small area.

5.3 Summary on Live Frequency Hopping Trial

The introduction of random baseband FH in a live network has been carried out. The aim has been to learn about frequency hopping from a practical side and find the advantages and disadvantages. Several observations, itemised below, were made.

Concerning the Functionality:

- When using FH, the intra-cell handover should be turned off, since there is little purpose in making intra-cell handovers in a FH network.
- The RXQUAL handover trigger should be switched from RXQUAL 4 to RXQUAL 5 in order not to increase the number of handover per call.

Concerning the Results:

- When introducing frequency hopping the RXQUAL distribution changes. Less RXQUAL 0 occurs, while more RXQUAL 1 to RXQUAL 6 has been measured. For the optimised network less RXQUAL 7 is seen for the FH case than for the non-FH case. That is, even when looking at RXQUAL the network has improved.
- By introducing frequency hopping an immediate decrease in the dropped call rate is seen. When optimising the handover algorithm, the rate decreases even further. In the SONOFON case the dropped call rate was decreased by one third from around 3.1% to around 2.1%, simply by introducing FH. Further network optimisation has over the following couple of months reduced the dropped call rate to approx. 1.3 %.
- A quality loss (measuring using RXQUAL) is seen, when comparing the raw BER before and after introducing frequency hopping. This is caused by the re-tuning and removal of the intra-cell handover functionality.

- It was found that a 2 dB handover hysteresis decreases the number of handovers. This is not only the case for FH, but also for non-hopping networks.

6. SUMMARY AND CONCLUSIONS

This chapter has dealt with the influence of frequency hopping on the system performance of a GSM network. At first the capacity limitations of a GSM system were described as coverage, soft and hard blocking.

Network simulation results, which will be used as reference for the rest of the book, have been performed. They showed that a 1/3 reused network is soft blocked, while a 3/9 reused network is hard blocked. It was also shown that a general increase in network quality is seen when going from fast to slowly moving mobiles. This is due to the macro diversity gain achieved at low speeds. Two different network topologies were compared. One with 90° antennas and another one with 65° antennas. It was shown that the network using the 65° antenna topology gives better quality than the network using the 90° antennas. For historical reasons it was chosen to use the 90° antennas topology in this book.

It was shown that the RXQUAL-bad percentage is quite a good indicator for dropped call rate in a non hopping network, but not for the frequency hopping network.

A frequency hopping solution for networks with a limited amount of spectrum was presented, called MAIO-Management. Simulations showed that this feature gives better quality than a 1/3 reused network and a non-hopping network.

Finally at the end of this chapter the results from a trial of the introduction of frequency hopping in a live non-hopping network was described. It was shown that some adjustments to the handover algorithm had to be made to get satisfactorily performance. The dropped call rate decreased from 3.1% in the non-hopping network to 1.3% after the introduction of frequency hopping, adjusting the handover algorithm and re-tuning the frequency plan.

Chapter 7

POWER CONTROL AND DTX IN A FH GSM NETWORK

The minimum effective frequency reuse can be determined from the level of interference and is directly proportional to the available network capacity. Traditionally networks are designed with an overall network strategy of allocating the available resources, where the distribution of the traffic requires it. In practice the fact that the environment and the traffic distribution is not the same for every square km leads to non-uniform cells. Omni-directional sites are used aside of sectored sites. Cells have different sizes, different antennas, use different tilting, are placed on different heights and so on. These network irregularities can cause problems in terms of how to handle the interference within the network. As illustrated in Chapter 2 one way of dealing with the interference problem is by using power control. Power control enables dynamic regulation of the transmitter output power.

Another effective way of lowering the level of interference is by switching off the transmitter when speech is not present. This functionality is known as discontinuous transmission (DTX). During a normal conversation, each person speaks, on average, for less than 50% of the time. Therefore, if it would be possible to exploit this on all users DTX could also lower the overall level of interference and thereby increase the capacity.

In this chapter power control and DTX as used in a FH GSM network is studied. In Section 1 a general introduction to power control is given, in terms of the basic functionality, previous work and potential gain from ideal power control. Section 2 briefly introduces DTX in the same way as with power control. In Section 3 the downlink GSM power control algorithm is treated. The basic functionality, the optimal settings and some performance enhancement proposals are investigated. Initially a simplified copy of a real (operating) power control algorithm is used to estimate the error committed by power control. Also network simulations have been carried out to see the influence of the power control algorithm in a complete network. This is a typical example of the of the CAND approach offered by developing CAPACITY. Section 3 describes the most essential results from a live test trial carried out of downlink power control as well as of DTX in a live FH network. Both features are compared to the simulation studies also carried out.

Section 0 treats DTX as it is implemented in GSM. The feature is studied on link as well as on system level. Especially the potential capacity gain as well as the data estimation inaccuracy is studied with DTX. The chapter ends with a summary on the essential conclusions on downlink DTX and power control in a FH GSM network, given in Section 5.

1. AN INTRODUCTION TO POWER CONTROL

In order to achieve a capacity gain from power control in systems like GSM, frequency hopping should be used, since this spreads the quality gain from power control over all users. If not, only some of the mobile stations and/or base stations will profit from power control and no overall capacity gain can be assumed.

Advanced mobile systems, as e.g. GSM or UMTS, offer the possibility of uplink as well as downlink power control. Depending on the specific type of system and level of interference, the uplink and downlink power control has different vitality.

For the uplink direction power control can improve the performance in three ways. By equalising the power received at the base station the "near-far" problem, arising from power level differences due to signal fading and path loss, is compensated for. This is primarily important for wide-band CDMA systems. Furthermore, uplink power control compensates the channel fading to enhance the co-channel interference protection and finally by lowering the output transmission power the power consumption of the mobile is reduced.

In the downlink direction the "near-far" problem literally never exist, since the desired signal and the interfering intra cell signals are typically exposed to close to identical distance path loss and fading [144]. Concerning the fading compensation for co-channel protection the downlink power control will enhance the connection. This becomes especially important for future systems where a lot of the traffic will be data on the Internet. In such situations the amount of downlink traffic will typically exceed the uplink by several times, making the downlink the capacity limiting direction.

For GSM the primary objective is therefore to choose the transmitter output power in up- and downlink such that a sufficient transmission quality is just maintained in all communication links. This will maximise the network capacity. The possibility to manage co-channel interference by means of power control, will be the essential topic of this chapter, in particular the effect in a GSM type of system and the power control algorithm used in GSM. Throughout the studies only the downlink is modelled, since it is believed to be the limiting link, in particular for networks with uplink antenna diversity [29,231].

Dimensioning a GSM network according to the level of interference has traditionally been done by ensuring a satisfactorily *C/I* at the cell border, whereas the remaining part of the cell is not considered. However, with power control the surplus quality created within the cell can be exploited. The mobiles closer to the base station will typically experience a better quality, why the output transmission power can be reduced. The idea of this is shown in Figure 71. The mobile experiencing a *C/I* of only 9 dB may have marginal quality while a *C/I* of 20 dB is more than enough.

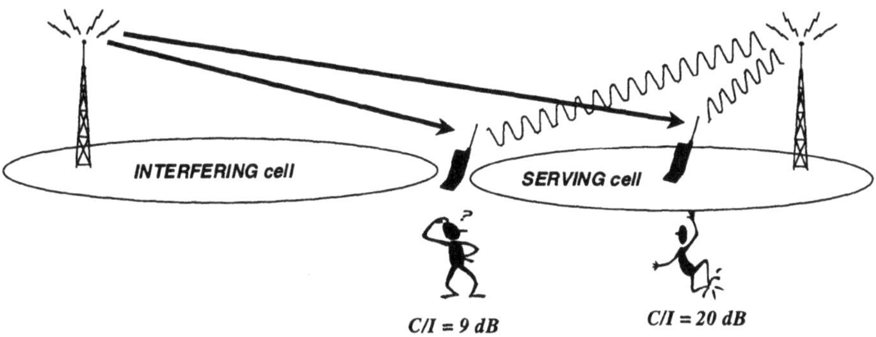

Figure 71. The surplus cell quality that can be reduced and exploited by power control.

1.1 Previous Work Concerning Power Control

Power control related research could be found as early as back to the 1950's in the planning radio broadcast networks. However, it was not until the 1970's, when satellite systems were treated, that analytical approaches were developed [138]. These initial studies where enhanced and generalised for cellular radio systems during the next couple of years. Based on the interest of the early analogue cellular systems several simulation studies of various power control schemes proved during the 1980's potential capacity improvements [69,155,204]. More theoretical studies have followed were an optimum centralised power control scheme, optimum in terms of minimising the interference probability, has been derived [252]. In the literature it is described how both centralised and distributed algorithms with continuous power levels and *C/I* objectives have been developed and their convergence properties have been investigated. Centralised algorithms require a central controller with complete knowledge of all radio links and their power levels in the system. During the last couple of years most of the power control studies have concentrated on the distributed implementation. Computational complexity makes the centralised approach impractical to implement in real time for large systems. All large scale operating commercial types of networks, as GSM, utilise distributed power control schemes, where the power regulation is based on measurements from each individual link. Correspondingly the GSM type of power control algorithm is especially powerful in treating local (in time and place) interference hotspot problems, while it also works on the overall network level. In this thesis only distributed power control algorithms are treated.

The appliance of power control in combination with other network features is also very relevant and has correspondingly been studied thoroughly. Power control along with resource management decisions is a typical combination, but also features like discontinuous transmission and frequency hopping has been combined with power control. A combination with mobile admission control has been treated in [19] whereas power control to enhance mobile removals is studied in [8]. Most of these studies have been of more theoretical character since they have not been linked to a specific system.

1.2 The Potential Gain from Power Control

Basically two different types of distributed algorithms for power control in cellular radio systems have been proposed in the literature. One is based on the principle that the power should be reduced as the path loss decreases. The simplest and mostly widespread of this type of algorithms keeps the received signal strength constant [224], i.e. the change in path gain is ideally fully compensated for. In the second type of algorithm the focus is instead on link quality. Since in situations, where capacity is needed, the users experience interference, it seems natural to use the quality to control the transmission power levels [5]. Furthermore, studies have shown that a large capacity gain can be obtained using a power control scheme trying to provide the same quality to all users [252].

In this section the maximum power control gain using ideal conditions for such two types of power control algorithms is estimated. Two kinds of power control algorithms are studied. Initially in Section 1.2.1 a power control algorithm that compensates fully for the path loss gain is studied, while in Section 1.2.2 a power control algorithm operating based on quality (*C/I*) is treated. The calculations are kept as simple as possible, but provide a comprehensive indication of the potential gain of power control in terms of reduction in output transmission power.

1.2.1 Ideal Gain from Power Control Solely Based on Carrier Path Loss

In this case a simple power control algorithm based on the received signal strength is described. Interference is ignored. The power control algorithm simply turns down the power, if the received signal strength is higher than required. Correspondingly this power control algorithm is completely load independent.

The dynamic power control range is specified to 30 dB (as in GSM) and the path loss model from CAPACITY has been used:

$$L\ [\mathrm{dB}] = K + 10 \cdot \log_{10}(d^{\gamma}) \tag{1}$$

In Equation (1) d is the distance from the base station to the mobile, γ is the path loss slope and K is a constant depending on among other the antenna heights and frequency [134]. The optimum power control functionality is defined to be when the base station transmitting output power is only at maximum on the cell border. If above this level the power should be reduced. In the calculations it is assumed that all links with mobiles within a distance of 1 km to the base station operates with minimum transmitting power, since the path loss model of Equation (1) is too simple to be used near the transmitter.

The situation is shown in Figure 72, where an omni-directional cell (for simplicity) with radius R is depicted, along with an illustration of the path loss.

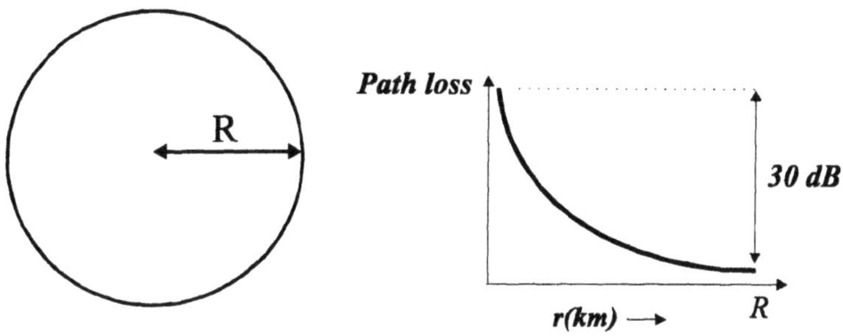

Figure 72. The situation considered with a cell radius, R, equal to the distance where the path loss is exactly 30 dB.

The calculated cell radius corresponding to this situation is depicted in Table 17 using different path loss slopes. It is seen that if e.g. a path loss slope of 3.5 is used, the path loss is 30 dB at a distance of 7.2 km from the base station.

Table 17. Ideal cell radii for different path loss slopes when doing the power regulation based on the simple carrier path loss.

PATH LOSS SLOPE (γ)	CELL RADIUS
2.0	31.6 km
3.0	10.0 km
3.5	7.2 km
4.0	5.6 km

It is assumed that the power control algorithm keeps the received power constant to a certain level. The gain from power control at a certain distance d from the base station is therefore:

$$\text{Gain\_at\_distance\_d} = \frac{1}{10^{-[10\cdot\gamma\cdot\log_{10}(R)-10\cdot\gamma\cdot\log_{10}(d)]/10}} \tag{2}$$

The gain is limited to 1000, since the dynamic power control range has been set to 30 dB. It can be seen that if d is equal to the cell radius R, the gain is equal to 1, which means no gain. If Equation (2) is integrated over the entire cell area (pixel by pixel), the average gain of power control for the entire cell can be found. Using the maximum ranges found in Table 17, the values in Table 18 are found for the average power control gain for the different cases.

Table 18. Ideal average power control gain within the cell, as a function of the path loss slope and using the power control algorithm based on path loss only.

PATH LOSS SLOPE (γ)	AVERAGE GAIN
2.0	3.0 dB
3.0	4.0 dB
3.5	4.4 dB
4.0	4.8 dB

1.2.2 Ideal Gain from Power Control Based on Interference

In this case the power control algorithm is not based purely on the received signal level, but on the experienced *C/I*. The algorithm allows an output power reduction provided the *C/I* is above a certain threshold. The gain is therefore load dependent. The *C/I* threshold is chosen to 9 dB, since this is the min limit allowed in the mobile station according to the GSM requirements [64]. The gain is found with and without shadow fading, starting without shadow fading. A frequency reuse of 3/9 is assumed. When taking only the first tier of interferers (6 interferers) into account, the *C/I* in every location within the cell can be calculated as:

$$\frac{C}{I} = \frac{C}{\sum_{i=1}^{6} I_i} \tag{3}$$

In Equation (3) I_i is the interference power coming form interferer i and C is the carrier power. The power control gain at distance d from the base station can be found as follows:

If $C/I_{distance\_d} > 7.94$ (9 dB):

$$Gain\_at\_distance\_d = \frac{1}{10^{-\left[10\cdot\log_{10}((C/I)_{distance\_d})-9\right]/10}} \quad (4)$$

else

$$Gain\_at\_distance\_d = 1, \quad (5)$$

$C/I_{distance\_d}$ is the C/I at distance d from the base station. The average cell gain of the power control is found from integration across the entire cell area.

The same calculation can be carried out with shadow fading, but here Equation (3) changes, since the interference from the 6 interferers can no longer be modelled as a single value. Here a log-normal distribution is used to describe the fading, see Chapter 5. The resulting C/I can be calculated with the help of [210] or it can be simulated using Monte Carlo simulations. The values shown in Table 19, are made using Monte Carlo simulations. The gain from power control based on interference without shadow fading is also seen in this table as a function of the different path loss slopes.

Table 19. Ideal gain of power control based on interference of a 3/9 network for different path loss slopes with and without shadow fading.

PATH LOSS SLOPE (γ)	GAIN WITHOUT SHADOW FADING	GAIN WITH SHADOW FADING (6 dB STD.)
2.0	0.5 dB	0.5 dB
3.0	2.7 dB	2.0 dB
3.5	4.8 dB	3.1 dB
4.0	7.3 dB	4.4 dB

When comparing the two different scenarios with and without shadow fading, it is seen that with large path loss slopes the effect of including the fading becomes more and more important.

2. A BRIEF INTRODUCTION TO DISCONTINUOUS TRANSMISSION

Idealised simulations, not including short term fading, of DTX, have shown a linear proportionality between the DTX factor and the improvement in network quality, when combined with random frequency hopping [105]. This implies that a

DTX factor of 0.5 ideally reduces the interference as much 3 dB (in *C/I*). As with power control, DTX has in GSM to be used in a FH network in order to convert the interference reduction into an overall network capacity gain. The gain from DTX is invariant to the network load and location of the mobile stations, which, as described above, is different from power control based on network quality. Partly therefore, but also since few configuration parameters are required for specification, the DTX feature is much simpler than power control. Furthermore, when considering frequency planning the interference reduction gain from DTX may, according to the load independence, be included in the planning procedure with less caution than power control.

Uplink DTX has like uplink power control also the ability of conserving the battery power in the mobile station why it is widely used, whereas only a few network operators currently use downlink DTX.

Despite the apparent simplicity of DTX, it has been found important to study the feature in relation to FH GSM networks due to some experienced trial result problems. More about this subject is found in Section 3.5, whereas the general GSM DTX functionality is treated in Section 0.

The research conducted within the subject of DTX is limited, probably due to the simple nature of the functionality. Few studies have investigated the potential effect for system specific networks as GSM. One example is however [208], where a frequency planning principle based on link quality instead of the *C/I* is described. A significant increase is found when including DTX as well as the cell traffic load. A statistical method used for analysing network quality by the *C/I* outage probability is, for generalised Ricean/Nakagami channels and environments [156], presented in [203]. Here an idealised study of the effect from different voice activity patterns has been studied using variable voice activity factors.

3. THE GSM POWER CONTROL ALGORITHM

This section is related directly to the power control algorithm used in GSM. That is, the distributed power control algorithm, with discrete power levels and a limited dynamic range with both signal strength and interference objectives.

3.1 Introduction

As stated in Chapter 3, the GSM system measures periodically the received signal strength (RXLEV) and quality (RXQUAL) of the active communication link. Information on the problem of only having RXLEV and RXQUAL as radio link measurement parameters is described in the following.

A diagram indicating the principle of why and when an increase/decrease of the transmission power takes place in the power control algorithm as a function of respectively the upper and lower thresholds of RXLEV and RXQUAL is shown in Figure 73. In the figure some fixed examples of thresholds have been used. The bold marked text (at two places) indicates which power control regulation cause is of highest priority when multiple different choices can take place. Note that in order for the algorithm to be able to increase the output power at all, it is of course assumed that the maximum output power has is not been reached yet.

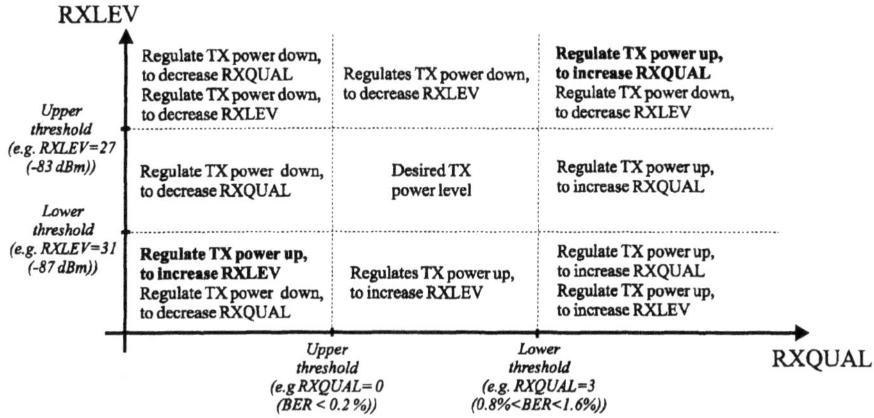

Figure 73. The possible up/down power regulations of the GSM power control algorithm.

Traditionally two different types of power control principles exist; open or closed loop power control. In the open loop type of algorithm one of the communication links is used to decide its transmission power based on the received signal strength or signal quality. Due to the used of network quality based regulation the open loop method cannot be used since the up- and downlink will not experience the same type of interference. One end of the communication link is not able to decide alone whether it should increase or decrease its transmission power based on its own signal quality measurements. The power control algorithm in GSM is a two-way closed loop power control algorithm. In Figure 74 it is shown how the uplink and downlink are regulated separately using a closed loop type of regulation. Principally, in GSM, the measurements reported at time instant t = [n–1] are used to regulate the transmitter output power at time instant t = [n]. Therefore a certain undesired time delay is introduced. For GSM the regulation is based on at least the previous 2 measurements, with one sample each 480 ms. This characteristic makes the GSM power control algorithm a slow type of power control algorithm. The same procedure is used in both uplink and downlink.

Research concerning the performance of the GSM power control algorithm and how to enhance the performance has lately taken place. An example is the stability analysis of the GSM algorithm presented in [119]. An actual enhancement of the GSM power control algorithm is e.g. presented by the introduction of a faster power control regulation [216]. Here an increase in regulation speed of a factor of approximately 100 has been proposed for a specific type of GSM base station, denoted a home base station. Instead of only allowing a regulation each SACCH frame the regulation takes place every burst.

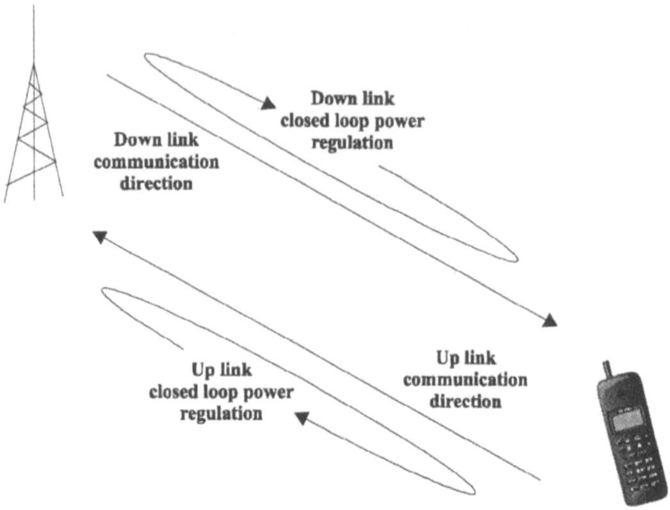

Figure 74. Principle of the two ways closed loop power control algorithm used in GSM.

As stated above, the downlink GSM power control algorithm uses the averaged values of the signal strength (RXLEV-AV) and quality (RXQUAL-AV), to regulate the output power of the base stations or the mobile stations [2]. The operator can specify the size of the averaging windows, within a certain range between 2 and 32 SACCH frames. Additionally the power is regulated according to the following 4 operator specified threshold combinations:

1. **Upper-threshold RXQUAL-AV**: If RXQUAL-AV are lower than the Upper-threshold RXQUAL-AV the output power of the base station is decreased.

2. **Lower-threshold RXQUAL-AV**: If RXQUAL-AV are higher than the Lower-threshold RXQUAL-AV, the output power of the base station is increased.

3. **Upper-threshold RXLEV-AV**: If RXLEV-AV are greater than the Upper-threshold RXLEV-AV, the output power of the base station is decreased.

4. **Lower-threshold RXLEV-AV**: If RXLEV-AV are lower than the Lower-threshold RXLEV-AV, the output power of the base station is increased.

If one of the thresholds in item 2 or item 4 are applicable, the power is always increased (if not transmitting with maximum output power), without looking at the other two thresholds. This corresponds to the bold marked regions in Figure 73.

Each operator can specify the size of the dynamic range, however no greater than 30 dB. The step size of the power increase and power decrease can furthermore be specified individually. The last important parameter is the time interval, i.e. a timer during which one power regulation step can take place.

A lot of other parameters are involved in the implementation of the power control algorithm, however for simplicity it has been chosen only to treat the most important ones. For a more detailed description see [2]. However, one additional functionality should be mentioned. A variable power change step (up) size functionality has been included in the downlink power control algorithm. This functionality allows larger power up regulations if needed. It can be useful in situations where the mobile currently transmitting at minimum output power moves away from line-of-sight and the received signal quality suddenly drops rapidly. The basic idea is then that the BSC can use this functionality to quickly regulate the power up. It is therefore only relevant in situations where the required power change is so large that it would require several power control commands.

3.2 The Simplified Power Control Algorithm

To illustrate the most important functionality's and limitations of the downlink power control in GSM, a simplified model of the algorithm has initially been made. In this model the power regulation is based on the received signal strength, i.e. the output power regulation in the real GSM power control algorithm based on quality has been ignored. A typical scenario, used for illustration throughout this part is shown in Figure 75. Using the path loss model from CAPACITY and transmitting with a base station output power of 34 dBm, the received level at the mobile is -81 dBm.

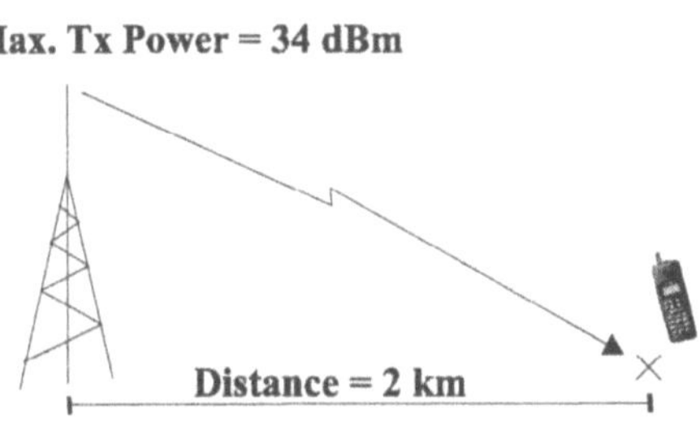

Figure 75. The scenario used as example to describe the simplified power control algorithm.

If applying log-normal fading of the same type as in all CAPACITY simulations, the resulting received signal strength could look like shown in Figure 76. Here the received signal strength both with and without power control is shown. The power regulation is based on the parameter settings shown in Table 20.

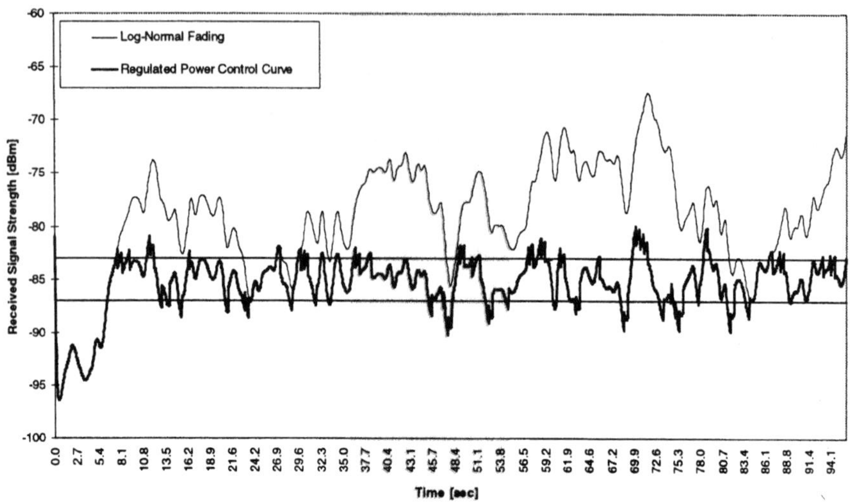

Figure 76. Example of the regulated output power of a typical fading signal as in CAPACITY using the simplified GSM power control algorithm.

Table 20. The parameter settings for the simplified power control algorithm.

PARAMETER	VALUE
Mobile speed	3 km/h
Log-normal fading correlation distance	110 meter
Log-normal fading decorrelation criteria	1/e[1] between samples of correlation distance
Log-normal fading standard deviation	6 dB
Upper Threshold	-83 dBm
Lower Threshold	-87 dBm
Upwards & downwards step size	2 dB
Dynamic output power range	30 dB
Regulation window size	2 SACCH frames

In Figure 76 it is seen how the base station output power is regulated so that the received signal strength at the mobile, is located quite well within the allowed 4 dB range. When the received signal strength at the mobile decreases below -87 dBm the power control increases the base station transmitting power if possible. In the same way if the received signal strength is above -83 dBm, the algorithm decreases the base station output power if possible. In this example it is also observed how the power regulated signal strength curve always lie below the non-regulated signal strength curve. The GSM power control algorithm cannot increase the base station output power to more than maximum, equivalent to no power control at all.

An equivalent 4-hour test drive, carried out on January the 28th 1998, measuring the downlink received signal level of a mobile station in the streets of Copenhagen, has been carried out. Around 16 min. of the signal strength measurements from the test drive is shown in Figure 77, along with the same signal exposed to the simplified power control algorithm described above.

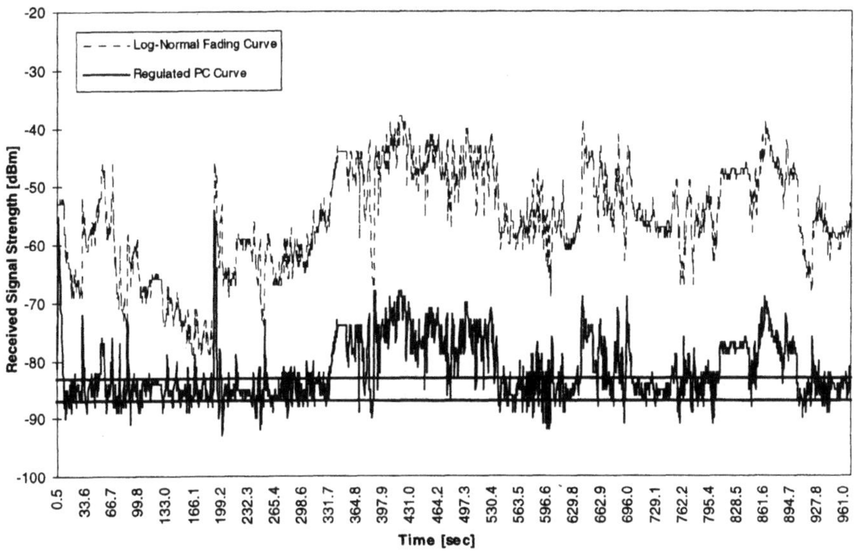

Figure 77. Live signal strength measurements exposed to the simplified PC algorithm.

It is seen how the reduction in transmitting power, equal to the area between the two curves, is very large (in most cases about 20 dB). This is due to the fact that the dynamic range of the algorithm is limited to 30 dB and sometimes, typically in urban areas, a larger range is needed. Therefore two factors limits the performance to be achieved from power control arise. The limited dynamic range and the non-perfect power regulation. The effect of only having a limited dynamic range is treated in Section 3, while the size of the non-perfect power regulation is treated in the following using the simplified power control algorithm.

3.3 Performance of the Simplified PC Algorithm

Concerning the simplified power control algorithm, a related but alternative way of describing the performance (compared to the one used throughout the remaining part of the chapter) is used. Here performance refers directly to how well the algorithm regulates the power. That is, how small is the error committed by the algorithm compared to the ideal perfect regulation, as a function of how large a reduction in output power is achieved.

To investigate theoretically how well the algorithm performs, an extra modification to the power control algorithm has at this point been made. An infinite dynamic range has been allowed, i.e. the power can be increased to more than the transmitting power if necessary. This modification has been made to see how well the algorithm can keep the received signal strength within the regulation range between the upper and lower thresholds. Again the parameter settings from Table 20 have been used, except for the speed of the mobile station, which is simulated for the case of 3 km/h as well as 50 km/h. The error committed by the power control algorithm, is found using the following method:

- The upper and lower level power control trigger thresholds are set equal. The error is described by the absolute difference between the power-regulated curve and the upper/lower level thresholds. With equal thresholds this error can be used to find the overall standard deviation of the error. By calculating the absolute error as the sum of squares of the difference of all samples (once each 480 ms), an estimate of the error variance can be found, if dividing this sum by the total number of samples [205, p.251]. Finally the square root is taken to get the standard deviation of the error (measured in dB) and it is possible to model the error.

Another method using different level trigger thresholds has been used in [164].

The interference reduction achieved by the power control algorithm is also found. It is calculated per sample as the difference between the received signal strength of the non-regulated transmitted signal and the received signal strength of the power regulated signal. It is therefore a measure (in dB) of the reduced transmitting power per sample. This value is summed for all samples and divided by the total number of samples to get one value describing the average interference reduction per sample.

Even though the power control algorithm allows different step sizes (up and down), only equal values are analysed here. Sometimes a greater step value for stepping up is used than for stepping down. The reason for this is that one wants to be sure that the power is turned up quickly enough (to avoid bad quality). However, since the variable power change step (up) size downlink functionality can save a call in those situations, it is more relevant to treat equal values.

Initially the slowly moving mobiles are studied, i.e. the situations where the power control algorithm is believed to perform relatively well.

3.3.1 CASE 1: A Mobile Speed of 3 km/h

Initially a variation of the step size is carried out. As stated above, with equal level triggers the error calculated is correlated to the standard deviation of the error, provided the probability density function of the error is a normal distribution. This study therefore describes the minimum standard deviation of the error (committed by the power control algorithm) for slow moving mobiles. The results are shown in Figure 78.

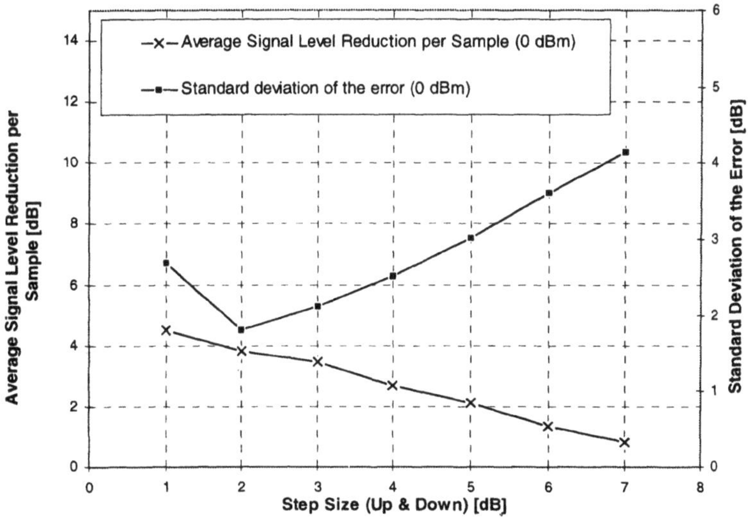

Figure 78. The signal strength reduction and standard deviation of the error when using an upper/lower threshold of -85 dBm and a mobile speed of 3 km/h.

As expected, the average signal level reduction per sample decreases with increasing step size. Furthermore, the error committed increases with increasing step size. Considering both criteria's the optimum situation is therefore found with a step size as low as possible (except for a step size of 1 dB which is not possible in the live network). 2 dB has therefore been chosen.

The probability density function of the error using a step size of 2 dB has been found and is shown in Figure 79. Also, based on the error samples a standard deviation and mean (assuming a Normal distribution) has been estimated. They are in this case found to be a mean of -0.004 dB and a standard deviation of 1.8 dB. This estimated Normal distribution is also plotted in Figure 79.

From the figure it seems like the error is very correlated to the normal distribution. To be sure that the assumptions are correct, the same data are plotted in a Normal distribution probability plot. In such a plot a normal distribution fits a straight line. The result is shown in Figure 80.

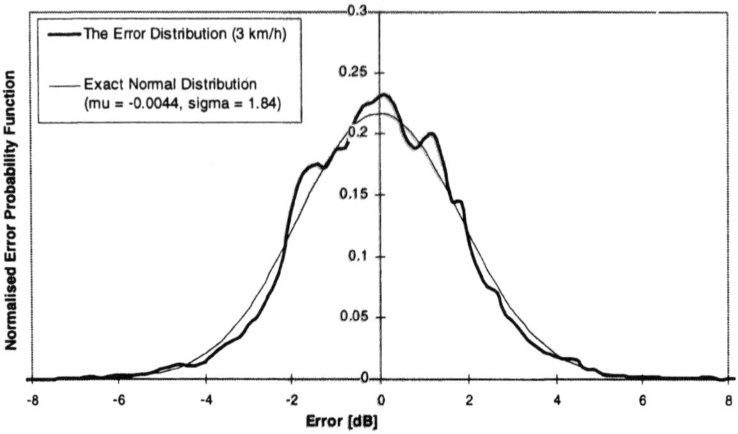

Figure 79. Error probability density function (3 km/h) and the equivalent Normal distribution.

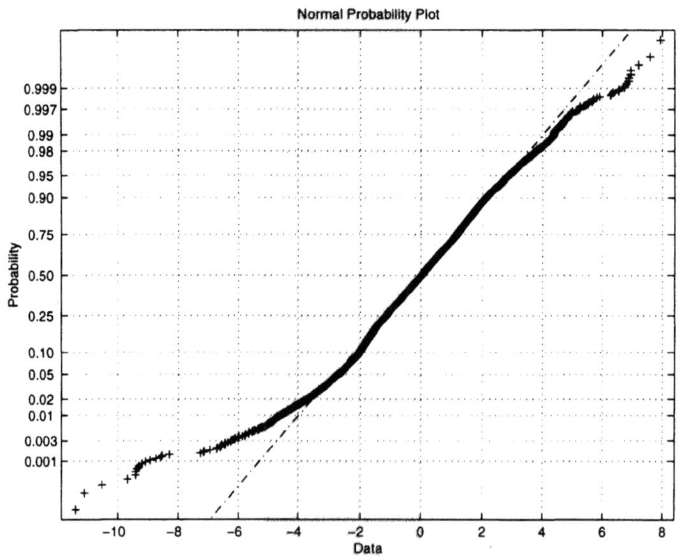

Figure 80. Normal distribution probability plot of the error with a mobile speed of 3 km/h.

In Figure 80 'Data' refers to the error measured in dB, as was the case in Figure 79. It is seen how the data fits a straight line within the probability between 2% and 98% quite well. It is therefore assumed that the error is normally distributed for the case of slowly moving mobiles. This means that the sample standard deviation along with a normal distribution of zero mean can be used as a performance indicator of the simplified power control algorithm, at last for a mobile speed of 3 km/h.

3.3.2 CASE 2: A Mobile Speed of 50 km/h

The faster moving mobiles have also been simulated. As was the case with the slow mobiles, the initial part concerns the step size to find the optimum for both the standard deviation of the error and the signal level reduction. The simulation results are shown in Figure 81. In this situation the standard deviation of the error is much larger. The minimum is found somewhere with a step size between 2 and 7 dB, but it is never below 4.7 dB. Compared to the case of 3 km/h it can be seen that the average signal level reduction per sample is significantly smaller. Based on the 3 km/h estimations, it has been chosen to continue with a step size of 2 dB, since the difference (in standard deviation of the error) in the 50 km/h case between a step size of 2 or 7 dB is quite small.

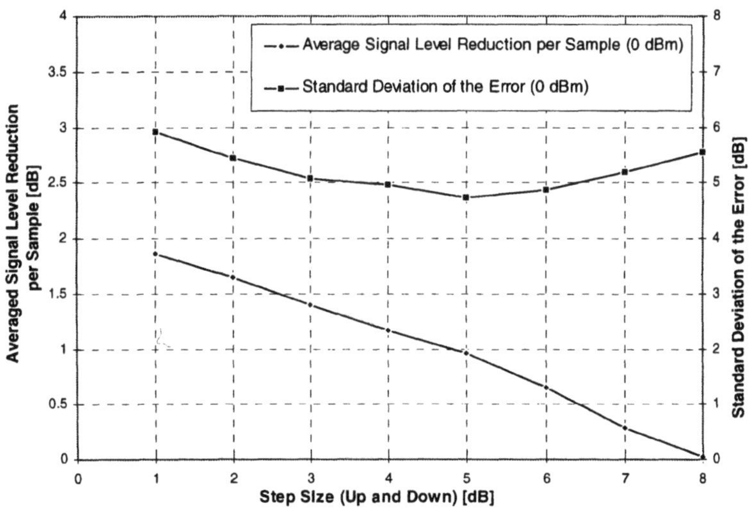

Figure 81. Signal strength reduction and standard deviation of the error when using an upper/lower level threshold of -85 dBm and a mobile speed of 50 km/h.

The probability density function of the error using a step size of 2 dB has been determined and is shown in Figure 82. Again, based on the error samples standard deviation and mean (assuming a Normal distribution) is estimated. They are in this case found to a mean of -0.003 dB and a standard deviation of 5.5 dB. This Normal distribution is also plotted in Figure 82.

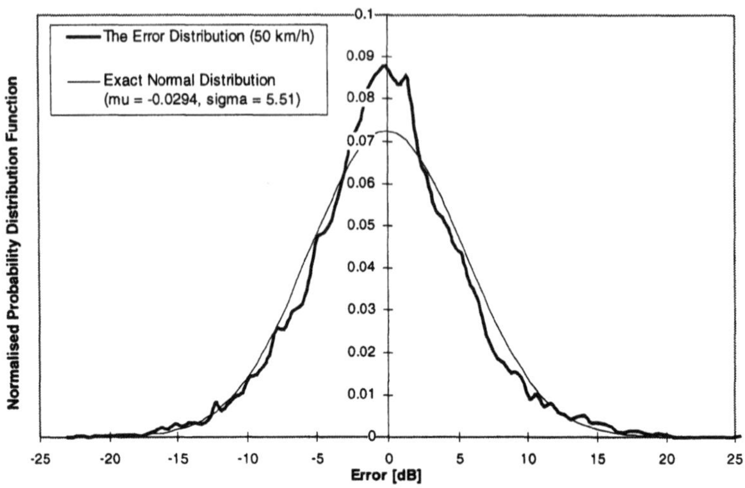

Figure 82. Error probability density function with equal level trigger thresholds and a mobile speed of 50 km/h and the corresponding Normal distribution.

Again it seems like the error distribution is highly correlated to the Normal distribution, however not as nicely as with the slowly moving mobiles. As with the slower mobiles the samples were plotted in the normal distribution probability plot and the results showed that within a probability from 10 % to 90 % the data fitted a straight line. It is therefore also assumed that the error samples are normally distributed for the case of faster moving mobiles, and the sample standard deviation along with a normal distribution (of zero mean) reflects the performance of the power control algorithm.

3.3.3 Performance of the Simplified Power Control Algorithm

Table 21. Summarised performance characteristics of the simplified PC algorithm.

Upper threshold [dBm]	Lower threshold [dBm]	Mobile speed [km/h]	Step size down & up [dB]	Max. signal level reduction per sample [dB]	St. dev. of the error [dB]
-85	-85	3	2	3.9	1.8
-85	-85	50	2	1.6	5.5

If looking at the worst case scenario it seems like the standard deviation of the error committed by the algorithm is as high as 5.5 dB. When comparing to the simulated log-normal fading with a standard deviation of 6 dB, the results are not to impressive. Only 0.5 dB has been compensated for. However, it should be remembered that these results are based on power regulations from signal strengths only, i.e. the GSM algorithm will probably perform better since it also exploits the received quality estimations.

3.4 Network Simulations of the GSM PC Algorithm

To perform a more detailed analysis of the GSM power control algorithm, a network level study is required. A model of the GSM power control algorithm has been implemented in CAPACITY. It therefore includes power control regulation triggers based on RXLEV as well as on RXQUAL and is correspondingly based on the received signal strength and experienced network quality.

The aim with these network simulations is twofold. First of all to configure the power control algorithm implemented in CAPACITY as well as possible and secondly to determine the potential gain from downlink power control.

Initially the various parameters required to configure the algorithm is described. For the simulation results using CAPACITY concerning the configuration of the parameters, the FER has been used as the performance measure. Since power control is studied for FH networks only the FER is believed the most reliable and powerful measure as described in Chapter 4 and 5. For the second aim, to determine the capacity, the *C/I* and RXQUAL soft blocking parameters have also been used.

3.4.1 Algorithm and Network Configuration Parameters

Several parameters have to be specified to operate the implemented power control algorithm in CAPACITY. The ones that has been focused on in order to optimise the settings are the 4 upper and lower quality and signal strength regulation thresholds. Also the step size (in dB) used for the regulation has been treated, however always with equal magnitude for step up and down. Finally the maximum dynamic range has also been treated.

Along with the power control settings, the corresponding handover settings are of course of great interest, since they have a direct influence on each other. All these parameters have also been kept unchanged, using the default handover parameters described in Chapter 8.

The simulated network uses a frequency plan with a frequency reuse of 1/3 and is highly loaded with an offered amount average network load of 30 %. 6 TRXs are allocated per cell in a network consisting of 48 sites, all with a cell radius of 3 km. A random FH network including DTX and no BCCH have been used.

Since all parameters in the input configuration file in CAPACITY influences directly on each other, the aim is initially to determine a default set of power control parameters, where the network works as well as possible. These parameters are then used as default power control parameters throughout the rest of the theses. To find the optimum settings for each possible simulated network would be far to time consuming, why this default-parameter-method has been used. The best parameters are found based on simulations of mobiles moving at 3 km/h as well as at 50 km/h.

3.4.2 Power Control Settings with a Mobile Speed of 3 km/h

Each of the parameters that are investigated is treated separately. In all cases the performance is now being illustrated using the cumulative distribution of the FER. In Chapter 6 it was explained how to understand this type of graph.

The Minimum Power Control Regulation Interval

As stated above it has previously been found that a minimum power control regulation interval of 2 SACCH frames was found optimum for the typical urban GSM specified propagation environment. Different sizes have been tried out, ranging from 2 to 16 SACCH frames [164]. In order for the power control algorithm to be able to react quickly enough it seems reasonable that the averaging should be as small as possible.

The Up/Down Step Size

The performance using an averaging of 2 SACCH frames is shown in Figure 83 for various step sizes. In all case the up and down step sizes are equal.

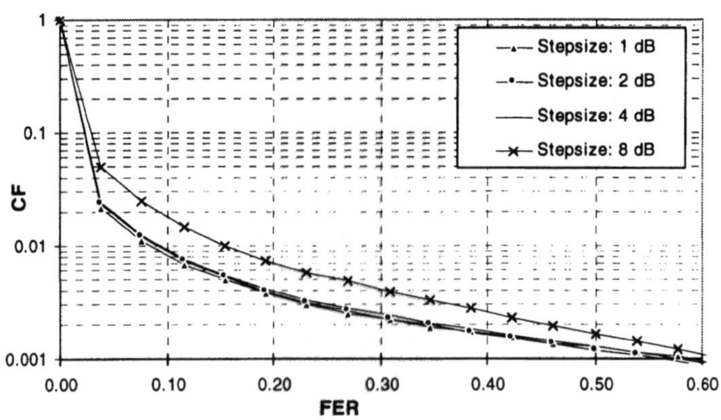

Figure 83. Cumulative frequency of the FER and a window size of 2 SACCH frames.

For this case, it is of little importance whether the step size is 1, 2 or 4 dB, however with 8 dB the performance is degraded significantly. It should here be mentioned that a step size of 1 dB is theoretical since in GSM the smallest possible step size is 2 dB. It is therefore chosen to continue the study of the power control algorithm with slow moving mobiles using a step size of 2 dB.

The Level Upper/Lower Thresholds

Also different triggers based on the received signal strength have been tried out ranging from –75 to –95 dBm. Using an averaging window of 2 SACCH frames, and a 2 dB step size, the different level threshold combinations shown in Figure 84 have been studied.

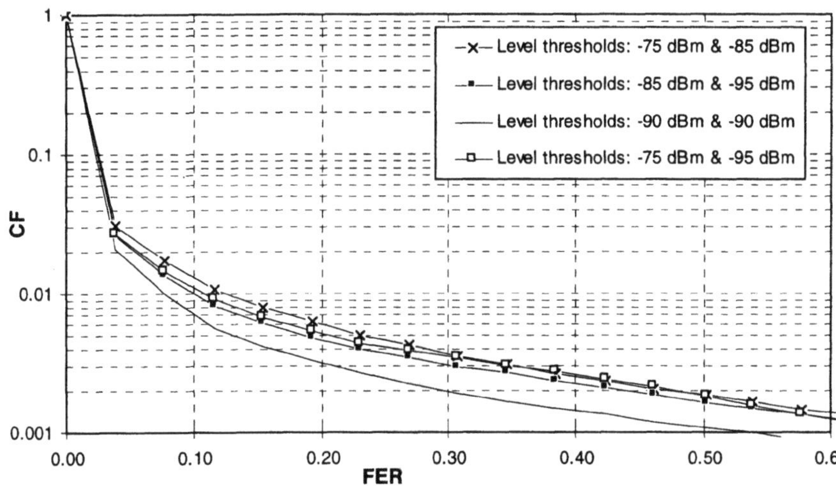

Figure 84. Cumulative frequency of the FER, with various upper and lower level thresholds.

It seems like for the 4 level threshold settings only little difference is found. The set with equal thresholds of –90 dBm might be a little bit better than the rest. The study will therefore continue with equal level trigger thresholds of –90 dBm.

The Quality Upper/Lower Thresholds

Using the settings found above, different quality trigger settings have also been investigated. Again the results are presented as the cumulative frequency of the FER, as shown in Figure 85. It seems like for the low FER values that an upper RXQUAL trigger of 1 and a lower of 4 is superior to the two other, which have therefore been chosen for the lower threshold.

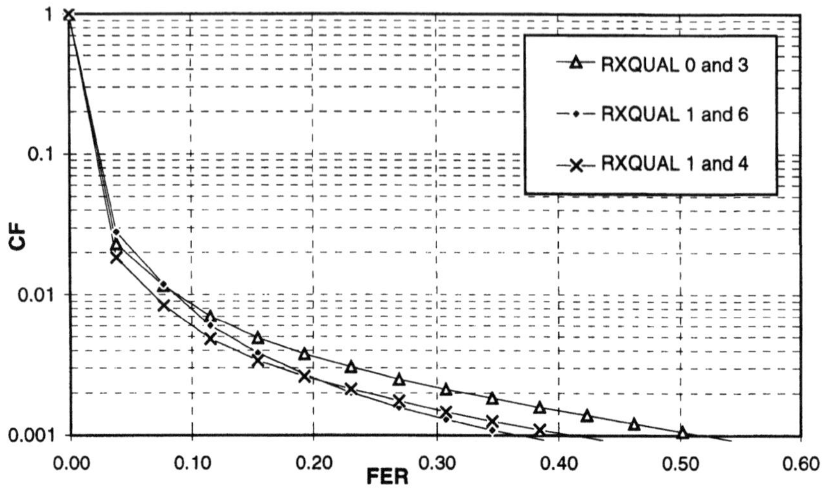

Figure 85. Cumulative frequency of the FER with various upper/lower quality thresholds.

The Maximum Dynamic Range

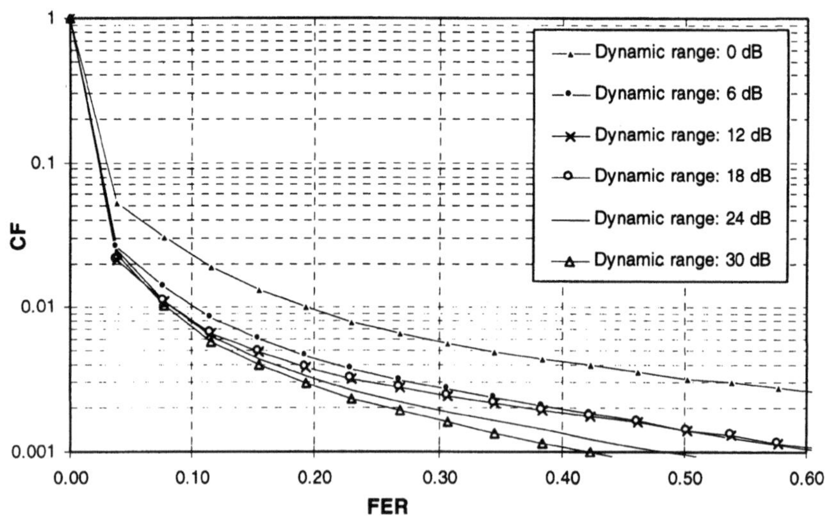

Figure 86. Comparison of the network quality with various dynamic power regulation ranges.

The last parameter to treat for the slow mobiles is the size of the dynamic range. As stated previously a maximum of 30 dB can be used. The issue of using smaller ranges is investigated. Using the default settings except for the level triggers, which have been set to –85 dBm and –95 dBm, the performance shown in Figure 86 were found. It is seen how already at a dynamic range of 6 dB a large part of the gain is achieved. There is little point in using a larger range than 12 dB.

The Transmission Output Power

Two simulations, one with and one without power control, were carried out using the setting found above for the power control. The probability distributions of the base station output power reductions are shown in Figure 87.

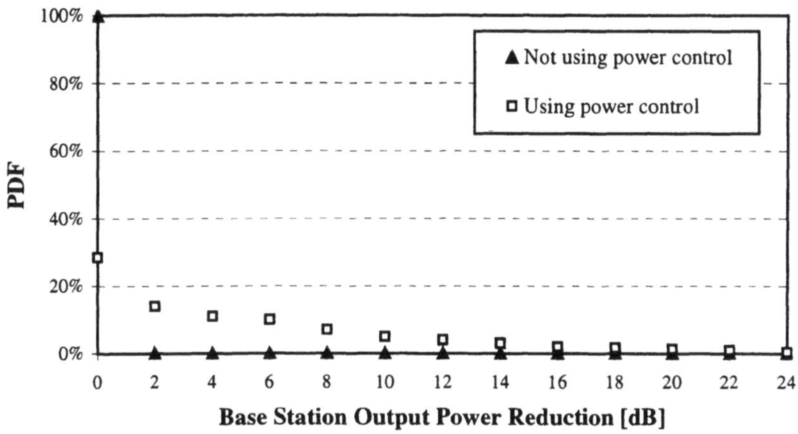

Figure 87. Probability distribution of the base station output power reduction, with and without power control.

It is seen how close to one third of the reported measurements are still based on full power from the base station.

3.4.3 Power Control Settings with a Mobile Speed of 50 km/h

The same set of simulations as for the slow mobiles has been carried out for the faster mobiles, in order to see if the settings found for he slow mobiles should be different when simulating the 1/3 reused network with 30 % load. Furthermore, the large speed dependency on the algorithm performance discovered when using the simplified level based power control algorithm, is also of interest.

The Minimum Power Control Regulation Interval

As with the slow mobiles, it has previously been found that a minimum regulation interval of 2 SACCH frames is optimum [164].

The Up/Down Step Size

The simulation results with an averaging using 2 SACCH frames for 4 different step sizes are shown in Figure 88. All step up and down sizes are equal.

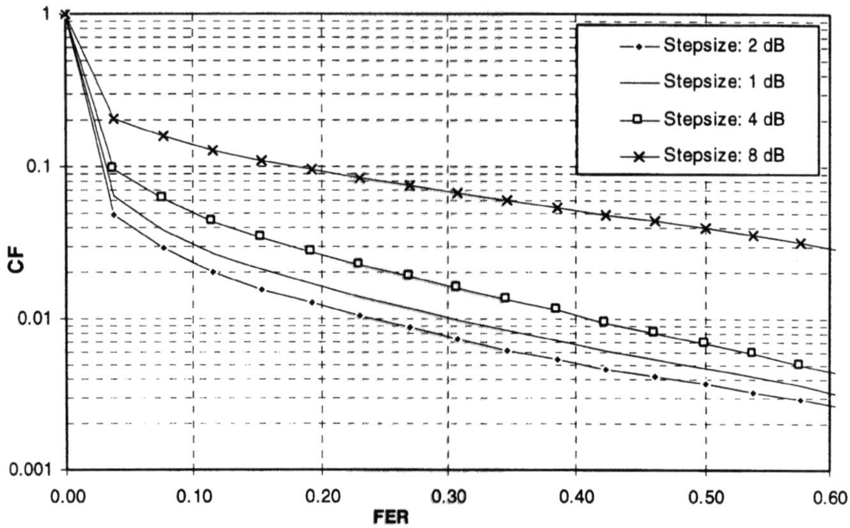

Figure 88. Cumulative frequency of the FER, with a window size of 2 SACCH frames.

It is seen how a large difference between each of the 4 sizes is now seen for these faster mobiles, when compared to the slow ones, see Figure 83. A step size of 2 dB is found to be the best. Already with a step size of 4 dB the performance is degraded significantly. In general, for all 4 simulations, the performance has degraded quite a lot compared to the case of mobiles moving at 3 km/h.

The Level Upper/Lower Thresholds

The same level trigger settings have been tried out as with the slow mobiles. The results are seen in Figure 89. The two best cases are the one with equal upper and lower thresholds of –90 dBm and the one with an upper threshold of –85 dBm and lower threshold of –95 dBm. In practice any of the 4 configurations could be used, but it was chosen to continue with the settings of –85 and –95 dBm.

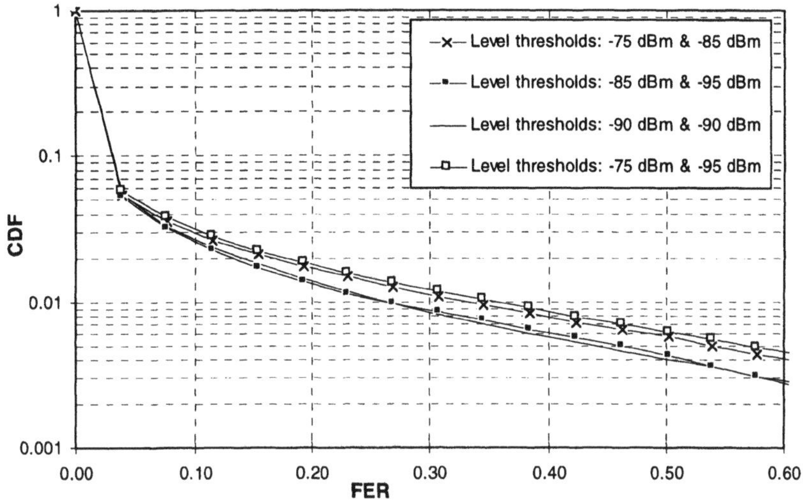

Figure 89. Cumulative frequency of the FER with varying signal level thresholds.

The Quality Upper/Lower Thresholds

Using the settings found above different quality trigger settings equivalent to the ones used for the slow mobiles have been investigated.

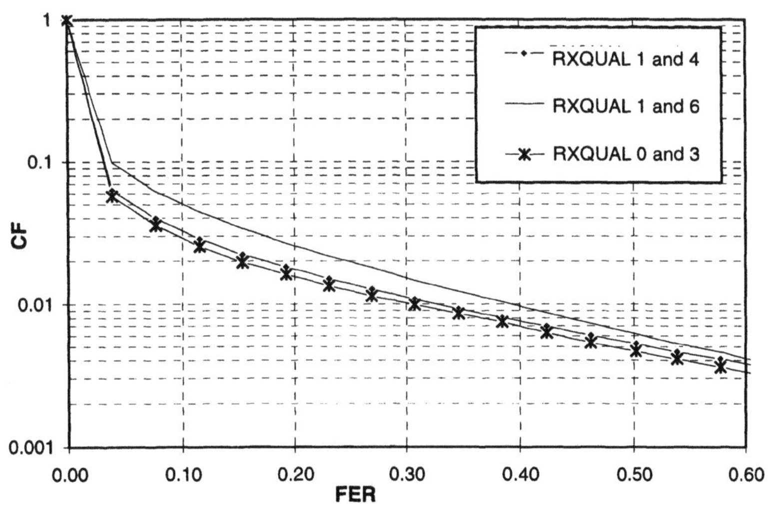

Figure 90. Cumulative frequency of the FER using various upper/lower quality thresholds.

Again the results are presented as the cumulative frequency of the FER, as shown in Figure 90. Using the thresholds of RXQUAL 1 and 6 degrades the performance compared to the other two. There is little different between using RXQUAL 0 and 3 or 1 and 4, why they could both be used as default settings. For the specific scenarios simulated it seems like when using RXQUAL 0 and 3 for the power regulations, the network performance is a bit superior. They have therefore been chosen as the default settings for mobiles with a speed of 50 km/h.

The Maximum Dynamic Range

Using the settings found for 50 km/h the effect of limiting the dynamic range of the base station output power is also studied, as shown in Figure 91.

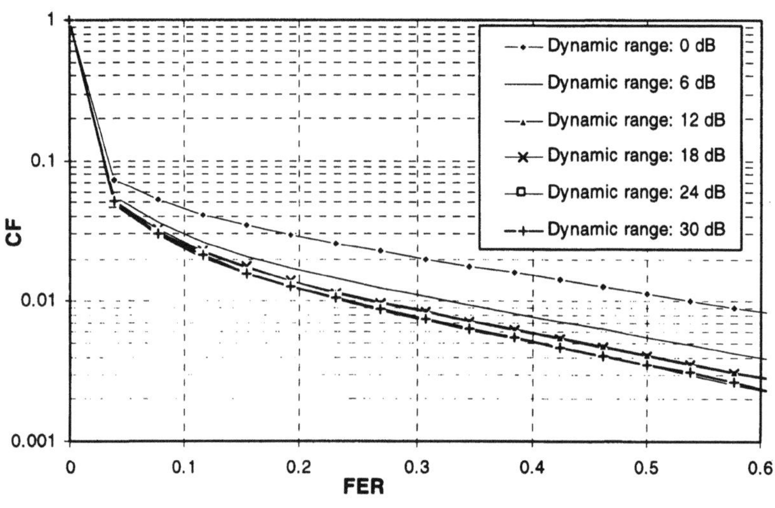

Figure 91. Network quality with various dynamic power regulation ranges.

Again it is seen how most of the gain is achieved already with a dynamic range of as little as 6 dB.

The Transmission Output Power

Two runs, with and without power control were carried out using the settings found above. The probability distributions of the base station output power reductions are shown in Figure 92. Again, as for the slow mobiles, it is seen how a lot of the reported measurements are still based on full power from the base station. However, the actual number is actually lower than for the slow mobiles. That is, from this figure it could seem like the algorithm performs better for the faster mobiles. However, as illustrated using the simplified power control algorithm, this conclusion should be drawn with great care. Reducing the base station output power is not necessarily enough. It also has to be the correct reduction.

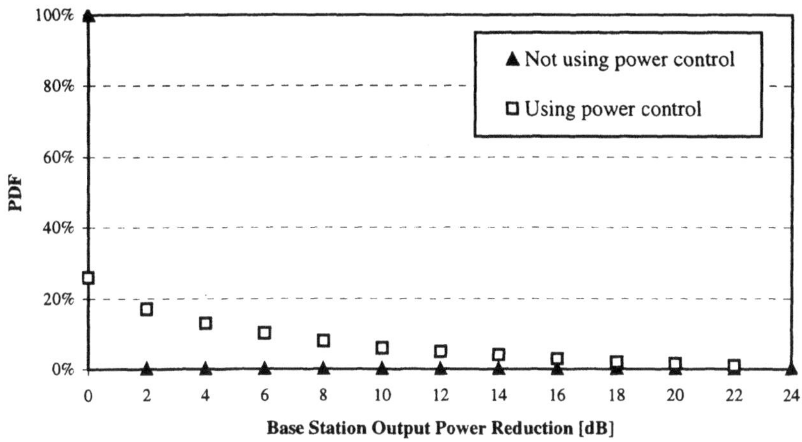

Figure 92. Probability distribution of the base station output power reduction at a mobile speed of 50 km/h.

3.4.4 Default Power Control Settings for CAPACITY

The best power control settings have been found for respectively a mobile speed of 3 and 50 km/h, when simulating a 1/3 frequency reused random FH network, including log-normal fading (as well as Rayleigh fading) and a 30 % network load. The DTX functionality was also included.

For both the fast and the slow scenario a minimum power control regulation interval as small as possible has been found optimum. The shortest possible period is 2 SACCH frames. A step size (of both up and down) as small as possible, has also been found to give the best results for both fast and slowly moving mobiles. The minimum is 2 dB. Concerning the power control triggers based in received signal strength (level), the algorithm can actually work with equal upper and lower thresholds. For the fast moving mobiles the GSM power control algorithm is equivalently good with an upper threshold of -85 dBm and a lower threshold of -95 dBm. These will reduce the number of power control regulations based on signal level, and have correspondingly been chosen. The power control triggers based on network quality, has for the low mobiles been found to RXQUAL 1 (upper) and RXQUAL 4 (lower). For the faster mobiles RXQUAL 0 and 3 or RXQUAL 1 and 4 could be used with almost equivalent performance. Therefore RXQUAL 1 and 4 are chosen as the default triggers. Finally the available dynamic regulation range has been treated. For both situations most of the gain is achieved with a dynamic range as little as 12 dB. A dynamic range of 30 dB is used for the default settings, however any value between 12 dB and 30 dB could be used with almost identical performance.

When considering both the fast and the slow mobiles, the default power control parameters shown in Table 22 were therefor chosen for CAPACITY.

Table 22. Default CAPACITY power control configuration parameters.

PARAMETER	VALUE
Power regulation interval	2 SACCH frames
Step size (up & down)	2 dB
Upper level threshold	-85 dBm
Lower level threshold	-95 dBm
Upper quality threshold	RXQUAL 1
Lower quality threshold	RXQUAL 4
Maximum dynamic range	30 dB

3.4.5 Simulated Capacity Gain from Power Control

To illustrate the potential gain of power control, the performance of a GSM network with and without power control and for fast and slow mobiles has been simulated. In this case, since a description of the potential gain from power control is attempted, the *C/I* and BER as well as the FER are shown. The simulated network has again been a 1/3 reused network with 30 % load as before. The accumulated *C/I*, BER and FER statistics are shown in Figure 93, Figure 94 and Figure 95.

It is seen that with the *C/I* statistics, at the 10 % outage little gain from power control is seen for both slow and fast mobiles. However, at the 1 % outage of the *C/I,* a gain around 1.3 dB for the slow and 1.9 dB for the faster mobiles is achieved from power control. Again looking at the BER, at the measurements worse than 6.4% (RXQUAL 6 and 7), it is not possible to see much gain.

Looking at the FER on the other hand, a clear gain from power control is seen for both fast and slow moving mobiles. The FER statistics also show a general network performance degradation for faster moving mobiles. If comparing the two curves without power control a significant difference is seen. The reason for this is believed to arise from the handover algorithm, as discussed in Chapter 8.

To emphasise this comment it could be mentioned that the dropped call rates from the simulations above are identical for the different speeds. Without power control the dropped call rate is approximately 0.9 % for mobiles moving at 3 km/h and 50 km/h, while with power control it is around 0.1 %.

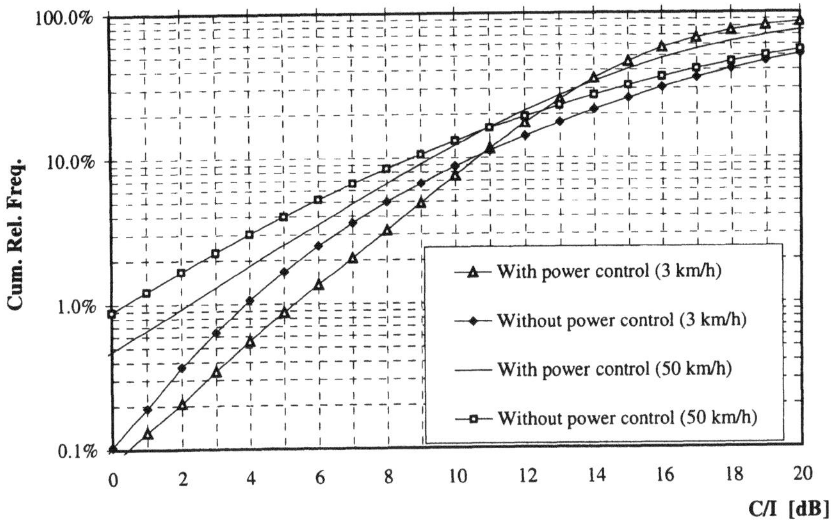

Figure 93. *C/I* statistics of a network with/without power control for fast and slow mobiles.

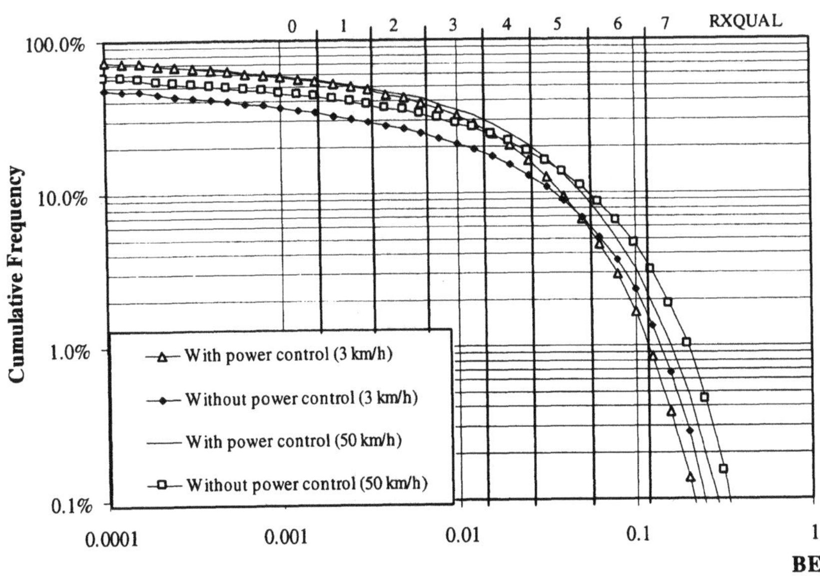

Figure 94. BER statistics of a network with/without power control for fast and slow mobiles.

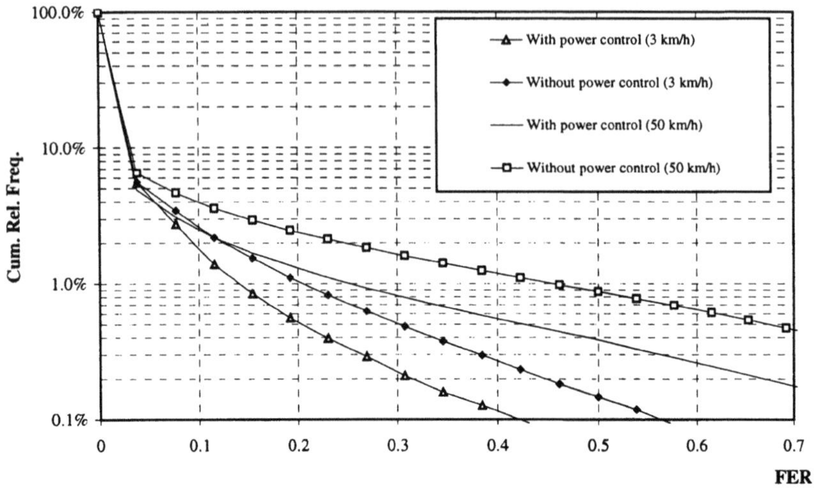

Figure 95. FER statistics of a network with/without power control for fast and slow mobiles.

Considering the two scenarios of fast and slow mobiles, it here seems like the power control algorithm performs almost equally well, which is contradicting to the conclusions drawn using the simplified algorithm. The reason for this is that with the default RXQUAL and RXLEV thresholds found above a relatively conservative regulation philosophy is chosen. This is especially important for the RXQUAL thresholds, where more aggressive settings like equal upper and lower thresholds of RXQUAL 4 have confirmed this. Furthermore, in the initial study using the simplified algorithm, the regulation was based on received signal strength exclusively, while for this study the estimated quality also determines the regulation. This will increase the performance of the power control algorithm in CAPACITY and thereby minimise the error committed.

Also, as stated, in the initial study only one link were treated, while with CAPACITY an entire network is simulated. Therefore in the latter case the effect from the handover algorithm as well as the network load is also reflected in the simulation results. The price of having the more conservative settings is a slight capacity reduction for the slowly moving mobiles, while the error committed for the faster moving mobiles is minimised.

3.4.6 Further Studies on the GSM Power Control Algorithm

The situation looked upon so far, has been where the offered network traffic has been equal to the maximum designed traffic load (according to Erlang B with 1 % blocking in case of a flat network). While the network load stays below this level the algorithm will work properly, but if for some reason the amount of cell traffic increase to more than this designed level the power control gain will decrease. This issue is treated in [110], where a continued study on the GSM power control algorithm is carried out as a function of the traffic load and frequency reuse. In reality this means that in situations where the gain power control algorithm is really important, it vanishes. One way of treating this problem could be by modifying the power control algorithm so that it uses dynamically load dependent RXQUAL thresholds.

3.5 Trial Results of Downlink Power Control and DTX in a FH Network

As described in the end of Chapter 6 a switch from using the intelligent underlay overlay (IUO) algorithm to using FH took place in SONOFON in August 1997. The introduction of FH, how it was done and what results were achieved, were described. A trial introducing downlink power control and DTX in the random baseband FH network was correspondingly carried out where the most significant results are described here.

A description of the most important results from the trial of FH, power control and DTX can be found in [177] or for more details, see the internal SONOFON documents [169,167, 168].

3.5.1 The Aim of the Trial

With FH, downlink power control and DTX in combination it should be possible to retain approximately the same network capacity as with IUO by itself [125]. Furthermore, when considering the daily traffic profile excluding the busy hour, the overall network quality might be increased compared to the IUO case, where the minimum dropped call rate (at last in the SONOFON network) was somewhat independent of the load. Two steps therefore describe the aim of this trial:

1. Configure the power control algorithm as good as possible using the simulation results from CAPACITY.
2. Determine if downlink power control and DTX (in combination with FH) in a live GSM network increases the network quality as expected.

The latter step is from the beginning a problem, since it is not a trivial thing to measure the network quality in a FH network, see Chapter 6.

3.5.2 Network and Parameter Configuration

The test area has been the same as was the case with FH, see Chapter 6. Correspondingly an area of approximately 200 cells (the large area) has been used. Out of these 200 cells, 61 cells in the centre have been used for the actual performance evaluation (the small area) as in the FH trial. That way all the cells in the small area are primarily interfered by other FH frequencies and can be included in the test measurements, as described previously. The most important characteristics, describing the network of the trial, are summarised here:

- A frequency plan with an average frequency reuse factor of approx. 13 has been used. This relatively high frequency reuse means that the frequency plan can probably operate even without FH.
- Concerning the frequency allocation, no splitting between the BCCH and TCH frequencies has taken place. See Chapter 10 for more information on this subject.
- Heuristic frequency planning, with non-grouped frequencies, has been carried out.
- The available bandwidth has been 8.8 MHz (44 frequencies).
- All cells in the small area have more than 2 TRX's per cell.
- Uplink DTX and power control has been enabled during the entire trial.
- Random baseband FH was used.

3.5.3 Test Procedure

Having already activated FH in the region, the introduction of downlink DTX and power control were done according to the following steps:

1. Enable downlink DTX and verify the functionality using a few sites.
2. Disable downlink DTX and enable downlink power control using the parameter settings found from the CAPACITY simulations, and verify the functionality. If necessary optimise the parameter settings of the algorithm even further.
3. Provided both features work well, enable downlink DTX as well as power control in the random FH GSM network using the settings found.

Remembering from the frequency hopping test, that great emphasis had to be put into not reducing the network quality during the entire trial, the trial was carried out. One of the ways of ensuring this was by starting out using a very limited dynamic range for the power control algorithm and then gradually increases the range.

3.5.4 Trial Results

The first step in the trial concerns the introduction of downlink DTX in the FH area. Ideally this should, as previously described, be quite simple since only one thing should be specified besides deciding whether it should be turned on or not. A feature allowing different weighting of the SUB and FULL values[19] can be used in the averaging process of the RXUQL and RXLEV samples for the handover and power control algorithms. The possibility of using different averaging factors has been allowed since the inaccuracy in the data estimation when using the SUB compared to the FULL is larger, refer to Section 4.2. This is one of the main issues in the study of DTX and will therefore be treated in further detail later.

Enabling downlink DTX on a limited number of sites, using a weight of the FULL twice as high as for the SUB, the RXQUAL statistics using the bad quality based on RXQUAL 6 and 7, looks like shown in Figure 96. Uplink DTX has been enabled throughout the entire trial.

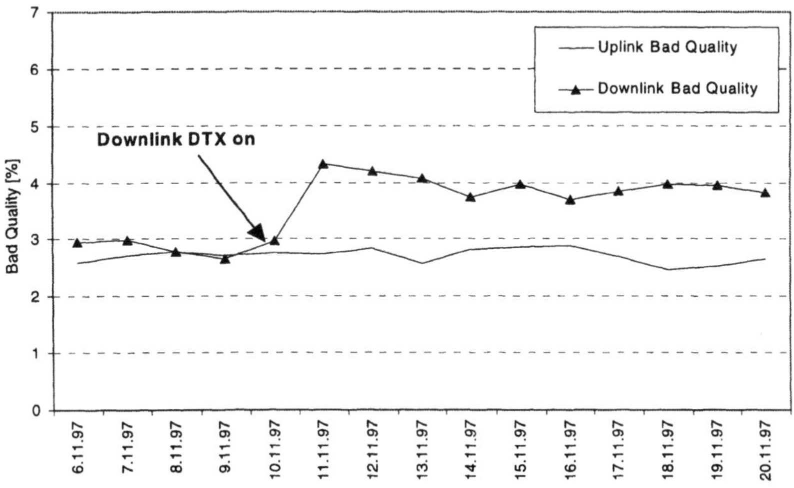

Figure 96. Measurements of the bad quality (%) when introducing downlink DTX in a FH GSM network.

Other types of statistics were measured before and after the enabling of DTX, such as the dropped call rate and the handover cause distribution. All of them were unaffected. To get a better indication of the experienced speech quality, several test drives exploiting a test mobile were carried out. The results from different of such

[19] The SUB and FULL refers to the samples when using respectively the samples from DTX mode and the samples from the non-DTX mode, see Section 4.1.

measurement methods were identical. The subjective speech quality was not degraded with DTX.

From other operators the same experience was reported with DTX. Due to the conflicting trial results when using respectively RXQUAL or the subjective speech quality (or the dropped call rate) to describe the network performance, it was decided to stall the use of downlink DTX until further studies had been conducted. To ensure nothing was wrong with the equipment and to clarify what was happening, a more thorough study of the feature was to be carried out.

Step two of the trial, involving the introduction of downlink power control, was therefore started. Again some operators had reported problems with the mobile stations when using this feature in combination with baseband FH. Therefore quite a conservative trial procedure was chosen. An initial dynamic output regulation range of 6 dB was used. The power control configuration parameters found from the network simulations using CAPACITY were used. When having the algorithm up running properly, the idea was to increase the dynamic range step by step as far as possible.

Initially several tests (on 3 cells) were carried out to ensure the right functionality. Having ensured that, the downlink power control feature was enabled in all cells within the entire large area. The measured busy-hour average downlink transmitting signal power reduction from the base stations after this introduction is shown in Figure 97.

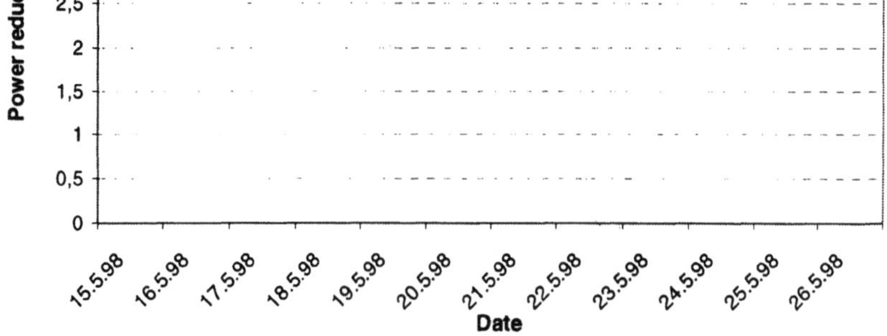

Figure 97. Average busy-hour signal level base station output power reduction using downlink power control with a maximum dynamic range of 6 dB.

When looking at the output transmission power from the base station, a relatively large average reduction around 4 dB was achieved with the small dynamic range of 6 dB. This seems reasonable when compared to the CAPACITY simulations. However, as shown in the simulations of the simplified power control algorithm the reduction in output power does not necessarily reflect an increase in performance. The output power of course has to be reduced with as few errors as possible when compared to the ideal regulation. Other statistics such as the RXQUAL probability distribution and the dropped call rate were measured, however despite the clear output power reduction, no improvement or degradation could be seen from these parameters.

The trial continued by increasing the dynamic range to 10 dB. For this situation an average base station output power reduction of 6 dB were achieved. However, in this case a slight increase in bad quality, as well as in dropped calls, was also seen. Therefore, in case of the SONOFON network, it was decided to configure the default maximum dynamic range to 6 dB.

4. DISCONTINUOUS TRANSMISSION IN GSM

Discontinuous transmission (DTX) should, as described in Section 2, be a straight forward network feature, where basically the only question is whether to turn it on or not. However, as indicated in the live network trial in the previous section, when network operators activate DTX in the live networks, the experience has surprisingly enough been quite bad. With DTX activated in a live network, the amount of bad quality (RXQUAL 6 and 7) typically increase. Primarily therefore only few operators have used downlink DTX, whereas with the uplink the experience with bad quality has been neglected in order to extend the lifetime of the batteries in the mobiles. However, as briefly indicated in Section 2 a significant interference reduction of ideally as much as 3 dB (in *C/I*) can be expected from DTX. Therefore it was decided to study the functionality of DTX in GSM to find the reason for this increase of bad quality. The measurement accuracy quickly turned out to play a key role in the explanation of the problem, and has become one of the primary issues of this section.

4.1 The Basic Functionality of DTX in GSM

Having enabled DTX in the network, the dedicated mobile can either be in DTX mode or not in DTX mode, i.e. both modes are experienced during the same conversation. In GSM when a mobile station is *not* in DTX mode (the subscriber is talking), 100 out of the existing 104 TDMA bursts per SACCH multi-frame can be used to estimate the RXQUAL (4 idle bursts are not used) [105]. When a mobile station on the other hand is in DTX mode (subscriber not talking), only 12 TDMA bursts are sent within the 0.48 seconds. The non-DTX mode is in GSM denoted the FULL, while the DTX mode is referred to as the SUB. Figure 98 show the burst transmission in DTX mode (the SUB) with the 12 TDMA bursts.

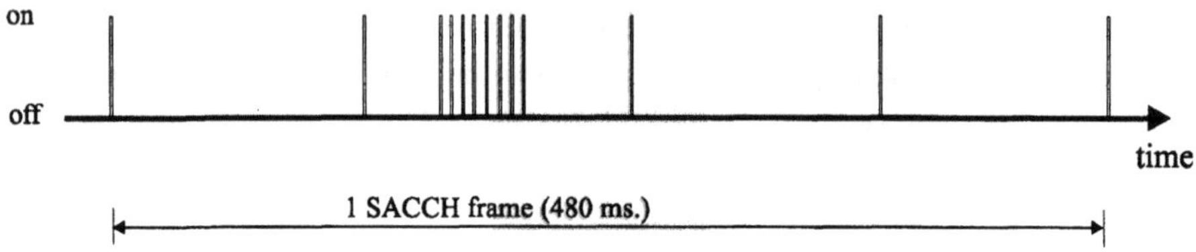

Figure 98. Schematic time-power diagram of a SACCH frame showing the SUB and no traffic channels [65].

Speech detection is carried out at the transmitting end by a voice activity detector (VAD). It distinguishes between speech superimposed on environmental noise and noise without speech being present. The output of the VAD is used to control the transmitter switch. If the VAD fails to detect every speech event, the transmitted speech will be degraded due to clipping. On the other hand, if the VAD identifies noise as speech too often, the effectiveness of DTX is diminished. Both of these factors result in degraded performance.

At the receiver end, the background acoustic noise abruptly disappears whenever the radio transmitter is switched off. Since switching can take place rapidly, it has been found that this noise switching can be very annoying. In very bad cases the noise modulation greatly reduces the intelligibility of the speech. This problem can be overcome by generating a synthetic signal known as 'comfort noise' inserted at the receiver whenever the transmitter is switched off. If the characteristic of the comfort noise matches the transmitted noise, the gaps between abrupt talk can be filled in such a way that the listener does not notice the switching during the conversation. Since the noise constantly changes, the comfort noise generator should be updated constantly [136].

Of the 12 bursts sent in DTX mode in GSM, 4 constitutes the SACCH frame used for signalling, see Chapter 3. The 8 remaining bursts, see Figure 98, contains the silence descriptor frame (SID frame) refreshing the comfort noise characteristics.

4.2 RXQUAL Estimation Accuracy with DTX

RXQUAL is measured by the mobile station for the downlink and by the base station for the uplink. Two factors can degrade the RXQUAL accuracy estimation with DTX. The fact that only 12 samples are available within the 480 ms. while for the FULL 100 samples are available for the same time period can cause some inaccuracy in the estimation. It is literally a down sampling that takes place. The other potential factor is the actual link level calculation of each RXQUAL value. Each of these issues is treated in the following.

4.2.1 Link Level RXQUAL Estimation Accuracy with DTX

The estimation of RXQUAL is done by calculating the BER prior to the decoding in the receiver (with a sample period of 480 ms.). Then this BER is logarithmically mapped to the linear RXQUAL scale ranging from 0 to 7, as shown in Chapter 3. The technique used for calculating the BER is not specified in the GSM recommendations. Therefore it is up to the vendor (of either the mobile station or the base station) to decide what principle to use. The only requirements specified concern the RXQUAL estimation accuracy. These requirements are shown in Table 23 [65].

Table 23. ETSI specified RXQUAL estimation accuracy requirements [65] for both the FULL and the SUB.

QUALITY BAND	PROBABILITY THAT THE CORRECT RXQUAL BAND IS REPORTED BY THE MS	
	FULL RATE CHANNEL	DTX MODE
RXQUAL_0	90 %	65 %
RXQUAL_1	75 %	35 %
RXQUAL_2	85 %	45 %
RXQUAL_3	90 %	45 %
RXQUAL_4	90 %	60 %
RXQUAL_5	95 %	70 %
RXQUAL_6	95 %	80 %
RXQUAL_7	95 %	85 %

Figure 99 show one way of estimating the BER, which has been implemented in the GSM link simulator. Other methods of estimating the BER and RXQUAL can be used, such as using the known training sequence, but they are not considered here. In Figure 99 frame *A* has been transmitted across the radio channel and has been de-interleaved but not yet decoded. If this frame is compared to the original encoded frame (here denoted frame *B*), we get the actual BER before decoding, here denoted the 'real BER', see Equation (6). This is possible in a simulator since the exact transmitted frame is known, but of course not in the real live situation.

$$real\ BER = (frame\ B) \oplus (frame\ A), \qquad (6)$$

The 'estimated BER' calculated in the receiver is found by encoding the received (and decoded) frame again (frame *C* in Figure 99) and then comparing it to frame *A*. Therefore frame *C* can be viewed as an estimate of frame *B*.

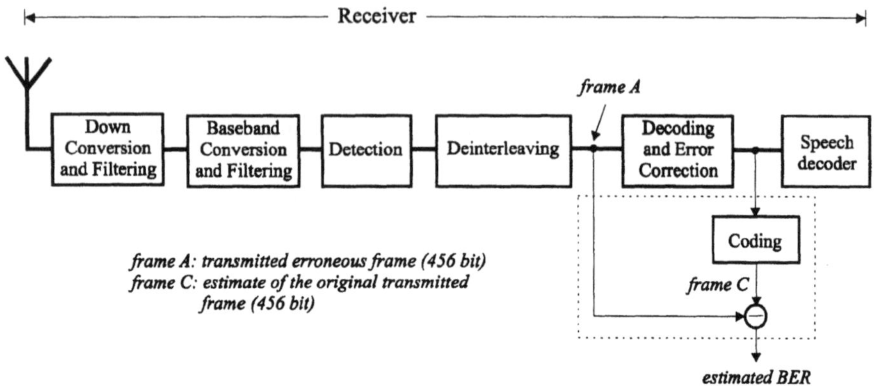

Figure 99. The technique studied to estimate the BER in the receiver.

If all the errors in frame *A* can be corrected by the decoder, frame *C* will be identical to frame *B* and the estimated BER will be completely identical to the real BER. However, if errors still occur after the decoding, the estimated and real BER might differ.

Figure 100 and Figure 101 show the difference between the real BER and the estimated BER as a function of the real BER for the FULL as well as the SUB.

The GSM link simulator has been used to generate the results. A typical urban channel (TU) [64] with mobiles moving at a speed of 50 km/h has been used, with random FH over 8 uncorrelated frequencies. All samples in the simulation concerning the SUB are averaged over 12 bursts, while 100 bursts are used for the FULL. It is seen that the error committed in the BER estimation is in general highest for high real BER. Also, the SUB has a higher error than the FULL, i.e. the variance of the SUB is higher than the FULL. This seems reasonable since the BER of the SUB is estimated using fewer samples. It is also seen how the error committed from the BER estimation is not symmetric. The average estimation error is positive. For the FULL no negative samples are found at all, while also for the SUB the error is clearly more positive than negative. This means that the estimated BER is lower than the real BER!

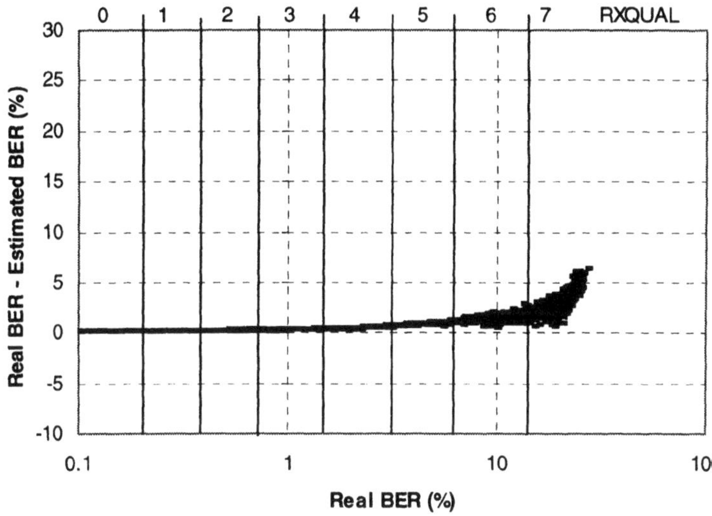

Figure 100. Error of the estimated BER as a function of the real BER, for the FULL.

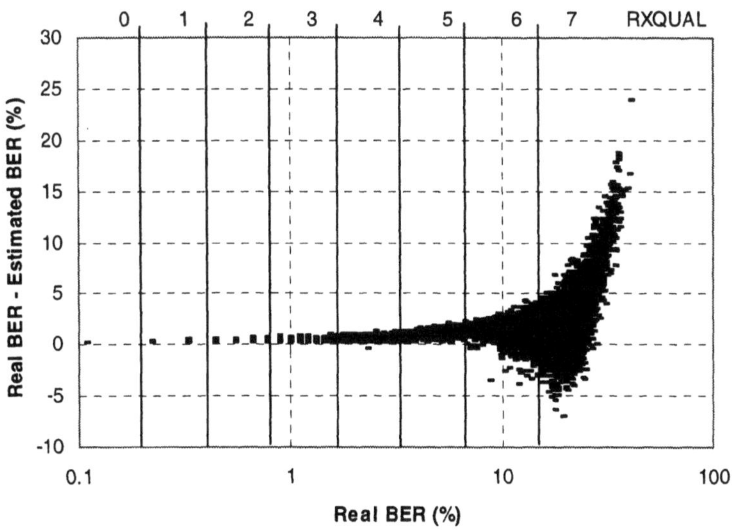

Figure 101. Error of the estimated BER as a function of the real BER, for the SUB.

An explanation can be found when looking to Figure 99. As long as all errors in frame *A* are corrected in the decoder, no estimation error will be made. However, if errors are still left after the decoding, this will introduce errors in frame *C*. These errors are not errors located in random places, since the error patterns in frame *B* and *C* are quite correlated. If e.g. the first 10 bit of frame *B* are corrupted and the decoder can only correct 5 of them. Then it is likely that some errors also will occur in the first 10 bit of frame *C*. Correspondingly the estimated BER will be lower, since some errors in frame *B* are also present in frame *C* and will therefore not be detected.

If the real RXQUAL (based on the real BER) is compared to the estimated RXQUAL (based on the estimated BER), only a small percentage of the values differ. If they are not equal, the difference is never more than one RXQUAL value. Therefore, when compared to the specifications shown in Table 23, they are fulfilled 100 %. In other words, the RXQUAL estimation method studied is sufficiently accurate.

4.2.2 System Level RXQUAL Estimation Accuracy with DTX

To model the effect of only having 12 bursts for the SUB, while 100 for the FULL when estimating the RXQUAL, both modes have been implemented in CAPACITY and system level simulations have been made. The default network parameters described in Chapter 5 have been used for the FH network with power control.

In the implementation in CAPACITY the estimated RXQUAL is 100 % accurate, i.e. the estimated RXQUAL is identical to the real RXQUAL. To find the relationship between the FULL and the SUB both values are calculated in parallel, which is equivalent to the live network when DTX is not switched on. Therefore, in the simulations the FULL values are used for the handover and power control algorithm.

Two different networks have been simulated, a 1/3 and a 3/9 reused network with mobiles moving at a speed of 3 km/h. The network frequency load in the simulations were respectively 23 % and 68 %.

Network Simulation Results

Figure 102 show the simulation results. The distributions of the FULL and the SUB are shown for both the 1/3 and the 3/9 reused network. It is seen that, while the network is exactly the same, the distributions of the SUB and the FULL for each of the 2 networks change. With the SUB more RXQUAL 0, 5, 6 and 7 occur. The percentage of RXQUAL 6 and 7 increase in the 1/3 case from 8.2 % to 11.0 %, while in the 3/9 case from 3.9 % to 6.0 %. In other words, the amount of bad quality can very well nearly double, when changing the estimation method from FULL to SUB. The change is simply caused by the higher variance of the SUB.

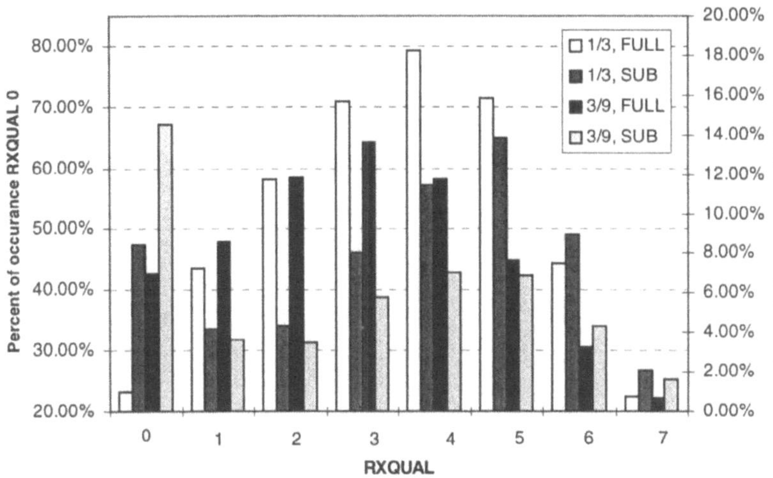

Figure 102. The SUB and FULL RXQUAL distribution from CAPACITY for a 1/3 and a 3/9 reused network.

By applying Bayes Rule [205] the probability of each SUB value can be expressed using the conditional probability of the SUB given a certain FULL as shown in Equation (7).

$$P(SUB=j)=\sum_{i=0}^{7}P(SUB=j\,|\,FULL=i)\cdot P(FULL=i) \qquad (7)$$

In Equation (7) *P(SUB=j)* is the probability of the SUB value being equal to RXQUAL *j*, *P(FULL=i)* the corresponding probability of the FULL value to be equal to *i* and *P(SUB=j|FULL=i)* is the conditional probability of the SUB being *j*, while the FULL is equal to *i*. The conditional probability *P(SUB=j|FULL=i)* is essential since it is through this parameter the error can be seen. This is shown in the following. Figure 103 show the estimated conditional probability of the SUB for different FULL values, *P(SUB=j|FULL=i)*, as found from the simulations of the 1/3 reused network.

An interesting thing is seen. The maximum value of a certain estimated probability distribution of the SUB does not necessarily lie at the corresponding FULL value. Looking at the estimated conditional probability, it is seen that the highest probability lies at 5, while also a value of 0 is very likely to occur. In other words, when the actual value is equal to RXQUAL 4, it will often happen that the reported SUB RXQUAL value is 5 or even 0. The reason why a value of 0 will often occur is because a lot of RXQUAL 0 in general occur in a network of high quality.[20]

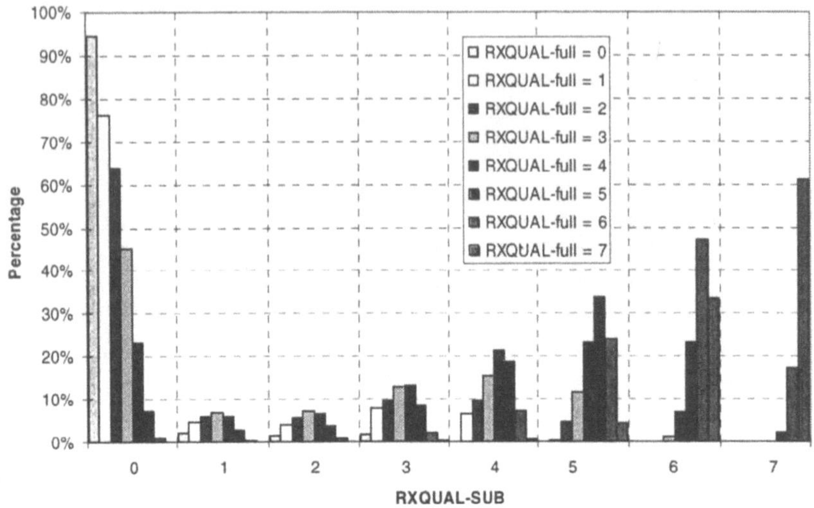

Figure 103 The estimated conditional probability of the SUB for different FULL values of the RXQUAL.

Live Network Measurements

Figure 104 show the distribution of the downlink RXQUAL SUB and FULL values of two cells measured from the live network using A-bis interface measurements. Both random baseband FH and downlink power control has been used. During the measurements the DTX feature was not enabled, so since both the SUB and the FULL are always reported, they can be compared directly.

If compared to the results in Figure 102, it is seen that the absolute distributions are different. This is of course due to the fact that the two scenarios differ in frequency plan, environment, power control, handover settings and so on. On the other hand, as with the simulations it is seen that more RXQUAL 0, 5, 6 and 7 occur for the SUB than for the FULL. Therefore the typical network operator will initially believe that the network has become worse after the introduction of DTX. In cell 1 the bad quality percentage increases from 1.15% to 1.81%, while for cell 2 it goes from 0.95% to 1.45%. All this while the network is exactly the same.

[20] This may not be correct if a very tight frequency reuse, synthesised FH and aggressive power control is used.

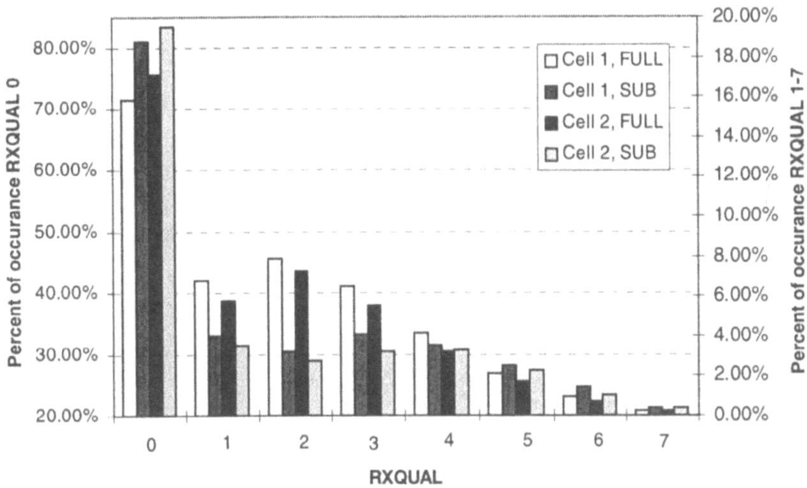

Figure 104. Downlink SUB and FULL RXQUAL distribution of two cells in a live network.

Figure 105 show the conditional probabilities for the SUB values for the live network measurements. Just as in the simulation results it is seen that the probability of having SUB values equal to 0 occurs relatively often. It can also be seen that the individual SUB value is spread out around the equivalent FULL.

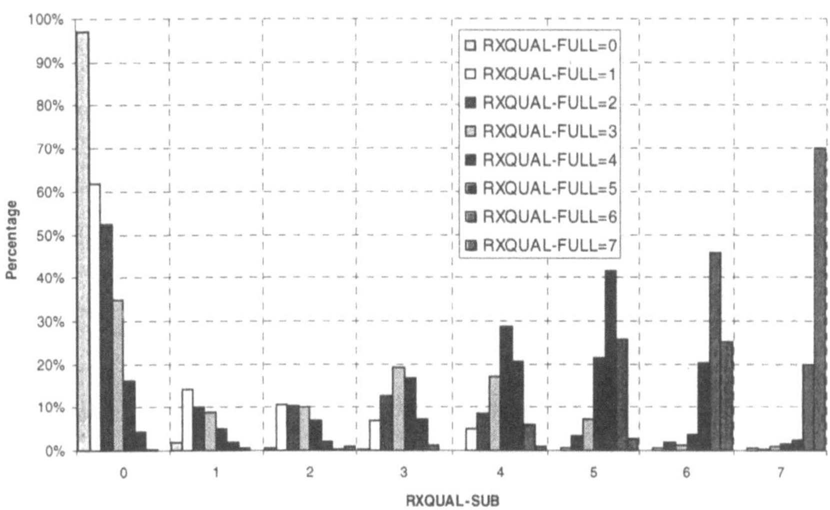

Figure 105. Conditional probability of the SUB for different FULL values (live network).

4.3 The Gain From DTX in a FH GSM Network

Detailed studies on the gain to be achieved from DTX in a random FH GSM network with and without power control have been carried out. The expected idealised 3 dB gain of DTX (with a DTX factor of 0.5) [105] has been verified for the network modelled more accurately with a burst-wise time resolution. In Figure 106 an example of a random FH network with a 1/3 frequency reuse, a 30 % average network load and a mobile speed of 3 km/h has been shown for simulations with/without DTX and with/without downlink power control.

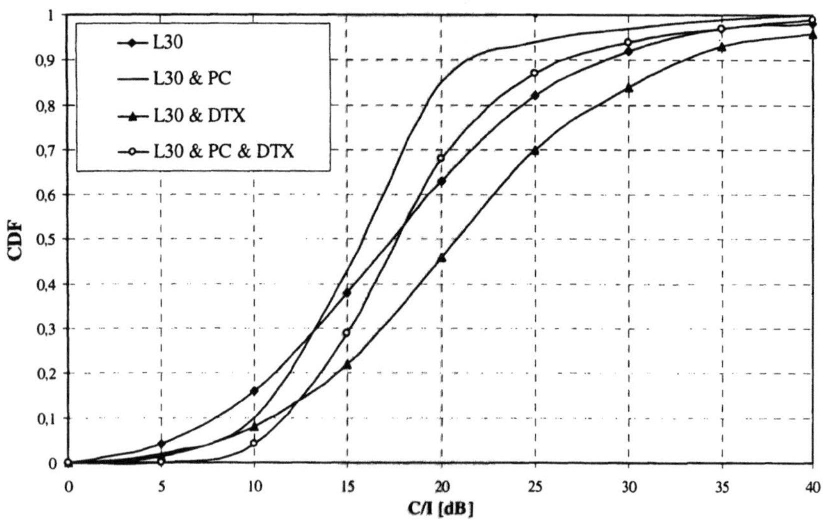

Figure 106. Example of the gain from power control and DTX in a random FH GSM network.

It can be seen how the gain (in *C/I*) from DTX is independent on what outage percentage is used. For power control on the other hand the gain depends on the outage used. The combined effect of using DTX in combination with power control is seen. At the low outage percentages a substantial gain can be achieved.

5. CONCLUSION ON POWER CONTROL AND DTX IN A FH GSM NETWORK

In this chapter the impact of using downlink power control and DTX in a FH GSM network has been described using different simulation models as well as live network measurements.

Simplified Studies of Downlink Power Control in GSM

Using a simplified version of the GSM power control algorithm, only able to regulate the base station output power based exclusively on the received signal strength, it has been found that the standard deviation of the error committed by the algorithm, is highly speed dependent. For slowly moving mobiles (3 km/h) a standard deviation of the error around 1.8 dB was found, whereas for the faster mobiles the corresponding value was 5.5 dB. That is, using a log-normal fading with a standard deviation of 6 dB, only 0.5 dB is in worst case compensated for by this level based power control algorithm.

Downlink Power Control in a Frequency Hopping GSM Network

A more realistic model of the algorithm has been implemented in CAPACITY, with the purpose of investigating, in detail, how well the GSM power control algorithm works when configured as well as possible. A typical macro cellular environment with a tight frequency plan of reuse 1/3 and a network load of 30 % has been simulated. As with the simplified study, the slow fading has been modelled by a log-normal fading with a standard deviation of 6 dB, a fading correlation distance of 110 meters and a de-correlation criteria of 1/exp(1) between the number of samples within the correlation distance.

A set of default settings have been found for the simulated network, see Table 22. It was discovered that already at a maximum dynamic range of 6 dB a large part of the potential gain is achieved and with 12 dB almost all of it. The achievable capacity gain from the power control algorithm has also been treated. Using the default settings, the effect from power control for both slow and fast moving mobiles was shown using the *C/I* and BER as well as the FER. For the simulated situation, a limited power control gain could be seen in the *C/I* (10 % outage) as well as the BER (RXQUAL 6 and 7) statistics. However, when looking at the FER, a significant gain from power control is seen for both fast and slow moving mobiles. Considering the two scenarios of fast and slow mobiles, the performance degradation found using the simplified level based algorithm, with increasing speed, could not be seen using the default GSM power control parameters. The somewhat conservative settings of the upper and lower RXQUAL thresholds means that the error committed is minimised. However, if stressing the algorithm with more aggressive settings, using equal upper and lower RXQUAL thresholds, the performance degradation were found for the faster mobiles.

Comparisons of the two FER curves (3 and 50 km/h) without power control, a significant difference was found. The fast mobiles experience worse performance. It means that a general network degradation arise when increasing the mobile speed. The reason for this should be found in the handover mechanism. During the period where a handover is attempted bad quality will typically be measured. When moving quickly away from the cell bad quality quickly occurs. This, on the other hand, does not necessarily introduce more dropped calls since the mobile has to experience bad quality for 10 sec. before it can be dropped. To emphasise this comment, it could be mentioned that the dropped call rates from the simulations above are identical for the different speeds. Without power control the dropped call rate is approx. 0.9 % for mobiles moving at 3 km/h as well as 50 km/h, while with power control it is in both cases around 0.1 %.

A live network trial implementing downlink DTX and power control in a baseband FH network has also been carried out. With downlink DTX, inconsistency between some of the performance indicators were encountered when measuring the network quality, why it was decided to analyse the feature before further use. Concerning the downlink power control a relatively large output power reduction was seen using the default power control settings found using CAPACITY. With a dynamic range as little as 6 dB, an average reduction of 4 dB was measured. However, the corresponding gain could not be seen from either the RXQUAL or the dropped call rate. It was, based on the trial, therefore not possible to determine the size of the gain from downlink power control.

DTX in a Frequency Hopping GSM Network

The introduction of downlink DTX in a FH GSM network has through a live network trial, as well as from other operators, shown an increase in RXQUAL 6 and 7, referred to as bad quality. Due to the potential gain to be achieved from DTX, the feature has been treated in detail to find and solve the problem. It has been shown how the problem arises due to the RXQUAL estimation accuracy with DTX.

One example of a well-known method of calculating the RXQUAL has been implemented in the link simulator. This method has been proven quite accurate and fulfils the GSM requirements 100 %. Also the effect of only having 12 samples with DTX instead of 100 without DTX for each SACCH frame has been studied. This down sampling has been shown to produce the increased amount of RXQUAL 6 and 7. An equivalent increase, as measured in the live network, has been found using CAPACITY. Therefore, with DTX, the parameter settings of the power control and handover algorithms have to be done more conservatively than without DTX. This will degrade the performance achieved from DTX.

It has been shown how the gain from DTX is directly proportional to the DTX factor and that the absolute size of the gain is independent of what outage is used. On the contrary, the gain from power control depends on the outage percentage. Therefore, when combining downlink DTX and power control, the utilised statistical outage percentage becomes important. Furthermore, it has been shown how the combined effect from DTX and power control increases the network performance when compared to using DTX or power control by them selves.

Chapter 8

HANDOVER ALGORITHMS IN A GSM NETWORK

Handover[21] is the mechanism that transfers an ongoing call from one cell to another as the mobile moves through the coverage area of a cellular system. As smaller cells are deployed to meet the increased capacity demands the number of cell boundary crossings and thus handovers increases. Each handover requires network resources to re-route the call to the new base station, so minimising the number of handovers minimises the switching load. Each handover has a certain change of failure. This has a direct influence on the quality of service (QoS), as a handover failure can lead to a dropped call. All this leads to the certitude that the handover is an essential part of a communication system.

In this chapter handover algorithms that can be implemented in a frequency hopping GSM system, are studied. It consists of two parts. First simulation and theoretical models are presented and evaluated. Secondly several handover features, which can improve the network performance are proposed and studied.

1. INTRODUCTION

In this section an introduction to handover algorithms is given. First some basics are presented in the Section 1.1, followed by a short overview of the literature on handovers in Section 1.2. This section concludes with the outline of this chapter.

1.1 Handover Basics

Some of the basic aspects and parameters of the handover procedure are presented here. Basically there are two kinds of handovers: hard handovers and soft handovers. A hard handover occurs when the old connection is broken before the new connection is activated [190], while in soft handover both connections are held simultaneously for a while. Soft handovers are most often used in CDMA systems, where the hard handover is mostly found in TDMA and FDMA systems. In this document soft handovers, are not treated, since CDMA systems are not considered. More about soft handovers can be found in [255].

[21] The two terms handover and handoff are used interchangeably within the literature.

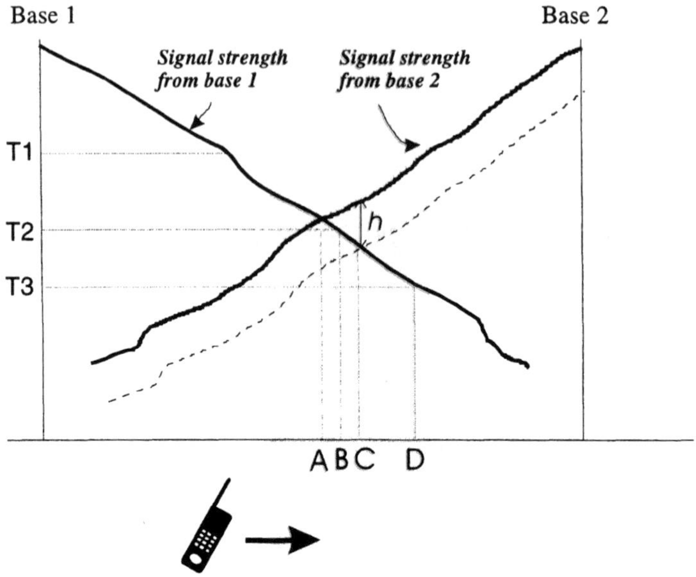

Figure 107. Example of the handover principles and parameters between two cells [190].

Figure 107 shows a mobile station moving from base 1 to base 2. The shown signal strengths are average signal strengths, so the fast fading is removed. The signal strength of base 1 decreases, whereas the signal strength of base 2 increases. It is clear that the mobile has to make a handover at some point. However there are several methods, which can give different points, where to make the handover:

- **Relative signal strength**. At all time the base station, from which the highest power is received, is chosen. This decision is based on the averaged output power. In Figure 107 the handover will occur at position A. This method has been shown to give too many unnecessary handovers, when the current base station signal is still adequate [84].
- **Relative signal strength with threshold**. A mobile station is only allowed to make a handover if its level decreases under a certain threshold level. For example, if the threshold is set at the level T1 in Figure 107, the handover is made, as in the previous method, in A. If the threshold is set to T3, then the handover is not made until the mobile arrives at D. This means that the mobile drives a long way into the new cell, causing cell stretching and more interference to the other users. Setting the threshold too low can also cause call drops, since a handover might come too late to save a call, which is suffering from bad quality. The process of setting the threshold can give some problems as well, since the optimum level might be different per cell.

- **Relative signal strength with hysteresis.** In this case a handover to another cell is only allowed, if the signal level from the other cell is sufficiently stronger than the own cell. A margin is defined, which also is called handover margin (HOMargin)[22] (h in Figure 107). The handover will occur in this case in point C. The hysteresis has a positive effect: it avoids the ping-pong effect, i.e. the repeatedly making handovers between 2 cells.
- **Relative signal strength with threshold and hysteresis**. This is a combination between the 2 previous methods.
- **Prediction techniques**. In this case it is tried to predict the signal strengths in the future, so that the optimal handover moment can be found. More about this can be found in [107].

The different methods are only one aspect of the handover process. A lot of questions remain, such as: How long should the averaging process be? How many and how should the neighbour base station be measured? The above description assumes only signal level as a reason for handover, but there can be many other reasons, such as quality, traffic reasons, etc., to make handovers. An operator has many parameters to control, which act directly on the handover algorithm.

All this leads to the conclusion that the handover process is complex. This chapter concentrates mainly on the handover in a frequency hopping GSM system, but it is tried to keep the scope as wide as possible.

1.2 Literature Study

Handover algorithms have been studied extensively. In this section an overview of the available literature on the handover process is given.

The initiation of a handover depends on a lot of different variables, such as the length and shape of the averaging window, the threshold level, and the size of the hysteresis. They are studied widely in literature, often as part of the analysis of new or existing handover algorithms.

A model for analysing handovers, based on signal strength measurements, made by mobile station in a log-normal fading environment, is presented in [228]. Handovers, which solely are based on signal level, are related to level crossings of the difference between the received signal strengths from two base stations. In [228] the performance of such an algorithm is derived by modelling the level crossings as Poisson processes with time varying rate functions. The relation between HOMargin, the averaging window (d_{av}) and number of handovers has been studied. Figure 108 shows mean number of handovers along the path between 2 bases as function of HOMargin. As expected, the number of handovers decreases, when HOMargin increases.

22 The term hysteresis is also used.

In [213,256] a model is presented, where an absolute as well as an relative level threshold is used. A handover is only made if the signal level falls under a certain absolute level. This way the number of handovers can be decreased.

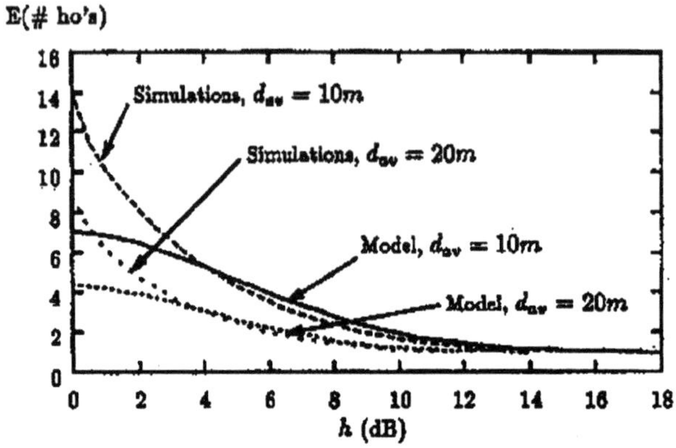

Figure 108. Mean number of handovers along the path between 2 base stations [228].

Choosing a new base station is one aspect of the handover algorithm, choosing a channel at the new base station is another. Some algorithms, called joint base station and channel assignment, make both decisions at once [190]. The signal to interference ratio (C/I) can be used to make this decision [34]. The disadvantage of this method is that it causes cell stretching and unnecessary high output powers. A method to prevent this is using the 'maximum power handover' [33] in which the mobiles constantly search for a combination of base and channel assignment that minimises the uplink transmitted power. It reduces the call dropping, but the number of handovers per mobile increases drastically. To avoid this a timer is introduced, which is the concept of 'maximum handover with timer' (MPHT) [33].

Various other topics in handover control include a discussion of priority handover schemes [214]. This work is motivated by the fact that the strategy of minimising the number of handovers may not work well from a tele-traffic point of view in the case of hot spots, areas of dense traffic. Users may be dropped due to an algorithm, which denies additional handover attempts that may save the call. The blocking of a call at handover is more critical as the blocking of a new call, since the first will possibly lead to a dropped call. Therefore, several priority schemes to reduce the chances of unsuccessful handovers have been suggested.

The simplest way of giving priority to handover calls is to reserve some frequency channels for calls being handed over into the cell. This scheme, which is called 'cut-off priority scheme' (CPS), is being studied in [180] and it is found that the results are satisfying, i.e. a sufficient Grade of Service (GOS) is achieved for the new calls and sufficient priority for the handover calls. In [86], free channels for handover are achieved by allowing queuing on new calls. In [96] a non-prioritised handover scheme is compared to 2 schemes using prioritising. The two prioritising schemes both use CPS. They differ in the fact that one allows queuing on the handover calls and the other does not. It is shown that the scheme using queuing gives a lower blocking on the handover calls, while there is no significant difference in the blocking of new calls between the two schemes, using prioritising. Also in [254] a CPS scheme is considered, but the reserved number of channels is divided into two parts: one part is reserved exclusively for the handover of fast moving mobiles, while the other part can be used for the handover by all kind of mobiles. One of the conclusions was that the handover blocking of the fast moving mobiles can be made smaller than the handover blocking of the slow moving mobiles by reserving an appropriate number of channels in each cell exclusively for fast mobiles.

In [186] the handover algorithm is studied for a micro cellular environment. Among others it was found that the rapid change in the received signal strength when a mobile turns a corner, 'the corner effect', is shown to affect the uplink more than the downlink in a micro cellular environment [82]. An important conclusion in [108] is that in a micro cellular environment there exist two contradictory goals. In the LOS case a HOMargin is useful to avoid useless handovers (the ping-pong effect), whereas in the NLOS situation, a handover must be done as quickly as possible due to the sudden signal drop as a mobile turns a corner. Possible solutions to these contradicting goals include he use of umbrella cells, macro diversity and switching to a mobile controlled handover. It should be noted that only the use of umbrella cells provides compatibility with the GSM standard.

Another area where the handover algorithm is very important is the multilayer network. An example of a multilayer network can be seen in Figure 109, where 3 layers are present. The first layer consists of macro cells, which are covering the whole area. They can for example be used as umbrella cells when the umbrella cell concept is used [120]. The next layer consists of smaller cells, micro cells, which are present in the heaviest loaded areas. Finally pico cells can be used as a third layer. They are typically used inside offices. The traffic is distributed over the different layers by the handover algorithm. An extensive study of one kind of multilayer networks can be found in [242], while also Chapter 9 of this report looks into this. A problem in multilayer networks is that the speed of the mobiles has to be estimated, since it is desirable to have a different strategy between fast and slow mobiles. Fast users should stay on the higher layer, which means in the macro cells, like is shown in Figure 109. Otherwise these mobile stations will make too many handovers. Slower mobiles should be directed to the smaller cells on the lower layers. Several techniques for speed estimation have been studied [13,21,102,107]. Also other techniques have been used to overcome this problem. In [195] for example a permanent positive offset on the received signal level of the micro cells has been given. This encourages stationary users to enter the micro cell, while fast moving users are discouraged from choosing the micro cell via a received signal level offset for a penalty time period.

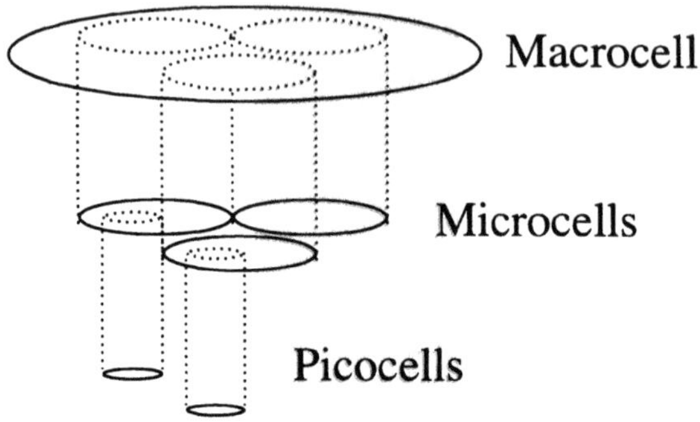

Figure 109. Multilayer network with macro, micro and pico cells.

There are different kinds of handover algorithms, which deal with multilayer networks. Examples are IUO [242] and umbrella handover [120]. They will not be dealt with here in this chapter, but IUO is the subject of the Chapter 9. The different layers in a multilayer layer networks can use a different air interface. In [72] for example the handover between a GSM and a DECT system have been studied, where the DECT cells are used as micro cells.

In [130] it is stated that the concerns, regarding handovers in smaller cells like micro cells [83,194] are exaggerated. This statement is based on results from field trials, where more then 145000 handovers between all types of cells were studied. It was found that in all cases the performance was satisfactory, since the handover failure rate was far below the 2 percent level commonly accepted. Field trials in three different cities with micro cells (Dublin, Hong Kong and Melbourne) have not shown any decrease in performance in terms of dropped calls for mobile stations, which are moving fast, compared to mobile stations with low mobility. A third aspect, that was found, was that the number of handovers, when installing micro cells does not increase drastically. This is due to the fact that micro cells are often installed in low mobility areas, so the number of handovers for the system as a whole does not necessarily increase more than marginally.

The application of non-standard approaches to handover control include neural networks, fuzzy logic, hypothesis testing and dynamic programming. Neural networks have been proposed as a possible tool to implement multi criteria handover algorithms [153]. In [157] it is shown that neural networks with the use of pattern recognition can decrease the number of handovers drastically, which has a positive effect on the signalling load.

1.3 Chapter Outline

The aim of this chapter is to show how the handover algorithm can be modelled and simulated in a frequency hopping GSM network and to show how enhancements to the handover algorithm influence on the performance of a frequency hopping GSM network.

A description of the handover algorithm used in CAPACITY is given in Section 2. Simulation results show the interaction and influence of the different parameters on the handover performance. The simulation results are verified by comparing them to network measurements. At the end of this section the influence of the introduction of frequency hopping in a non-hopping network on the handover performance is shown.

In section 3 a model is developed to find the minimum number of handovers per call. Simulations verify the model. Furthermore models, using Markov processes, are developed for a single and multiple cell scenario.

Several handover enhancements are presented and studied in Section 4. Channel reservation for handovers, queuing, traffic reason handover, dynamic HOMargin and combinations of these four enhancing methods are looked at. This chapter concludes with a summary in Section 5.

2. THE SIMULATION MODEL

In this section a description of the model used for the implementation of the handover algorithm in CAPACITY can be found in Section 2.1. Section 2.2 contains simulation results, showing the influence of different parameters on the handover performance In Section 2.3 measured data from a GSM network can be found, which is used to verify the handover performance in the simulator. Finally the last section the effects on the handover performance of introducing frequency hopping in a non- hopping network can be seen.

2.1 Modelling and Implementation in CAPACITY

As mentioned before the handover process in GSM is a very complex process. In this section it is described how it is modelled and implemented in CAPACITY.

In GSM the responsibility for the handover process lies entirely in the radio resource management (RR) layer, as this layers role is to establish and release connections between mobile stations and an MSC and maintain them despite user movements.

A natural way of implementing the different functions within the handover process can be seen in Figure 110. The mobile station (MS) and BTS make the link measurements, while most of the actual handover process is done in the BSC. The measurements are processed and the comparison with different thresholds is done here. If the threshold comparison indicates that a handover should be made, target cells are evaluated in the BSC. If a handover should be made between cells, which belong to different BSC's, the handover decision and comment is done in the MSC, while otherwise is stays at the BSC, except for certain kinds of more advanced handovers, such as the traffic reason handover, which are done at MSC level.

One advantage of having the possibility of doing handover at MSC level is that if BSC's of different manufacturers are used, the handover is performed inside a MSC of one manufacturer. This is guaranteed to work by the manufacturer, while there might be problems by making a handover between 2 BSCs of different manufacturers.

In CAPACITY the MSC is not simulated, so all handovers are performed at BSC level. The implementation in CAPACITY is described in Section 2.1.1, while the neighbouring cell measurements, needed for the handover algorithm, are the subject of Section 2.1.2.

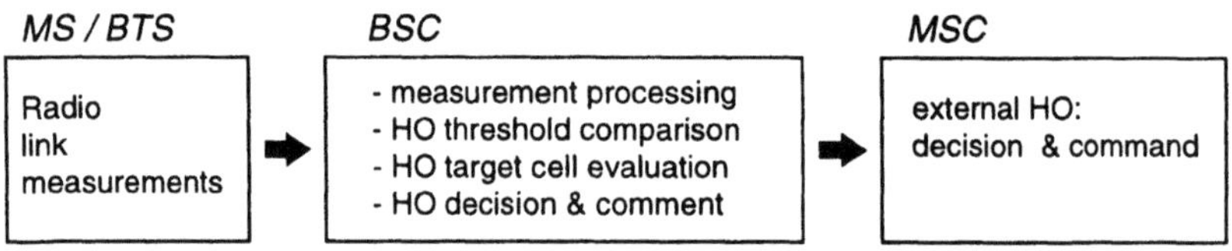

Figure 110. Different handover tasks and the physical location, where they are performed.

2.1.1 Handover in CAPACITY

While an active mobile station is driving around in a GSM network it is measuring the signal level of the neighbouring cells constantly, while also the own signal level and quality are monitored. Also the base station is measuring the level and quality of the active connection. According to the specifications [64] the following measurements have to be made in every SACCH multiframe:

- **RXQUAL-DL**, the downlink quality, measured by the MS.
- **RXQUAL-UP**, the uplink quality, measured by the BTS.
- **RXLEV-DL**, the downlink signal level, measured by the MS.
- **RXLEV-UP**, the uplink signal level, measured by the BTS.
- **Timing advance**, indicating the distance between the MS and the BTS.
- **RXLEV_NCELL**, the signal level of the 6 strongest neighbouring cells, measured by the MS.

These measurements can be used in the handover and power control algorithm, after possibly being weighted and averaged. In CAPACITY, not all of them are available. Only the downlink is simulated, so the uplink measurements are missing. The RXQUAL and RXLEV measurements are described in Chapter 3, while a description of the neighbouring cell measurements can be found in the next section.

In CAPACITY the handover evaluation scheme shown in Figure 111 is used.

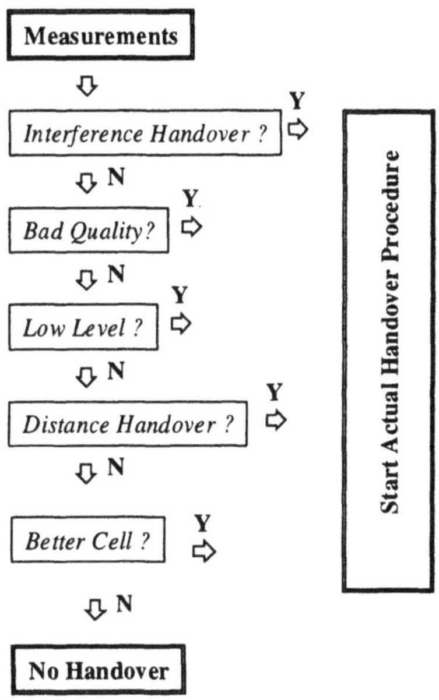

Figure 111. Handover evaluation scheme with prioritising of each handover trigger.

The handover evaluation is based on a row of triggers:

- **Interference.** When the averaged received signal level (AV-RXLEV) of a certain mobile station is satisfactorily high but the averaged quality (AV-RXQUAL) bad, a handover can be triggered due to interference. Typically, when triggered by the interference criteria, an intra-cell handover attempt (handover inside the own cell) is prioritised above an inter-cell attempt (handover between two cells), if frequency hopping is not being used.
- **Quality.** When the quality (AV-RXQUAL) of a mobile connection becomes worse than a certain threshold, a handover attempt is made to improve the quality of the mobile station.
- **Signal Strength.** When the received signal strength (AV-RXLEV) is lower than a certain threshold, a handover attempt is made to increase the level.

- **Distance.** When a mobile station is further away from its BTS than a specified threshold, a handover attempt is made. In reality this kind of handover is hardly used.
- **Power budget (better cell).** Constantly it is evaluated if the signal level of neighbouring cells is higher than the received signal level of the serving cell (to which the mobile station is connected). When the level of the neighbouring base station is more than HOMargin better than the level of the own cell, a handover attempt is made.

When one of these triggers indicates that a handover should be made, it is evaluated if any of the neighbouring cells fulfils the following two conditions:

1. The measured signal level (RXLEV) of the candidate cell has to be higher than a certain threshold, for example –104 dBm.
2. The power budget has to better than HOMargin. This means that the following equation has to be fulfilled:

$$\overline{L(j)} - \overline{L(i)} > \text{HOMargin} \qquad (1)$$

where *L(j)* is the level (RXLEV_NCELL) of candidate cell *j*, measured by the considered mobile station, and *L(i)* is the received level (RXLEV) from the own cell (*cell i*). So HOMargin indicates how much better the candidate cell has to be, before a handover to that cell is made.

If the 2 conditions, mentioned above, are fulfilled, the handover execution starts. In reality this execution contains a lot of different actions to be done by several parts in the network. This is not implemented in CAPACITY, since it is assumed that the handover execution is perfect. So when a handover is triggered and a candidate cell with a free channel is found, the handover is made instantaneously. No delay or information loss is suffered. In CAPACITY only the processes before the actual handover execution, such as the triggering, neighbouring cell measuring are simulated. More information about the exact handover execution in a GSM network can be found in [151].

In CAPACITY no handover failures occur, but there are blocked handovers. They occur when no free channels are available in the candidate cell. The mobile simply remains in the serving cell.

In CAPACITY handover prioritising is being used. Handovers have higher priority than new calls. Before a new mobile station is initialised in a cell, there is checked if there are any mobile stations, which want to make a handover to that cell. If this is the case, the handover is performed before the new call. When there still is a free channel left after all desired handovers have taken place or if there are no mobile stations that want to make a handover, the new call can be initiated.

Furthermore the priority of each individual handover is prioritised so that the mobile station that needs the handover the most has the highest priority, i.e. performs the handover first.

2.1.2 Neighbouring Cell Measurements

As described in the previous section, the mobile station makes measurements, which are used in the triggering of the handover and in the evaluation of the handover candidate cells. This makes the measurements an essential part of the handover process.

In order to make efficient handovers, the rate at which measurements are refreshed should be as high as possible. As mentioned in the previous section a mobile station has to measure not only its own level and quality but also the level of the neighbouring cells. The measurement reporting from the mobile station is carried on the signalling channel associated with each TCH, called SACCH. One SACCH frame, consisting of 184 bits is sent per SACCH multiframe (0.48 sec.). Hence the bit rate is 383 bit/sec. In CAPACITY the SACCH channel is error free, i.e. the measurement reporting is assumed perfect.

Due to the TDMA structure in GSM, it is possible for a mobile station to use the periods between transmitting and receiving, for listening and measuring other BTS's. Each mobile station receives a list of frequencies, which it has to measure, from its own BTS. These frequencies are BCCH frequencies, which among others carries the synchronisation and frequency correction channels (SCH and FCCH). The SCH contains the BSIC (Base Station Identity Code), which makes it possible for the mobile station to know where a certain BCCH signal comes from. A mobile station not only measures the signal level of the neighbouring cells, but it also acquires synchronisation with all cells, on which they report measurements. Hence, a mobile station is always pre-synchronised with all cells, it potentially is handed over to.

In CAPACITY all cell are synchronised, so each mobile station automatically is pre-synchronised with all other cells. In the simulator each BCCH of the neighbouring cells is measured over a period of 0.48 seconds, not a fraction of this period as in reality. This gives more perfect measurements in the simulator, since the fast fading is completely removed, whereas in reality there still may be rests of fast fading present in the measurements. This leads to potentially less handovers in the simulator, compared to reality, since in reality the measurements vary more, possibly leading to ping-pong handovers.

In GSM one measurement message, send from a mobile station to the BTS every 0.48 sec., contains the measured signal level of up to 6 neighbouring cells. However, a mobile station may pre-synchronise with more than 6 neighbour cells. If this is the case, then only those measurements corresponding to the 6 cells it receives best are reported to the BTS.

In CAPACITY a mobile station measures the signal level from all the cells from the 6 neighbouring sites plus from the two other cells of the own BTS. One cell and all its neighbours can be seen in Figure 112. The mobile station reports the measurements of all these 20 neighbours back to the BTS, instead of the maximum of 6 neighbours reported back in a real GSM system. Only a few of the 20 neighbouring cells are potential handover candidates, so this is assumed not to have great impact on the performance. In reality the handover candidates among the 20 cells will be measured in a real GSM network, if the neighbour cells are defined correctly.

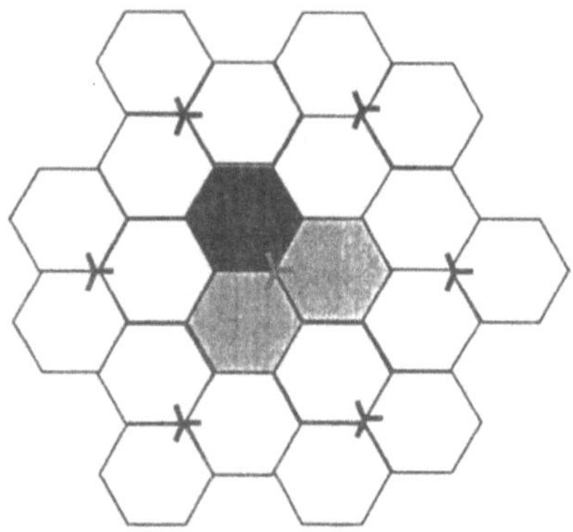

Figure 112. A cell (black) and all the cells measured as neighbouring cells in CAPACITY.

It should be noted that when discontinuous transmission is used, the measurements become more inaccurate, as reported in [241]. To avoid problems, the power control and handover settings should therefore be set more conservative. This results in a reduction in the gain from power.

2.2 Simulation Results

The general aim of designing an optimum handover algorithm is twofold: the quality of the network has to optimised and a mobile station should perform as few handovers per call as possible without degrading the quality There are two reasons for the latter. Every handover demands traffic capacity on the control channels in the GSM network. The FACCH channel is used among others during the handover. This channel steals capacity from the traffic channels by just 'throwing' data from the traffic channels away, thus degrading the quality, experienced by the user. This is the reason for the appearance of frame erasures on the traffic channel around the handover moment. Furthermore the call quality can be reduced dramatically if the handovers are performed while experiencing bad quality. When a handover is somehow triggered, it is because some parameter is not as desired. Consequently a handover is performed when the link is working in its "worst-case" conditions. This will cause a certain risk for a handover failure, which can lead to a dropped call.

In this section, simulation results, related to the handover algorithm, can be found. 4 different scenarios are used, by combining 2 speeds (3 km/h and 50 km/h) with 2 reuses (3/9 and 1/3). In all cases 3 frequencies are used per cell, giving 24 channels. The networks are loaded to 20%, while a cell radius of 3 km is used. The power control algorithm, described in Chapter 7, is being used with the optimal settings. The handover parameters have been optimised and they can be seen in Table 24.

Table 24. Handover parameters and the optimal settings used in CAPACITY.

Window RXLEV averaging	10 SACCH frames (no weighting)
Window RXQUAL averaging	1 SACCH frame (no weighting)
Window RXLEV neighbouring cells averaging	10 SACCH frames (no weighting)
Level Threshold	1 of the last 1 AV-RXLEV values < -110 dBm
Quality Threshold	3 of the last 4 AV-RXQUAL values >= 5
Interference Threshold	1 of the last AV-RXLEV > -85 dBm & 3 of the last 4 AV-RXQAL values >= 5
Power Budget	When level of a neighbour cell is more than 3 dB better than the level of the own cell.
Power Budget Interval	Every 6 SACCH frames (6 seconds)
Distance Handover	Disabled
Evaluation rule 1	The RXLEV of the candidate cell at least -104 dBm
Evaluation rule 2	Level of a candidate cell is more than 3 dB better than the level of the own cell.
Timer Successful Handover	5 SACCH frames
Timer Unsuccessful Handover	1 SACCH frame

'Power Budget Interval' is the interval in which the power budget trigger is checked once. So once per 3 seconds the power budget trigger is checked, while the other triggers are checked every half-second. 'Timer Successful Handover' and 'Timer Unsuccessful Handover' are the times until the next handover attempt after respectively a non-successful (blocked) handover and successful handover.

In the next subsections, AV-RXQUAL and AV-RXLEV statistics can be found. The handover reasons are studied as well as the output powers when a handover occurs.

It should be noted that the power control algorithm and the handover algorithm are running independently, but they are highly depending on each other. This should be kept in mind when setting the thresholds for these two algorithms. For example letting the power control algorithm turn down power at a RXQUAL value of 4, while a handover is triggered when the RXQUAL value becomes 5 or worse, will cause a lot of handovers, since the power control will cause a lot of mobiles to have a RXQUAL equal to 5.

2.2.1 RXQUAL Statistics

Figure 113 shows the statistics of the averaged RXQUAL values during the simulations. If the RXQUAL value of 0 is ignored, the distribution seems to be Gaussian around RXQUAL 3. This is due to the power control settings. The handover quality threshold is set at 5, which means that if 3 out of the 4 last AV-RXQUAL values are worse than 4, a quality handover is made. This means that about 1-7% of the total traffic will try to make a quality handover (assuming that the RXQUAL value does not change much during 3 SACCH frames), since between 2 and 20 % of the RXQUAL values equal to 5 or worse (since 3 'bad' RXQUAL values trigger a quality handover, this number has to be divided by 3). It should be noted that this fraction also includes the interference handover, since this is actually a kind of quality handover.

It can also be seen that the RXQUAL values of the 1/3 network are a bit worse than the RXQUAL values of the 3/9 network. Speed seems to have only little influence. More about this is found in Chapter 6.

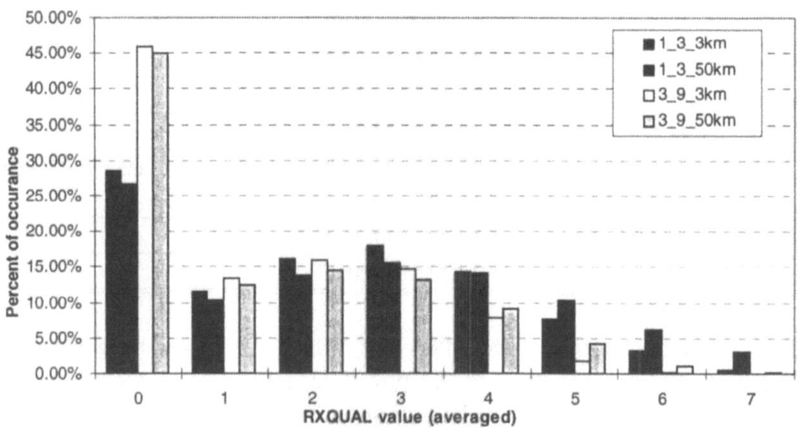

Figure 113. Averaged RXQUAL distribution for 1/3 and 3/9 reuse at 3 km/h and 50 km/h.

2.2.2 RXLEV Statistics

Figure 114 shows the distribution of the averaged received powers in the network during the simulations. As it can be seen there are no mobile stations, which have a received power below the -110 dBm, so no handovers based on level are made. This is as expected since the coverage is close to 100% in the simulator.

The received powers in the 1/3 network are higher. This is mainly caused by the power control, which does not turn down power as much in the 1/3 network as in the 3/9 network (see Chapter 7). An extra effect is that mobile stations can drive further out of the cell (cell stretching) in case of the 3/9 network, since the quality of a connection stays longer good.

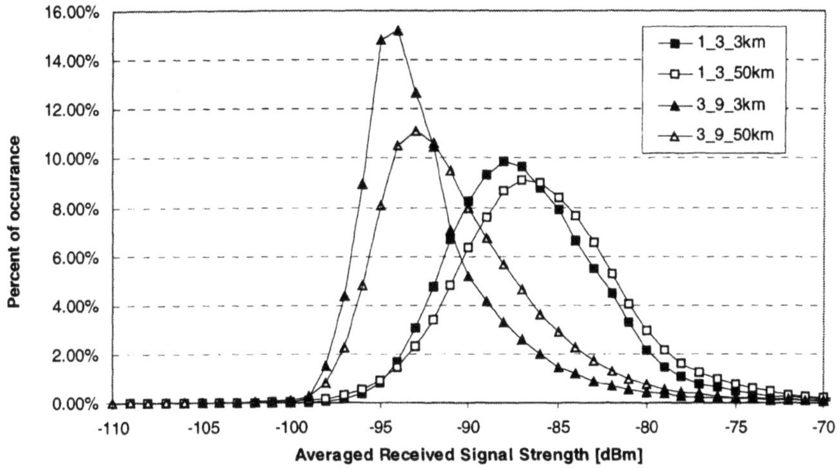

Figure 114. Averaged RXLEV distribution for the 1/3 and 3/9 reuse at 3 km/h and 50 km/h.

2.2.3 Handover Type - HO message

Figure 115 shows the distribution of handover reasons in the 4 different networks. As was concluded in the previous section no level handover are made. In the 3/9 network and in the 1/3 network with low speed most handovers are due to power budget, while in the 1/3 network with high speed a lot of quality handovers occur. This is not a desirable situation, since the failure rate for handovers is highest for quality handovers. It also can be seen that very little interference handovers occur, which is caused by the low load in the system.

It should be noted that by changing the quality threshold from RXQUAL 5 to 6, this distribution changes. The number of quality handovers will decrease while the number of power budget handovers will increase. The total amount of handovers and the quality in the network will remain the same. The quality threshold is chosen to be equal to RXQUAL 5.

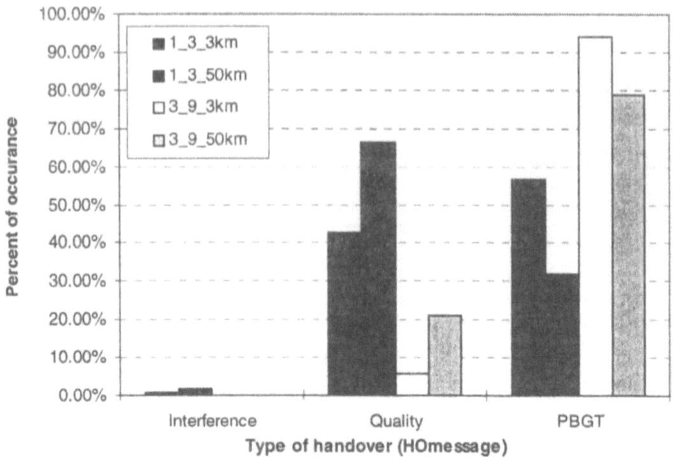

Figure 115. Handover causes for the 1/3 and 3/9 reused network at 3 km/h and 50 km/h.

2.2.4 BTS Output Power at Handover

The output power of the mobile stations at the handover is an important parameter. In the simulations all mobiles send at full power when they are making a handover. This is desirable in a macro cellular network, since the power control algorithm is designed in such a way that the power is maximum at the cell border. In a micro cellular environment this might however be different, due to cell overlaps. But that's not treated here. More about power control can be found in Chapter 7.

2.3 Live Network Measurements

In this section some handover statistics of a live GSM network are shown are compared to the simulation results. The area considered is Copenhagen, which can characterised as an urban area. Figure 116 shows the RXQUAL distribution for up and downlink on 2 different days. When compared to the simulation results in Figure 113, it is clear that the results from the live network are much better, i.e. there are more RXQUAL values equal to zero and less high RXQUAL values. The reason for this can be many, as e.g. the reuse is different, the traffic load differs or the environment is different than the environment modelled in CAPACITY.

It can also be seen that the quality on the downlink is slightly worse than the quality in the uplink.

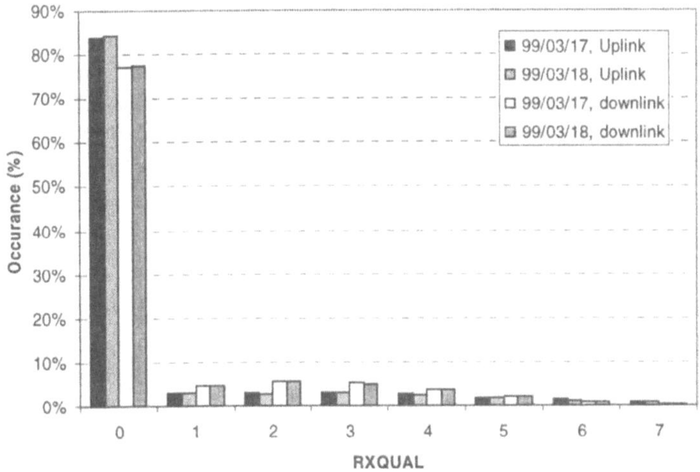

Figure 116. RXQUAL distribution of the considered area for up-/downlink on 2 days.

In the area, which is used for the measurements frequency hopping is used, so the interference handover is turned off (the reason for this will be explained in the next section). Figure 117 shows the distribution of the handover reasons in the area.

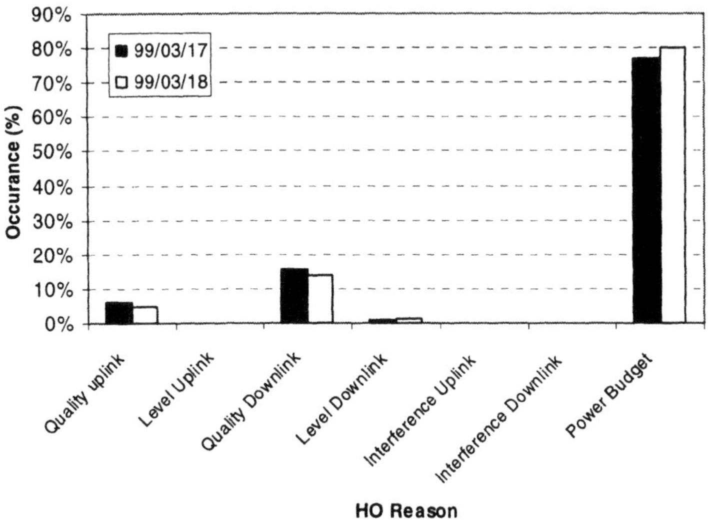

Figure 117. Distribution of handover causes in the considered area.

It can be seen that the mean reason for making a handover is power budget, while it is followed by quality. More handovers are made due to bad quality in the downlink compared to the number of handovers made due to bad quality in the uplink, which corresponds with the fact that the quality is worse in the downlink. Only a very few handovers are made for signal level reasons.

When compared to the simulation results in Figure 115, the results from the 3/9 reused network with a mobile speed of 50 km/h seems to fit quite nicely with the results from the measurements.

2.4 Frequency Hopping in Relation to Handovers

In Chapter 6, the introduction of frequency hopping in a live GSM network was described. Here again there is looked at this trial, but the focus is now on the handover statistics. A description of the complete trial can be found in Chapter 4.

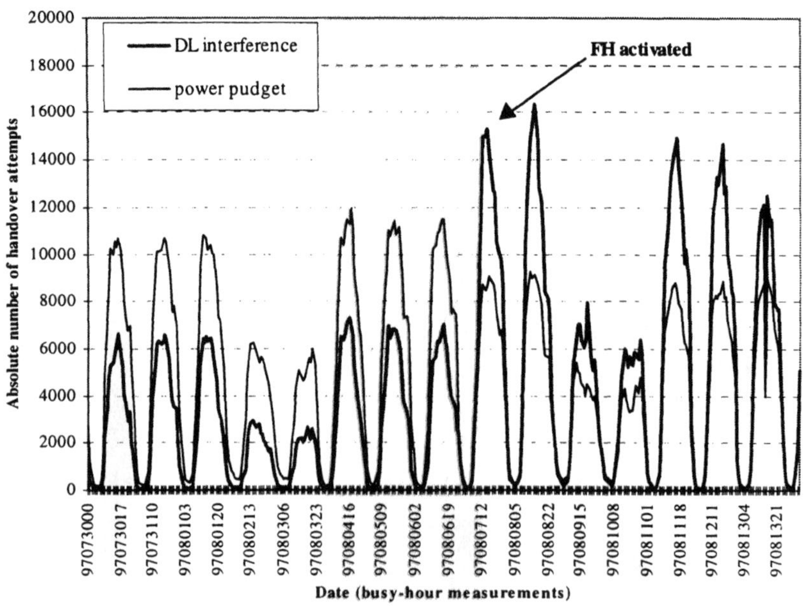

Figure 118. Handover causes of entire small area with/without frequency hopping. No other changes have taken place

The number of handovers, due to downlink (DL) interference and power budget during the trials is shown in Figure 118. The other handover reasons are not shown, since they do not change significantly. From the figure above it is obvious that after the introduction of frequency hopping, the distribution of handover causes has changed dramatically. Instead of using primarily power budget, as the SONOFON handover philosophy prescribes, the primary handover is now downlink interference. The absolute number of handovers is increased by 25%.

As described in Chapter 6, two things were done at this stage:

- The intra cell handover was turned off. In practice this means that interference triggers (uplink and downlink) are deactivated.
- A 2 dB HOMargin was introduced.

The effect from doing this is shown in Figure 119. The point were the changes were made is indicated with the label 'Changes'. The measurements are from 1 cell only.

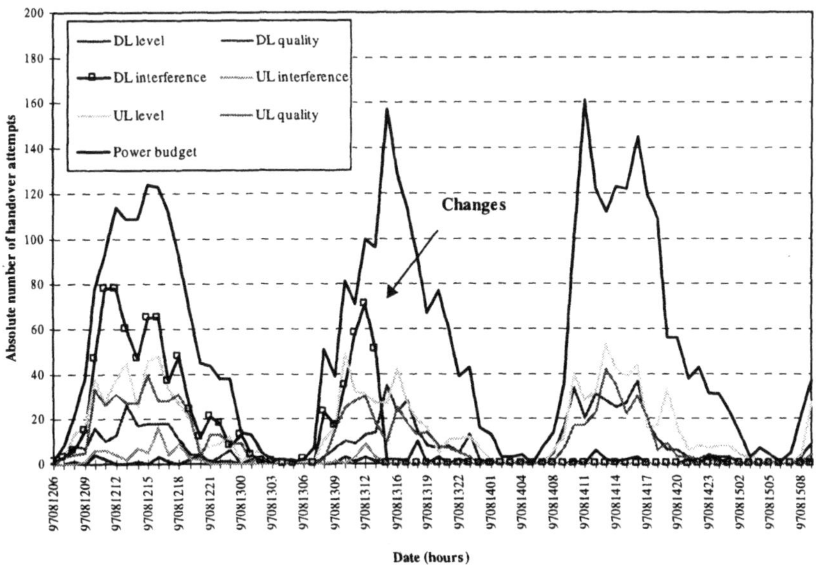

Figure 119. Handover statistics (one cell) during period of changes in handover settings.

It is seen how the number of handover attempts based on interference is simply removed. The other triggers do not seem to have been affected. What was also seen, was that the total number of handover attempts where reduced by 19 % after the change. That is, the absolute number of handover attempts is nearly back to the original value before introducing frequency hopping.

To investigate the effect of the quality trigger in the handover algorithm, the handover quality trigger is initially reduced from RXQUAL 4 to 5 on one site (004). This had the effect that the number of quality handovers decreased, while the quality, measured in dropped calls and RXQUAL was maintained. This change in quality handover was introduced in the whole network.

During the whole field trial the number of dropped calls was decreased, as was described in Chapter 6. The number of handover and distribution over the causes after introducing frequency hopping and optimisation is the same as for the non hopping network. The following was learnt from the field trial in relation to the handover algorithm:

- Intra cell handover should be turned off in a frequency hopping network, since it has no purpose and only causes more handovers.
- The quality threshold can be shifted, since a higher RXQUAL value is tolerated with frequency hopping than without frequency hopping. So if the threshold was at RXQUAL 4 before introducing hopping, then it can be set to RXQUAL 5 when using frequency hopping. The threshold could possibly also be set to RXQUAL 6, since RXQUAL 6 is representing partly good and partly bad quality, when using frequency hopping. This is however not tried in the trial.
- The HOMargin should be set to a positive number to avoid ping-pong handovers. This is not so much an effect of frequency hopping, but more a general finding for the handover algorithm.

3. THEORETICAL HANDOVER MODELLING

In this section different theoretical models to study handover algorithms are presented and evaluated. First in the next subsection, a simple model is developed to calculate the minimum number of handovers per mobile connection. Section 3.2 introduces the birth-death model for one cell, while this model is extended to multiple cells in Section 3.3. Section 3.4 studies the influence of mobility on the blocking statistics with the help of the developed models.

3.1 Simple theoretical analysis of handover probability

To relate different parameters, such as the speed of the mobile and the size of the cell, to the rate of handovers per call, the minimum probability of a handover has been analytically calculated using trigonometry and probability theory. The calculation is done using one cell of a 3-sector base station. The assumed cell coverage area is shown in Figure 120.

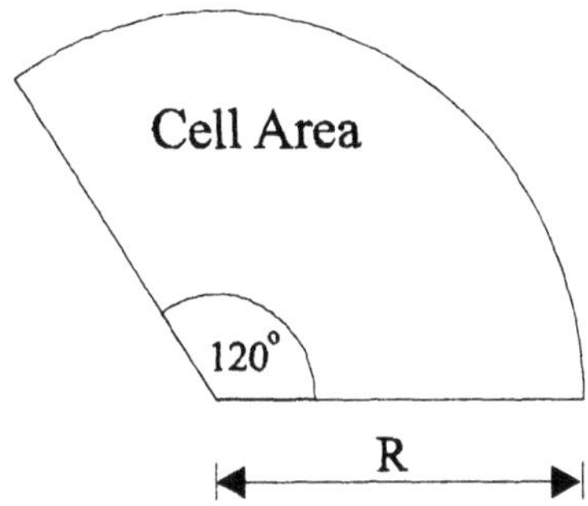

Figure 120. Cell area of the modelled sectored cell.

Handovers can occur for different reasons, like quality, level, interference, etc. The minimum number of inter cell handovers performed is equal to the number of mobile stations driving out of the cell area, shown in Figure 120. A parameter called, user mobility, μ_i, characterising the distance, which mobile station i drives during the mobile connection, is defined as:

$$\mu_i \equiv \frac{V_i T_i}{R} \qquad (2)$$

where V_i is the speed of the mobile station i, T_i is the speech time of mobile station i and R is the cell radius, as shown in Figure 120. The probability that mobile station i has to perform an inter cell handover is:

$$P_{hi} = P(\mu_i > d_i) \qquad (3)$$

where d_i is the normalised distance of mobile station i to the cell border. It is assumed that the mobile stations are placed randomly in the network and drive in one randomly chosen direction. The distribution of d_i for random placement and random directions can be seen in Figure 121.

This distribution can be approached by a normal distribution with mean $\mu = 0.25R$ and standard deviation $\sigma = 0.5R$, which also can be seen in Figure 121. A normal distribution is characterised by [219]:

$$p_x(x) = \frac{1}{\sigma\sqrt{2\pi}} e^{\frac{-(x-\mu)^2}{2\sigma^2}} \qquad (4)$$

where $p_x(x)$ is the chance that the outcome is x.

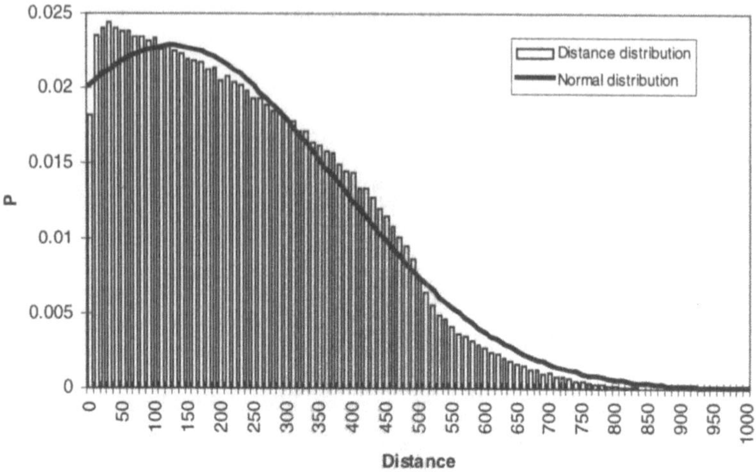

Figure 121. Distance distribution to the cell border d_i and the equivalent normal distribution.

Taking into account that the distance can not be negative, the distribution $p_n(x)$ of the distance to the cell border d_i can be written as

$$p_n(x) = \frac{2}{0.6915 \cdot \sqrt{2\pi}} e^{-2(x-0.25)^2} \qquad (5)$$

Since the speed of the mobile stations is constant, the distance, which a mobile station drives during one call, is only depending on the call duration. The call duration $p_e(x)$ is assumed to be negatively exponentially distributed according to:

$$p_e(x) = \frac{1}{b} e^{-\frac{x}{b}} \qquad (6)$$

where b is the mean. Using (6) and filling the mean call duration of 80 sec., the distribution of μ_i becomes:

$$p_\mu(x) = \frac{R}{80V} e^{-\frac{xR}{80V}} \qquad (7)$$

Using (3), (5) and (6), the intercell handover probability becomes:

$$P_h = \int_{x_n=0}^{\infty} \int_{x_\mu=x_n}^{\infty} p_\mu(x_\mu) p_n(x_n) dx_n dx_\mu \qquad (8)$$

Equation (8) can be simplified to:

$$P_h = \frac{2}{0.6915 \cdot \sqrt{2\pi}} \int_{x_n=0}^{\infty} e^{-2\cdot(x_n-0.25)^2 - \frac{x_n \cdot R}{80 \cdot V}} dx_n \qquad (9)$$

Figure 122 shows the handover probability as a function of the cell radius for a mobile speed of respectively 3 and 50 km/h.

In Figure 123 the average number of handovers per call is shown as function of cell radius R for two different speeds in the case of no blocking at access and handover. Also shown in this figure is the average number of handovers found by simulations in CAPACITY. All handovers were made based on distance only, i.e. a handover was made each time a cell boundary was crossed.

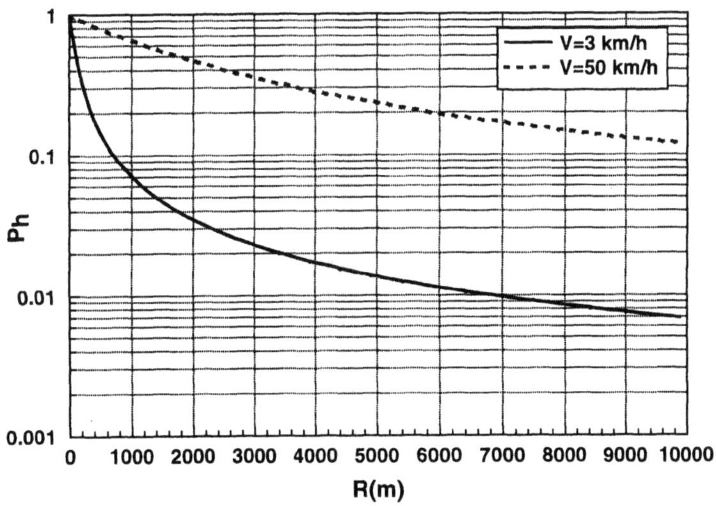

Figure 122. Handover probability as a function of *R* for a mobile speed of 3 and 50 km/h.

It can be seen that this average number of handovers per call is quite low. It should be remembered that it is a theoretical minimum, for example fading is not included in the calculations. It also seems that the theoretical model fits very nice with the points found by the simulations. The small differences are caused by the fact that a handover in the simulations is not independent of previous handovers. This has especially great influence at small cell sizes.

In fact this model would be true if no fading did exist and a handover was only made based on signal level. However the reality is much more complex, as the next sections will show.

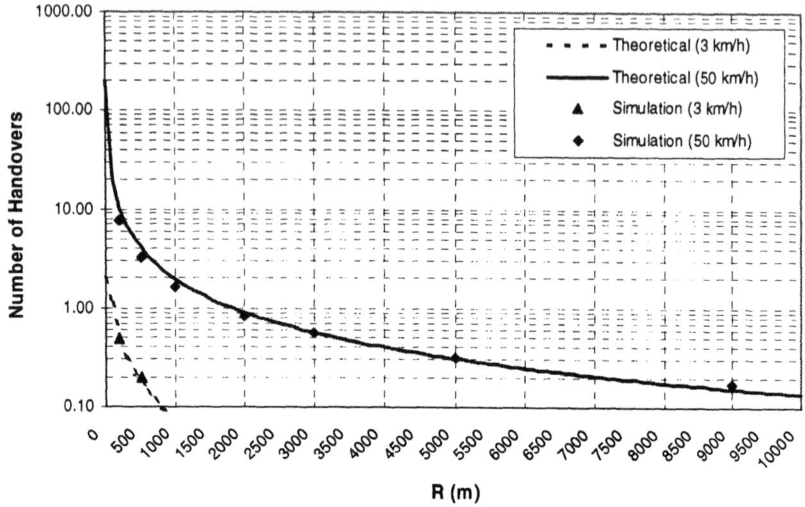

Figure 123. Average number of handovers per call as a function of the cell radius, R, for a mobile speeds of 3 and 50 km/h in the case of no blocking at access and handover.

3.2 Birth Dead Model

If a single cell with N cells is considered, a model, like is shown in Figure 124, can be made. As can be seen there are N states, corresponding to the number of channels in the cell. When there are k calls ongoing, then the cell is in state k. This kind of model is called a birth-death model [210]. The arrival process of calls is modelled by λ_i. λ_i reflects the probability of an arrival of a new call when the system is in state i, i.e. there are i calls ongoing in the cell. Similarly μ_i gives the probability of a call departure, when the system is in state i.

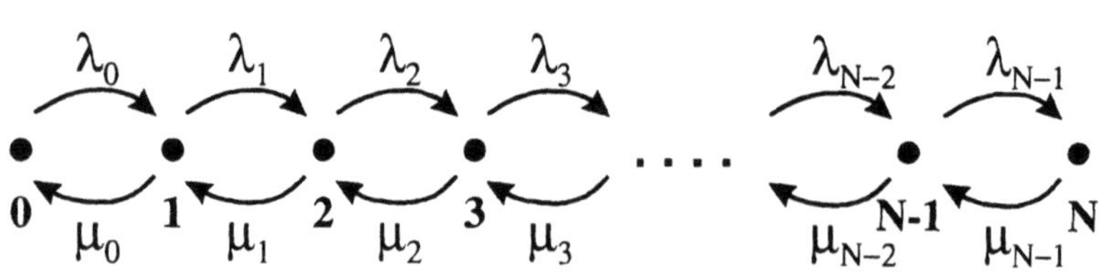

Figure 124. Birth-death process of a single cell with N channels.

A special case is when $\lambda_i=\lambda$, a constant. This means that the arrival process is independent of the number of calls in the cell. This is quite realistic in a mobile communications network, since it is not known how many calls there are in the system by the users. Similarly the departure process can be modelled as $\mu_i=i\mu$ (when more calls are in the system, the probability that more finish is higher). The arrival process is also called the birth process, while the departure process can be called death process. Four different processes can be found in a single cell in a mobile network:

- Arrival of new calls
- Handover away from the cell
- Handover into the cell
- Call finishes.

Those processes or events that cause a change of state in the cell under consideration are called transition events. It should be noted that some physical events that are pertinent to system performance, such as the arrival of a new call when all channels are occupied, do not cause a state transaction.

The arrival of new calls is normally modelled as a Poisson process [102], which can be described by:

$$p(k) = \frac{(\lambda T)^k e^{-\lambda T}}{k!} \tag{11}$$

where $p(k)$ is the probability of k arrivals in time T. λ is the arrival rate. The call finishing process is related to the call duration. The call duration is, as in [197], modelled as a negative exponential process with mean μ.

Each active mobile station in the network has a probability of making a handover p_h, so the handover rate out of a cell at time t is given by:

$$r_h(t) = p_h s(t) \tag{12}$$

where $s(t)$ is the state of the cell, i.e. the number of mobile stations. The handover away from a cell can be modelled by a Poisson process [198] with the handover rate $r_h(t)$. The same way handovers to a cell can be modelled by taking the amount of traffic in the neighbouring cells into account. The probability of making a handover p_h is an indication of the mobility in the system. High mobility means higher handover probability, while slow moving mobiles, for example pedestrians, are more associated by a low handover probability.

3.3 Multiple cells scenario

The model of the previous section is extended in this section to a multiple cell scenario A group of 7 cells is considered as can be seen in Figure 125. Cell 1, which will be called centre cell, has *N* channels, while the other cells have *cellN* channels. In the Figure two different kind of traffic streams can be seen:

- $A_{new}(n)$: this is the new traffic in cell *n*. As mentioned this is assumed to be Poisson distributed, depending on arrival rate λ.
- $A_{HO}(m,n)$: this is the amount of traffic, which makes a handover from cell *m* to cell *n*. Cells outside the 7 cells, i.e. the second tier, are not simulated, but the handover from those cells is. This handover traffic is denoted as $A_{HO}(o,n)$ and $A_{HO}(n,o)$, where *o* represents the second tier cells.

Some assumptions have to be made on this last handover (from first to second tier and vice versa), since normally the handover rate depends on the number of active mobiles in a cell (see equation (12)). It is not known how many mobiles there are active in the second tier cells, so it is assumed that a total of *CellN* mobiles are active in the second tier and possibly want to make a handover to the first tier.

The mean call time after a handover is equal to the mean of the whole call time μ, since the call duration of a call is assumed to be exponential distributed. The exponential distribution is memoryless and therefore the chance of a call finish at a certain time is independent of the actual call length at that time [103].

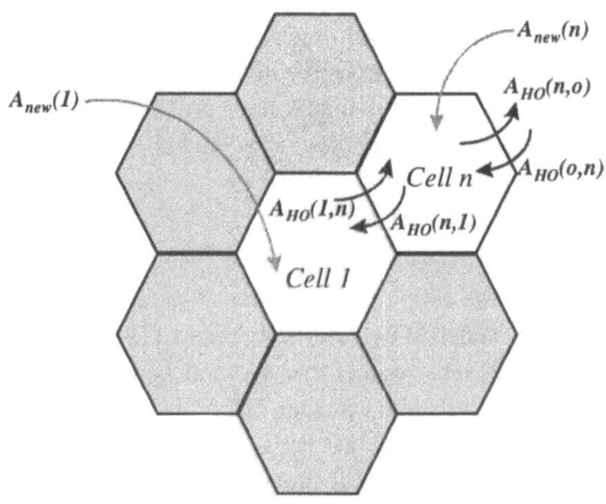

Figure 125. Considered scenario with 7 clustered cells.

Handover prioritising is being used. The meaning of this is that handovers have higher priority than new calls. Before a new mobile station is initialised in a cell, there is checked if there are any mobile stations, which want to make a handover to that cell. If this is the case, the handover is performed before the new call. When there still is a free channel left after all desired handovers has taken place or if there are no mobile stations that would like to make a handover, the new call can be initiated. Furthermore different handovers have different priorities. The order of handovers performed is as follows:

1. Handover from cells in the first tier to cells in the second tier.
2. Handover from cells in the first tier to the centre cell.
3. Handover from the centre cell to cells in the first tier.
4. Handover from cells in the second tier to cells in the first tier.

All this means minimises the blocking on handover coming from the centre cell, which is the cell of interest. It also should be noted that there are no handovers between the cells in the first tier. This is a simplification, which is made, since the main interest is in the centre cell.

The parameters for the cells in the first tier are kept the same, so the following is valid:

$$A_{new}(n) = A_{new\_1tier} \qquad \text{for } cell\ n \in 1 \text{ tier} \tag{13}$$

$$A_{HO}(1,n) = A_{HO\_to1tier} \qquad \text{for } cell\ n \in 1 \text{ tier} \tag{14}$$

$$A_{HO}(n,1) = A_{HO\_from1tier} \qquad \text{for } cell\ n \in 1 \text{ tier} \tag{15}$$

$$A_{HO}(o,n) = A_{HO\_from2tier} \qquad \text{for } cell\ n \in 1 \text{ tier} \tag{16}$$

$$A_{HO}(n,o) = A_{HO\_to2tier} \qquad \text{for } cell\ n \in 1 \text{ tier} \tag{17}$$

The parameter settings can be seen in Table 25.

Table 25. Parameter settings used for the multiple cell scenario.

Call duration	80 sec.
Simulated time	10000 sec.
P_h ***(from second tier cell to first tier cells)***	0.004
P_h ***(from first tier cells to second tier cells)***	0.01
P_h ***(between center cell and first tier cells)***	0.002
N	10
CellN	10
λ_{new} **(all cells)**	0.06

The handover probabilities need a bit more explanation. The handover probability between the center cell and the first tier cells is given as the handover probability between each of the two cells. Therefore the total handover probability from the centre cell to the first tier cells is 6 times this handover probability (since there are 6 first tier neighbours). The handover probability between the first and second tier represents all the handover activity between the first tier cells and all other cells except cell 1, so in a regular grid this means 5 cells. That is the reason for P_h *(from first tier cells to second tier cells)* to be 5 times greater than P_h *(between centre cell and first tier cells)*. Furthermore P_h *(from second tier cell to first tier cells)* is lower since it is assumed that there are always *CellN* mobiles active in the second tier cells. The value of 0.004 was found by tuning it, so that the net number of handovers between second tier cells and first tier cells was about zero. With these settings the results in Table *26* were achieved.

Table 26. Results from the simple general blocking simulations.

Load *Cell1* and cells in first tier	50%
Number of new calls in centre cell	619
Number of new calls in the first tier cells	3597
Number of handovers from the centre cell to cells in the first tier	512
Number of handovers from first tier cells to the centre cell.	509
Number of handovers from first tier to second tier	2593
Number of handovers from second tier to first tier	2226
Blocking new calls centre cell	2.1%
Blocking new calls first tier cells	1.1%
Blocking HO from first tier cells to the centre cell.	1.8%
Blocking HO from the centre cell to first tier cells	0.2%
Blocking HO from second tier cells to first tier cells	0.9%
Total blocking ingoing traffic to centre cell (HO and new calls)	1.9%
Total blocking cells in first tier(HO and new calls)	1.0%

It is well known that if it is assumed that blocked calls are cleared then the blocking probability P_B can be approximated by the Erlang B formula [210]:

$$P_B = \frac{A^N / N!}{\sum_{l=0}^{N} A^l / l!} \tag{18}$$

where N is the number of channels and A is the offered traffic in Erlang. Tables exist that describe the blocking for combinations of number of channels and load [109]. By the use of Equation (18) the blocking can be calculated. First the average call time within one cell has to be determined. This can be done with the following equation:

$$Call\_in\_cell\_time = \left(\frac{N_{new} + N_{HOin} - N_{HOout}}{N_{new} + N_{HOin}} \right) \cdot Call\_time \tag{19}$$

where N_{new} is the number of new calls in the cell, N_{HOin} is the number of handovers into the cell, N_{HOout} is the number of handovers out of the cell and *Call_time* is the average call time per mobile. So if there are 100 new calls in a cell and 100 handovers in and out of the cell, the average call time within the cell is half of the total call time. This can easily be understood since the total traffic stays the same in the cell, while the amount of incoming traffic (from handover and new calls) is doubled, if compared to the situation where there is no handover.

Applying this to the results in Table 26 gives a *Call_in_cell_time* 44 sec. and 41 sec. for respectively the centre cell and the first tier cells Now simply the offered traffic per cell can be calculated and the Erlang-B equation can be applied. The offered traffic for the centre cell and the first tier cells can be seen in respectively Table 27 and Table 28.

Table 27. Offered traffic in the centre cell.

Offered traffic from new calls	619*0.44/1000 = 2.7 Erlang
Offered traffic by handover from first tier cells	509*0.44/100 = 2.2 Erlang
Total	4.9 Erlang

Table 28. Offered traffic in the first tier cells.

Offered traffic from new calls	3597*0.41/(100*6) = 2.5 Erlang/cell
Offered traffic by handovers from second tier cells	2226*0.41/(100*6) = 1.5 Erlang/cell
Offered traffic by handovers from the centre cell	512*0.41/(100*6) = 0.3 Erlang/cell
Total	4.3 Erlang

Using Erlang-B tables a blocking of 1.7% and 0.8% is found for respectively the centre cell and the cells in the first tier. This fits quite well with the simulated values in Table 26.

3.4 Mobility Dependency

In the previous section a handover probability, p_h was defined. This is a measure of the mobility of a mobile station relative to the cell size, so:

$$p_h \cong \alpha \frac{V}{R} \tag{20}$$

where V is the speed of the mobile station, R the cell radius and α a constant.

The influence of the mobility in the network can be studied by varying the handover probability. Under the condition of blocked calls cleared the blocking can be calculated by:

$$P_B = E_{1,N}(A) \tag{21}$$

where $E_{1,N}(A)$ is the result of Equation (18) with N channels and A Erlang offered traffic. The offered traffic can be calculated (using (19)) as follows:

$$\begin{aligned} A &= \left(N_{new} + N_{HOin}\right)\left(\frac{N_{new} + N_{HOin} - N_{HOout}}{N_{new} + N_{HOin}}\right) \cdot Call\_time \Rightarrow \\ A &= (N_{new} + N_{HOin} - N_{HOout}) \cdot Call\_time \end{aligned} \tag{22}$$

where N_{new} is the number of new calls in the cell, N_{HOin} is the number of handovers into the cell, N_{HOout} is the number of handovers out of the cell and *Call_time* is the average call time per mobile. If it is assumed that the system is in balance, i.e. the number handovers out of the cell is the same as the number of handovers going into the cell is equal, (22) can be simplified to:

$$A = N_{new} \cdot Call\_time \tag{23}$$

This leads to the conclusion that the mobility has no influence on the blocking statistics, if the handover is in balance.

The model described in the previous section has been used to study the dependency of the mobility. The handover probability between the centre cell and the first tier cells is varied from 0 to 0.02, while the number of channels in all cells is set equal to 100. The other handover probabilities are scaled, so that the handover probability between 2 cells is always equal to p_h and the handover traffic is the same in the two directions, i.e. the system is in balance.

Figure 126 shows the overall blocking versus offered traffic. Also the theoretical Erlang-B curve is drawn. At offered traffic values higher than 75 Erlang, the curve fits the results quite well, but at lower amounts of traffic the results do not fit well with the Erlang-B curve. This is believed to be caused by the way the handovers and new calls are prioritised. First the handover into the centre cell are performed, followed by the handovers out of the centre cell, after which the new calls are started. This lead to a higher blocking on the handovers into the centre cell, since the mobile stations, wanting to make a handover out of the cell, are still there. This will especially be noticeable for high handover rates, which is valid for the points in the Figure, which do not fit the Erlang-B curve well.

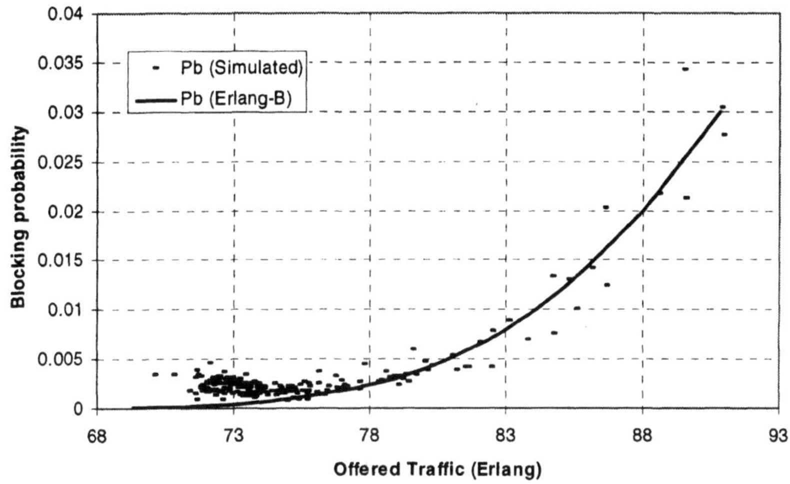

Figure 126. Overall blocking versus offered traffic.

4. HANDOVER IMPROVEMENTS

In this section a couple of enhancements, which can improve the performance of the handover algorithm are presented and studied. There are three kinds of improvements. First quality improvement, which simply means that the quality in the network is improved. Secondly the trunking efficiency can be improved, so the blocking on handovers and/or new calls decreases. The third improvement is a decrease in the number of handovers per mobile call.

In Section 4.1 channel reserving for handovers is studied, which is a method of decreasing the blocking on the handover traffic. Section 4.2 deals with a combination of queuing and channel reservation, while traffic reason handover is the subject of Section 4.3. Section 4.4 deals with a feature called dynamic HOMargin, which can improve the quality and decrease the number of handovers.

4.1 Channel Reservation for Handover Traffic

As mentioned in the introduction the blocking of a handover can lead to a dropped call. This means that a blocked handover may be worse than a blocked new call, since the users experience a call drop worse than a blocked call. A way of decreasing the number of blocked handover calls is to reserve some channels especially for the handover traffic.

Concentrating on the centre cell of the model presented in Section 3.3, it is known that there are two arrival processes. The handover arrival process and new call arrivals. Let there be N channels in the cell. Then N-m channels are available for both arrival processes, while only the last m channels are reserved for handover purposes. The two service processes have the same service rate μ, but different arrival rates. The new calls have arrival rate λ, while the handovers have arrival rate γ. This gives the state diagram as shown in Figure 127.

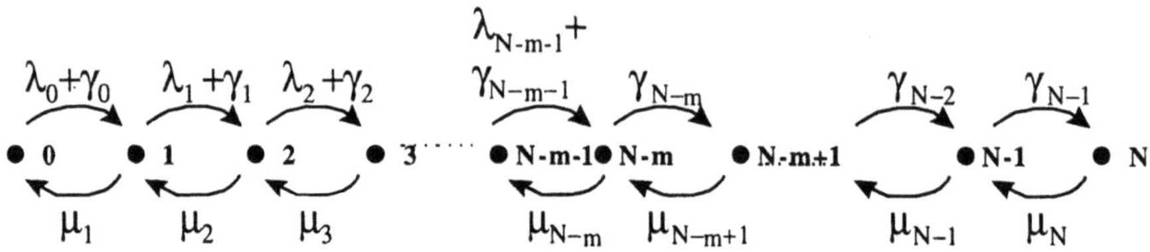

Figure 127. State diagram of a system with channel reservation.

The blocking probability of the handover and new calls is easily found by the use of balance equations:

$$P_{B\_HO} = \frac{A^N (1-r)^m / N!}{\sum_{l=0}^{N-m} (A^l / l!) + \sum_{l=N-m+1}^{N} \frac{A^l (1-r)^{l-(N-m)}}{l!}} \tag{24}$$

$$P_{B\_new} = 1 - \frac{\sum_{l=0}^{N-m-1} A^l / l!}{\sum_{l=0}^{N-m} (A^l / l!) + \sum_{l=N-m+1}^{N} \frac{A^l (1-r)^{l-(N-m)}}{l!}} \tag{25}$$

where N is the total number of channels, m is the number of channels reserved for handovers, r is the fraction of handover arrival, compared to the total traffic, and A is the total traffic offered in Erlang. So $A=(1-r)A$ is the amount of traffic (measured in Erlang) coming from handovers, while the amount of new traffic is equal to rA Erlang.

The blocking probabilities can be seen in Figure 128 for a different number of reserved channels and different r. The offered traffic is equal to 8 Erlang with 16 channels available.

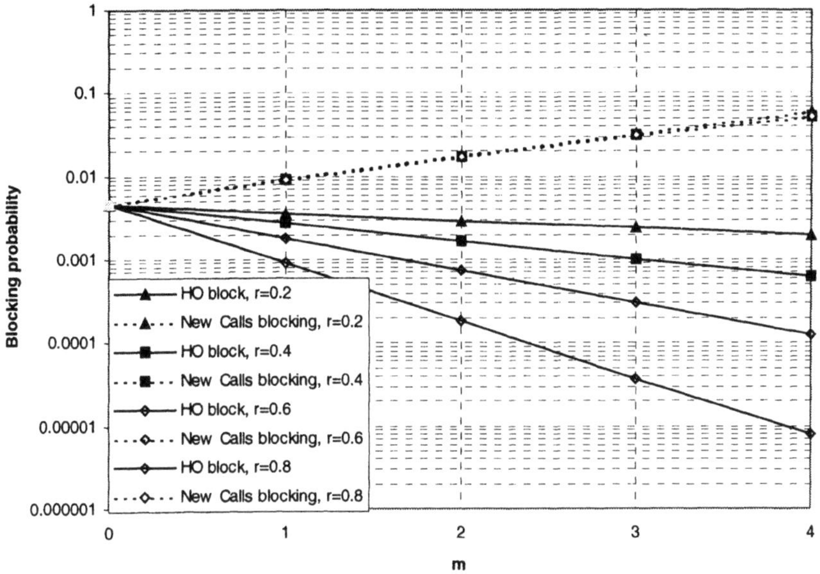

Figure 128. Theoretical blocking of new calls and handovers with channel reservation.

The situation with reserving timeslots has also been simulated with the model, which was presented in Section 3.3 and the simulated and theoretical results can be seen in Figure 129. It can be seen that there is a slight mismatch between the theory and simulation results. This caused by the fact that handovers have higher priority than new calls. This causes the new call blocking in the simulations to increase, while the handover blocking in the simulations decrease.

The results show that reserving some channels for handover can decrease the handover blocking. This lead however to more blocked new calls. So how many channels should be reserved for handover purposes? This depends on the costs, associated with blocked handovers and blocked new calls.

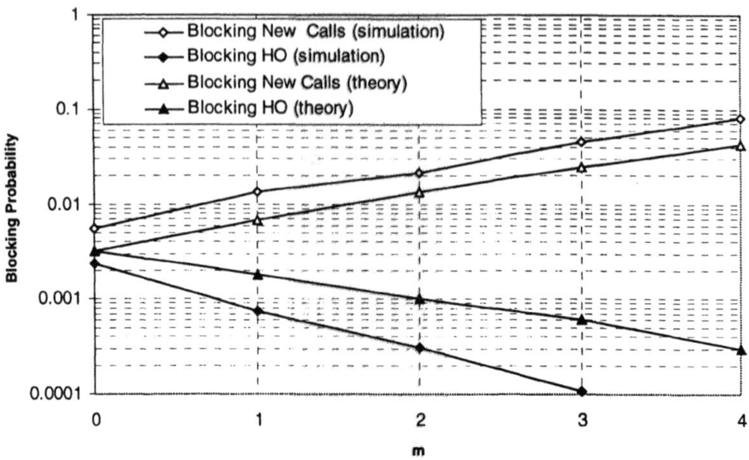

Figure 129. Theoretical and simulated blocking of new calls and handovers for $r = 0.45$ and $A = 7.7$ Erlang, $N = 16$ and varying m.

The total cost of blocked calls can be expressed as:

$$C = A \cdot r \cdot P_{B\_new} \cdot C_{B\_new} + A \cdot (1 - r) \cdot P_{B\_HO} \cdot C_{B\_HO} \tag{26}$$

where C_{B_new} and C_{B_HO} are the costs of the new call blocking (per Erlang) and handover call blocking (per Erlang). Figure 130 and Figure 131 show the cost as a function of m, while $N = 16$, $a = 8$ and the C_{B_new}/C_{B_HO} ratio and r are varied.

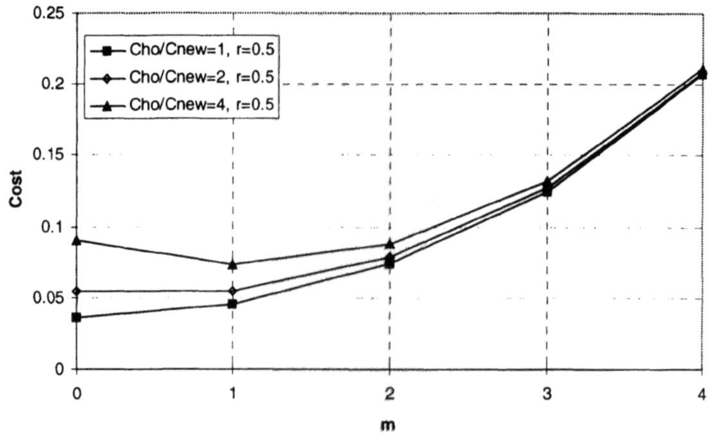

Figure 130. The cost as function of m for different C_{ho}/C_{new} ratios, $N = 16$, $a = 8$ and $r = 0.5$.

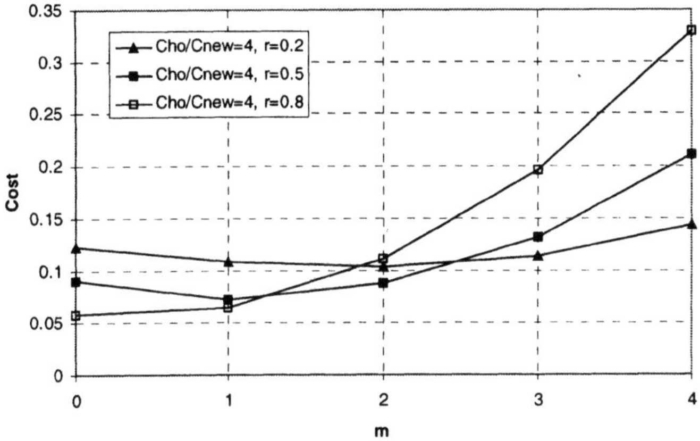

Figure 131. Relative cost as a function of m for different r, $C_{ho}/C_{new} = 4$, $N = 16$ and $a = 8$.

The figures shows that handover reservation is a good idea, when the costs associated with the blocking of handover traffic are higher than the costs associated with the blocking of new calls. Also at high handover rates, which leads to a higher r, channel reservation is useful. Such a situation may for example be found, when micro cells are used in a GSM network. In Figure 132 the cost is shown for a system with 100 channels per cell and an offered traffic equal to 80 Erlang. It can be seen that also in this situation it is a good idea to use handover reservation when the cost associated with a blocked handover is higher than the coast associated with a blocked new call. In general it can be concluded that reserving relative a few channels often gives the optimal situation.

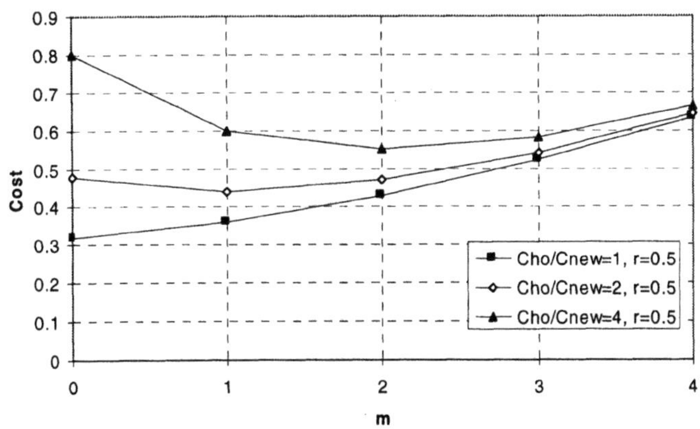

Figure 132. Cost as a function of m for different C_{ho}/C_{new} ratios, $N = 100$, $a = 80$ and $r = 0.5$.

4.2 Channel Reservation Combined with Queuing

The previous section showed that channel reservation for handover can decrease the blocking on handovers at the prize of a higher blocking at the new calls. Furthermore, queuing can decrease the blocking on new calls why the combination is studied in the following.

4.2.1 Analytical Model

Using queuing on new calls can decrease the blocking. The flow diagram for such a system with reserving channels for handover and queuing on new calls can be seen in Figure 133. Each state now consists of two numbers i_1 and i_2. The first one indicates how many calls are standing in the queue, while the second indicates how many channels are in use. It is assumed that both kinds of traffic, new calls and handover calls have the same service time and that their arrival rates are γ and λ respectively. By solving the balance equations the blocking for the handover calls and the average queue length can be found [86]. The blocking of the handover calls is given by Equation (27), (28) and (29).

$$P_{B\_HO} = P_{0,0}\left(\left(\frac{\lambda+\gamma}{\gamma}\right)^{N-M}\frac{(\gamma/\mu)^N}{N!}-\frac{\lambda/\mu}{N-M}\frac{(N-M)!}{(\gamma/\mu)^{N-M}}\frac{(\gamma/\mu)^N}{N!}\left(\sum_{l=0}^{N-M-1}\frac{\left(\lambda+\gamma/\mu\right)^l}{l!}\right)\right) \tag{27}$$
$$+\frac{\lambda/\mu}{N-M}\frac{(N-M)!}{(\gamma/\mu)^{N-M}}\frac{(\gamma/\mu)^N}{N!}$$

$$P_{0,0}=\left(1-\frac{\lambda/\mu}{N-M}\frac{(N\_M)!}{(\gamma/\mu)^{N-M}}\sum_{i_2=N-M}^{N}\frac{(\gamma/\mu)^{i_2}}{i_2!}\right)\cdot\Delta^{-1} \tag{28}$$

$$\Delta=\left(\sum_{i_2=0}^{N-M-1}\frac{(\lambda+\gamma/\mu)^{i_2}}{i_2!}\right)\left(1-\frac{(N-M)!}{(\gamma/\mu)^{N-M}}\frac{\lambda/\mu}{N-M}\sum_{i_2=N-M}^{N}\frac{(\gamma/\mu)^{i_2}}{i_2!}\right) \tag{29}$$
$$+\left(\frac{\lambda+\gamma}{\gamma}\right)^{N-M}\left(\sum_{i_2=N-M}^{N}\frac{(\gamma/\mu)^{i_2}}{i_2!}\right)$$

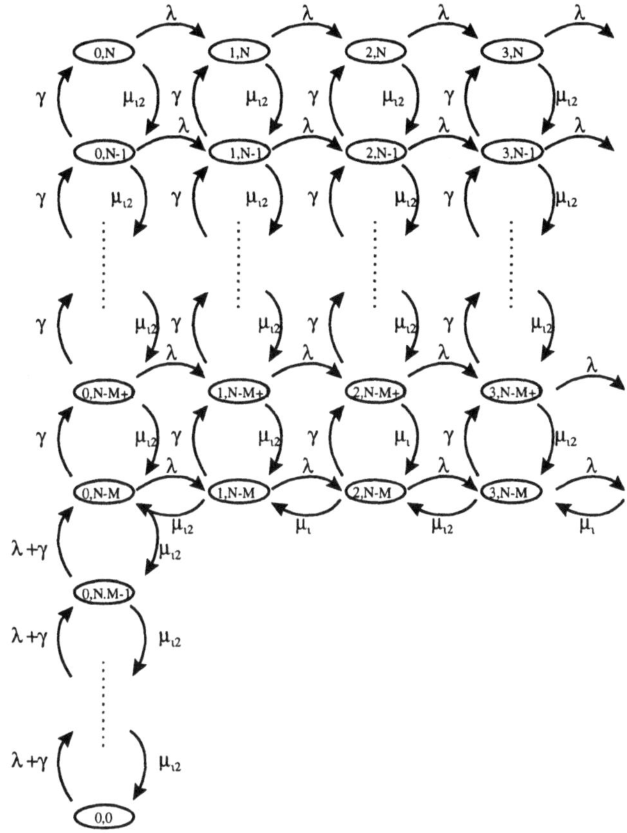

Figure 133. Flow diagram of system with channel reservation and queuing.

The probability of having new calls being delayed is equal to:

$$P_{Delay} = 1 - \left(\sum_{l=0}^{N-M-1} \frac{\left(\gamma + \lambda \big/ \mu \right)^{l}}{l!} \right) P_{0,0} \tag{30}$$

Figure 134 shows the blocking probability (handover) and the probability of having delay (new calls) for $N = 16$, $r = 0.45$ and 7.7 Erlang of total offered traffic. The results are very similar to the ones seen in Figure 129, except for the fact that the new calls are now delayed, not blocked.

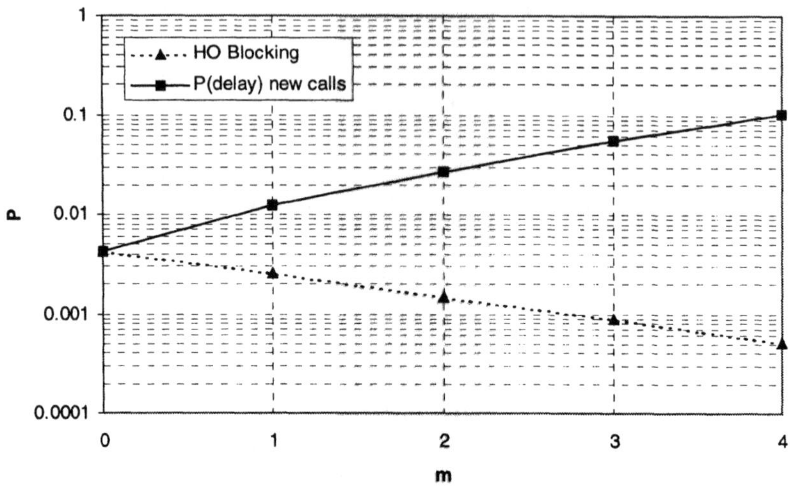

Figure 134. Blocking and probability of delay for $N = 16$, $r = 0.45$ and $A = 7.7$ Erlang.

4.2.2 Queuing in GSM

In GSM queuing can be introduced, since the specifications have no constraints in this area. Manufacturers are free to implement it.

There are several strategies to allocate a traffic channel, which can be grouped under the three following categories:

- **Very Early Assignment (VEA)** consists in allocating a TCH/x at initial assignment, when probable that the requested service will need such a channel.
- **Early Assignment (EA)** consists in allocating a SDCCH initially, then subsequently allocating a TCH as soon as it is known for sure that this type of service will be required.
- **Off Air Call Set Up (OACSU)** consists in allocating a SDCCH initially, then waiting until the called party has answered before attempting the subsequent assignment of a TCH/x.

Each of these strategies has its advantages and disadvantages in terms of capacity and grade of service. The VEA gives the shortest call setup, but is most capacity inefficient, since it can use a TCH/x when not necessary, for example at MS location update. The main drawback of EA compared to VEA is that the setup time is unnecessarily increased if a TCH/x is needed.

OACSU is the most efficient of the three, if looked at the capacity, since no TCH/x is allocated before the called party has answered the call. However there might be a noticeable delay in establishing a speech connection. The grade of service of this delay is by many GSM operators considered to be unacceptable [151]. In terms of grade of service the VEA and EA are nearly identical, whereas the EA and OACSU are more alike in terms of channel allocation strategy.

For the EA assignment it is possible to introduce a small queue period on the SDCCH channel when all TCH channels are being used (it is assumed that the SDCCH is congestion free). The call attempt will be on hold during the queuing time before blocking the call. The effectiveness of this queuing lies in parameters like: the maximum queuing length, the number of TCH channels and the mean call time in the cell. Queuing is called very efficient when [66]:

$$Queuingperiod \geq \frac{Mean\_calltime\_in\_cell}{Traffic\_Channels / cell} \tag{31}$$

This means that the queuing period should be at least as long as the average time in which one channel becomes free in a cell. When the mean call time in a cell is about 40 seconds and there are 24 channels per cell, the queuing length should be at least 1.7 seconds to be efficient.

4.2.3 Simulations

Queuing was implemented in CAPACITY. A 3/9 reused network, loaded 65%, has been simulated with the default settings (Chapter 5). Figure 135 shows the estimated PDF and CDF of the queuing time.

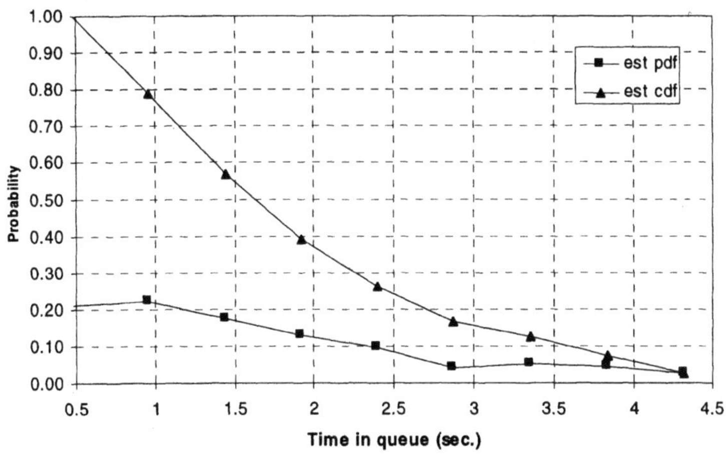

Figure 135. Estimated PDF and CDF of the time spend in queue before entering the network. Only the mobiles in queue are used. Maximum queue length has been 4.4 sec.

This is only for the mobile stations in the queue, i.e. mobile stations not coming in the queue are left out. The maximum length of the queue was 4.4 sec., which means that after 4.4 sec. in the queue the mobile stations (after trying to access the network once more) are blocked. About 2 % of the mobile stations come in the queue, i.e. 98 % of the mobile stations are accessing the network directly without coming in the queue.

It can be seen that in 38% of the cases the queuing period is longer than or equal to 2 seconds and about 43% of the mobile stations in the queue are there only for less one second. So only a small portion of the mobile stations comes in the queue (about 2%) and most mobile stations in the queue are only in the queue for a short period.

Figure 136 shows the delay and blocking probability of new calls as function of the number of channel reservations for handover, *m*. It can be seen that the probability of blocking and of delay increases with *m*, due to the fact that less channels become available for new calls. It can also be seen that longer queues give a lower blocking, since the mobiles have more chance to come into the network.

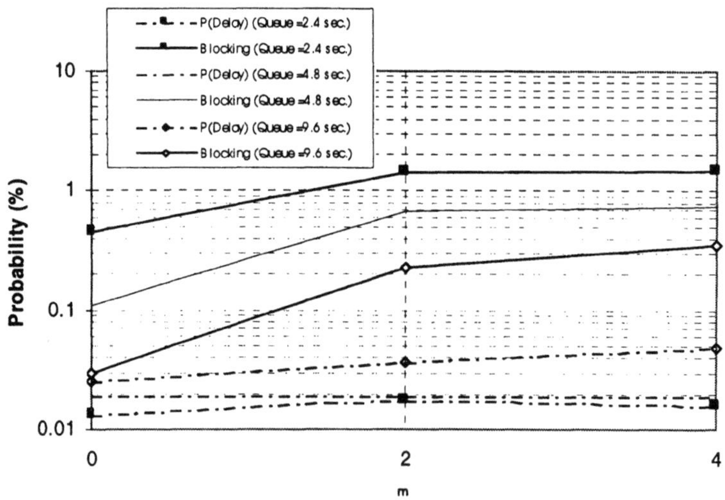

Figure 136. Delay probability and blocking probability of new calls as a function of the number of reserved handover channels (m).

Figure 137 shows the average waiting time for mobiles, which have been in the queue as function of the queue length. It can be seen that longer queue give longer average waiting times, while channel reservation also leads to a longer waiting time.

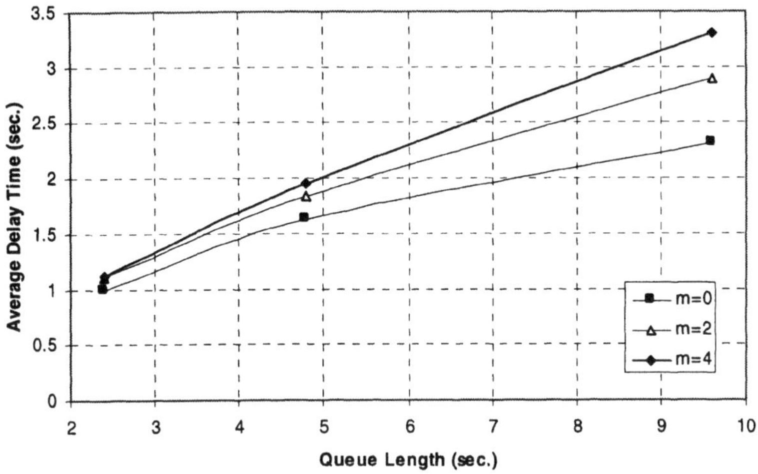

Figure 137. Average wait time as a function of the length of the queue for a different number of reserved channels for handover.

The following conclusions can be drawn:

- As already was concluded in the last section, channel reservation is often a good idea. It lowers the blocking on handover traffic. Relatively a few reserved channels often are enough to lower the handover blocking drastically. The prize to pay is that the blocking of new calls increases.
- Queuing can be used to lower the blocking on new call arrivals. The queue length in a network using effective queuing depends on the number of traffic channels and the average time in a cell. In GSM network a queuing length of a few seconds will decrease the blocking remarkably.
- Queuing on new calls and channel reservation for handovers can be used together to lower the blocking handover traffic and to minimise at the same time the blocking on new calls.

4.3 Traffic Reason Handover

In the previous section it was shown that queuing on new calls can decrease the blocking on new calls. Another way of doing is the use of traffic reason handover (TrHO), which is sometimes referred to as load sharing [232]. Traffic reason handover is a functionality that tries to equalise the traffic over the cells in the network. This can in some situations lead to less blocking and thereby improve the QoS.

Traffic reason handover is studied in this section. First in Section 4.3.1 the potential gain of traffic reason handover is analysed. Section 4.3.2 contains the description of the traffic reason handover algorithm, which is simulated. In Section 4.3.3 the simulation setup and results can be found, while the conclusions are in Section 4.3.4.

4.3.1 Potential gain from Traffic Reason Handover

In order to improve trunking efficiency it is advisable to equalise the traffic load over the cells. Directed retry and traffic reason handover make use of this principle. The first allows a new call to be served if the receiver is able to hear a transmitter of a neighbour cell [56] and is not considered here. Traffic reason handover allows calls to be served if the receiver or any other receiver in the cell is able to hear a transmitter in a neighbour cell. Traffic reason handover can be used to reallocate traffic from one cell to a neighbouring cell when close to congestion. The feature is based on the idea that neighbouring cells have an overlapping serving area.

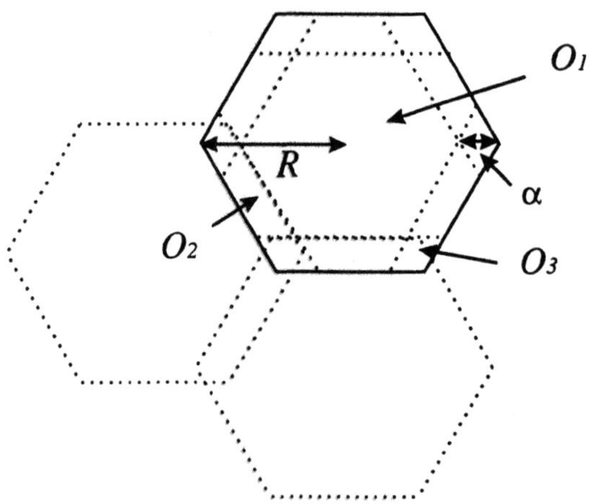

Figure 138. Cell with overlap area's O_1, O_2 and O_3.

A cell with a typical overlapping area can be seen in Figure 138. O_1 is the area where there is no overlap, while O_2 and O_3 are respectively the areas with overlap by 1 and 2 cells. R is the cell radius and α is the parameter, which indicates how large the overlap is.

If the load in all cells would be a constant, there would not be any gain from traffic reason handover. But since the load between cells is not 100% correlated, cells with a momentarily high load can direct traffic to neighbour cells, which have a momentarily lower load. This is the origin of the gain from traffic reason handover. The traffic in a cell with overlap, like in in Figure 138 , can be split into three kinds:

- The traffic in area O_1. P_1 is the probability of mobile station being in this area. These mobiles only have the possibility of accessing the channels of the own cell, N. The blocking probability for this traffic is called P_{b1}.
- The traffic in area O_2. P_2 is the probability of mobile stations being in this area. These mobiles have the possibility of accessing the channels of the own cell, but also the channels of one neighbouring cell. The blocking probability for this traffic is called P_{b2}.
- The traffic in area O_3. P_3 is the probability of mobile stations being in this area. These mobiles have the possibility of accessing their own cell, but also the channels of two neighbouring cells. P_{b3} is the blocking probability for this traffic.

Figure 139 shows a simplified overflow model, which can be used to find the gain from traffic reason handover. A is the offered traffic in a cell, while N is the number of channels per cell.

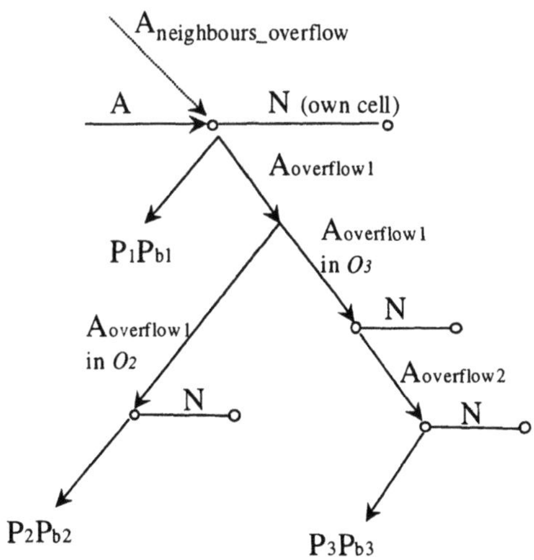

Figure 139. Simplified overflow model for traffic reason handover.

It is assumed that the traffic in the areas with overlap only is allowed to use a part of all available channels of the own cell, called m, since this traffic has the possibility of accessing the channels from the other cells. In other words the traffic in the overlap areas has lower priority than the traffic in the areas with no overlap.

The traffic in O_1 can only access N channels. This leads to blocking P_{b1} for P_1 of all mobile stations. The traffic in the areas with overlap tries to access m channels in the own cell first. The blocked traffic is treated as overflow traffic to the neighbouring cells ($A_{overflow1}$). This overflow traffic can be split into two parts:

- The traffic is in area with overlap of 1 cell ($A_{overflow1}$ *in* O_2). This traffic tries to access one of the N channels in a neighbour cell. The resulting blocking probability is P_{b2} for P_2 of all mobile stations
- The part of the traffic in the area with overlap of 2 cells ($A_{overflow1}$ *in* O_3). This traffic is offered to N channels and the blocked traffic is treated as overflow again ($A_{overflow2}$) to another N channels (since there is an overlap of two cells). The resulting blocking probability is equal to P_{b3} for P_3 of all mobile stations.

Since neighbouring cells may also be using traffic reason handover, a certain amount of traffic is handed over back from them. This traffic is called $A_{neighbours\_overflow}$. in the Figure. This traffic is assumed to be on average equal to $A_{overflow1}$, when the average load in all cells is the same. However in time they vary.

With this the total blocking, P_B, becomes:

$$P_B = P_1 P_{B1} + P_2 P_{B2} + P_3 P_{B3} \tag{32}$$

It is assumed that the overflow traffic, like new call arrivals, is Poisson distributed, which according to [194] gives quite accurate and meaningful results. The best result is achieved, when all channels of the neighbouring cells are available, i.e. there is no traffic generated in the neighbour cells. This means that there is not coming any handover traffic back from the neighbouring cells, so $A_{neighbours\_overflow}$. is equal to zero. With the help of [210], this leads to the blocking probabilities P_{b0}, P_{b1} and P_{b2}, shown in Equations (33), (34) and (35).

$$P_{B1} = \frac{A^N (P_0)^m / N!}{\sum_{l=0}^{N-m} (A^l / l!) + \sum_{l=N-m+1}^{N} \frac{A^l (P_0)^{l-(N-m)}}{l!}} \tag{33}$$

$$P_{B2} = \frac{A_{overflow1\_in\_O2}{}^N / N!}{\sum_{l=0}^{N} A_{overflow1}\_in\_O_2{}^l / l!} \tag{34}$$

$$P_{B3} = \frac{A_{overflow2}{}^N / N!}{\sum_{l=0}^{N} A_{overflow2}{}^l / l!} \tag{35}$$

The results for an overall blocking probability of 2% in the case of different N and α, while m has iteratively been optimised, can be seen in Table 29. Also the cell overlap can be seen as fraction of the cell, so when α is equal to 0.1R, then the area's O_2 and O_3 cover 9% of the whole cell. Not only the results, for when all channels in the neighbour cells are available, can be seen, but also the results in the case that there is just as much traffic in the neighbour cells as in the considered cell. This traffic is assumed to be Poisson distributed, and the handover traffic between the cells has iteratively being balanced, so that the handover traffic on average is the same in both directions. The gains are relative to the situation without traffic reason handover.

It can be seen that traffic reason handover can increase the maximum capacity of a cell drastically. The highest gains are achieved, when all channels in the neighbour cells are available, i.e. their load is equal to 0%. But also with the neighbour cells having the same load as the centre cell, the capacity of the centre cell is increased. This is due to the fact that the traffic in all cells is Poisson distributed: in time the traffic streams are not correlated, which makes that the centre cell often can successfully redirect traffic to channels from the neighbour cells and vice versa.

It should be noted though that the results with the load in the neighbour cells being equal to zero, are very optimistic, since in real life this situation hardly will occur. The results with the neighbouring cells loaded just as much as the considered cell, on the other hand, are the worst case results. The gain of traffic reason handover relies on traffic variations, so the gain will be highest in case of hotspots, i.e. for different loads in neighbouring cells.

Table 29. Maximum cell capacity at 2% blocking for different α with and without TrHO in the case of 0 % and 100 % load in the neighbouring cells.

N	α (cell overlap as a fraction of the cell)		No TrHO [Erl]	Tr1HO with neigh-bour cells load = 0% [Erl]	Gain	TrHO with equal cell load [Erl]	Gain
8	0	(0.00)	3.6	3.6	0 %	3.6	0%
8	0.1R	(0.09)	3.6	4.7	31 %	4.0	11%
8	0.2R	(0.36)	3.6	6.3	75 %	4.3	19%
8	0.5R	(0.75)	3.6	20.8	478 %	5.4	50%
16	0	(0.00)	9.8	9.8	0 %	9.8	0%
16	0.1R	(0.09)	9.8	12.5	28%	10.6	8%
16	0.2R	(0.36)	9.8	16.5	68%	11.3	15%
16	0.5R	(0.75)	9.8	51.1	421%	13.0	33%
24	0	(0.00)	16.6	16.6	0%	16.6	0%
24	0.1R	(0.09)	16.6	21.1	27%	17.8	6%
24	0.2R	(0.36)	16.6	27.6	66%	18.8	13%
24	0.5R	(0.75)	16.6	83.1	401%	21.0	27%

A large overlap gives more capacity than a small overlap, but even by just having a small overlap a considerable gain is achieved. The overlap of 0.1R (overlapping area is equal to 9 % of the cell area) gives a gain of at least 6 %, whereas if the overlap is equal to *0.5R* (overlapping area is equal to 75 % of the cell area) the capacity gain is 27 %.

4.3.2 The Implemented Traffic Reason Handover Algorithm

Traffic Reason handover was implemented in CAPACITY. It was added to the existing handover algorithm, which was described in Section 2.1. As was described the parameter HOMargin is used in the power budget trigger and in the handover candidate cells evaluation. HOMargin is kept the same for the whole network during time. With traffic reason handover a certain offset in level per cell is introduced. This offset (Δ (load (cell))) is depending on the load in the cell. Equation (1) becomes:

$$(\overline{L(j)} - \Delta(load(j))) - (\overline{L(i)} - \Delta(load(i))) > \text{HOMargin} \qquad (36)$$

where *Δ (load(k))* is the offset depending on the load in cell *k*. Some examples of possible sets of offsets can be seen in Table 30.

Table 30. Offset (Δ (load (cell))) for different cell loads.

LOAD	SET1	SET2	SET3
≤50%	0 dB	0 dB	0 dB
>50%	1 dB	0 dB	0 dB
>60%	2 dB	1 dB	0 dB
>70%	4 dB	2 dB	1 dB
>80%	8 dB	4 dB	2 dB
>90%	16 dB	8 dB	4 dB

When set 2 is used, and a certain mobile, mobile A, in cell i wants to make a handover. Cell j, which is loaded only 20%, is the only candidate. The load in cell i is 91%. At the same time there is a mobile station, called mobile B, in cell j, wants to make a handover to cell i, which in this case is the only candidate. Mobile A will make the handover if the averaged measured level of candidate cell j is HOMargin-8dB higher than the averaged received level in the own cell. So if a HOMargin of 6 dB is being used, the averaged measured level of the candidate cell can be 2 dB below the averaged received level of the own cell. For mobile B to make a handover however the averaged measured level of candidate cell i has to be HOMargin+8dB higher, so 14 dB in case of a HOMargin equal to 6 dB. In other words it is much harder to come to a cell with a high load than to a cell with a low load.

The overlap between two cells is dependent on the difference in offsets in both directions. If this difference is equal to zero, then there is no overlap, while when it is 14 dB, as in the example above, then the overlap has the size of the area covering this 14 dB. The size of the overlap is constantly changed with the load, but α (see

Figure 138) never becomes greater than 0.22R (corresponding to the overlap areas covering 39% of the cell) with these settings.

4.3.3 Network Simulations of Traffic Reason Handover

The investigated network structure is based on cells with a radius of 3 km and a regular base station grid with 48 3-sector base stations, corresponding to 144 cells. Each cell has 3 TRXs and frequencies, so 24 channels are available per cell. The frequency reuse is 3/9 and the BCCH is neglected. The mobile speed is 50 km/h and random hopping is used. HOMargin is set to 6 dB, while the load in the cells is averaged over 5 seconds before the correction factors are calculated (Table 30). The other simulation parameters can be found Chapter 5.

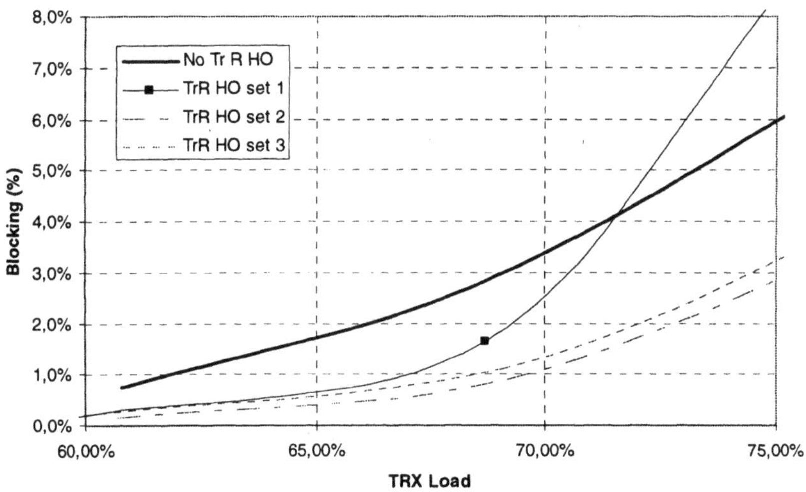

Figure 140. Blocking probability without traffic reason handover and for the 3 TrHO cases.

Figure 140 shows the blocking results of the simulated network. The results of no traffic reason handover and with the 3 different traffic reasons handover settings from Table 30 can be seen. It can be seen that with the use of traffic reason handover the blocking decreases. The network without traffic reason handover could be loaded up to 66 %[23], when the blocking limit is 2 %. With traffic reason handover the blocking is decreased to about 0.5 % (set 2). If a 2 % hard blocking limit the network with traffic reason handover can be loaded up to 73 %, so a gain of about 10 % in capacity is achieved. The capacity gain from traffic reason handover will be

[23] This is the amount of carried traffic.

higher for a lower number of channels per cell, since the trunking efficiency is already quite good for 24 channels per cell.

Set2 and set3 give better results than set1 of the traffic reason handover settings. This is due to the fact that set 1 is too aggressive: Δ can become much greater than the HOMargin.

The capacity gain is slightly lower than the capacity gain in Section 4.3.1 with $\alpha = 0.2R$. This is caused by the fact that a single cell and its neighbours are considered, as in the theoretical model, but an interacting dynamic network is simulated.

It should be noted that the simulations give worst case results, since the traffic is uniform distributed. So this gain should be viewed upon as minimum gain. In a real life network with hotspots, the gain will be higher.

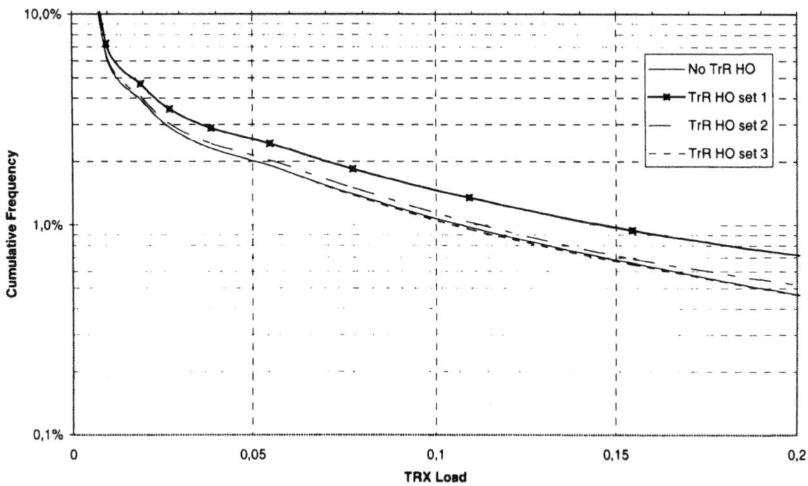

Figure 141. FER distribution (estimated CDF) without TrHO and for the 3 TrHO cases.

Figure 141 shows the quality of the network for the different settings. The estimated CDF of the FER samples (all mobile stations generate a FER sample for every 0.48 seconds) in the case of no traffic reason handover and with traffic reason handover is depicted. The load in all cases is about 77%. It can be seen that in all cases 10% of the FER samples have are worse than 1%. It also can be seen that the network quality is maintained while using traffic reason handover with this overlap and reuse, except for set 1, which worsens the network quality slightly.

The average number of handovers per mobile station slightly increases. In case of set1 only 1%, but in case of set 2 and 3 respectively 16 and 6%. This is the prize of traffic reason handover.

4.3.4 Conclusions on Traffic Reason Handover

Traffic reason handover was studied analytically and by simulation in this section. The theoretical model showed that the capacity gain from traffic reason handover depends on the size of the overlapping area between cells, the number of channels per cell and distribution of the traffic. When the cell overlap only covers 9% of a cell, and the neighbouring cells carrying just as much traffic as the considered cell, a capacity gain of 6 % was achieved. Higher gains were found in other situations.

Simulations of a 3/9 reused network with uniform distributed traffic show that the blocking can be decreased from 2% to about 0.5% when using traffic reason handover, while the quality of the network remains the same. It was also shown that a capacity gain of about 10% was achieved by using traffic reason handover. The simulations give a worst case result, since the traffic was uniform distributed. In a real life network with moving hotspots, the gain will be higher. The prize to pay for the gain of traffic reason handover is that the average number of handovers per call increases.

4.4 Dynamic HO Margin

HOMargin is an important parameter in the GSM handover algorithm, as can be concluded from the previous sections. In Section 2.1 it was explained that the power budget handover trigger and the second candidate cell evaluation rule looks like:

$$\overline{L(j)} - \overline{L(i)} > \text{HOMargin} \tag{37}$$

where L(j) is the level (RXLEV_NCELL) of candidate cell j, measured by the considered mobile station, and L(i) is the received level (RXLEV) from the own cell (cell i). So HOMargin indicates how much better the candidate cell has to be, before a handover to that cell is made.

HOMargin has a large influence on the number of handovers. The network simulator CAPACITY was used to study this influence. A 3/9 reused network was simulated with 30% load. Random frequency hopping was used, while the speed of the mobiles was distributed uniformly between 3 and 50 km/h. The handover settings are as described in Table 24, while the power control settings can be found in Chapter 7 and the other settings are as described in Chapter 5.

In Figure 142 the average number of handovers per mobile station can be seen as function of HOMargin. As expected the number of handovers decreases with increasing HOMargin. Due to the shadow fading, mobile stations start making handovers forth and back between two cells, when the HOMargin is small. This is called the 'ping pong effect'. The location of the first handover will lie relatively far away from the cell border if the HOMargin is high.

The HOMargin influences the quality in 2 ways. First of all, every handover has a negative effect on the QoS, since during every handover the control channel FACCH steals some capacity from the traffic channel, which leads to errors in the traffic channel. Furthermore there is a chance of a handover failure, which can lead to a call drop. There is a positive effect as well, since the QoS improves when a mobile is handed over to a better cell. In Figure 143 the normalised frequency of having no frame erasures or having less than 2 frames erased during a SACCH multiframe can be seen. These probabilities should be as high as possible, since they represent good quality.

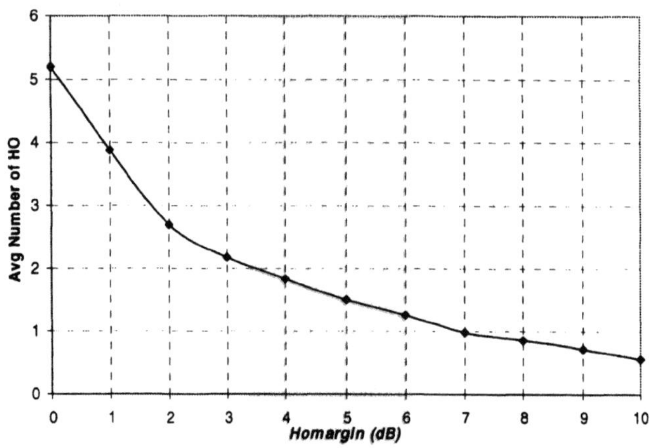

Figure 142. Average number of handovers per mobile station as function of HoMargin.

Figure 37 shows that a low HOMargin is preferable. Around 2-3 dB seems to be optimal. Lower values do not giver better quality. The probability of making a wrong decision, when the HOMargin is equal to 0 or 1 dB, is higher than when the HOMargin is equal to 3 dB. When making a wrong handover decision it always will take some time before a handover back can be made (minimum time between handovers in the simulations is set to 2.5 sec.). During this time bad quality is 'collected'.

On the other hand is a too high value of the HOMargin not good. Then a mobile is able to drive a long way past the cell border, causing cell stretching. The highest reasonable value seems to be 6 dB, since with higher values the quality decreases fast.

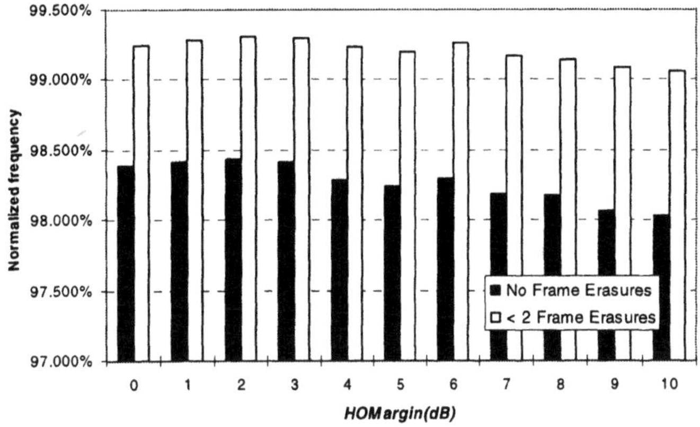

Figure 143. Normalised frequency of having no frame erasures or less than 2 frame erasures during a SACCH frame.

The results lead to the following dilemma:

- A low HOMargin leads to many handovers, but better quality. The better quality is desirable, but the many handovers are not, since each handover comes with the risk of a handover failure.
- A high value of HOMargin leads to cell stretching, but decreases the number of handovers.

The optimum balance between these 2 effects has to be found. Another possibility is making the HOMargin dynamic. Depending on the quality, experienced by the mobile, the HOMargin is set. When a mobile station is experiencing bad quality the HOMargin is set low, so that handover is made, while when the mobile is experiencing good quality, the HOMargin is set high, leading to a very low probability of a handover.

This dynamic HOMargin has been simulated by changing Equation (1) to:

$$\overline{L(j)} - \overline{L(i)} > \text{HOMargin}(RXQUAL) \tag{38}$$

The 11 different sets of HOMargin(RXQUAL) combinations that have been tested, all depending on the experienced RXQUAL value, can be seen in Table 31. The RXQUAL value, which dictates the HOMargin value used in the handover algorithm corresponds to the latest reported RXQUAL value of the mobile station. A 1/3 network has been simulated with 25% load.

Table 31. The different sets of HOMargin's [dB] as they depend on the RXQUAL value.

RXQUAL	S 1	S 2	S 3	S 4	S 5	S 6	S 7	S 8	S 9	S 10	S 11
0	3	6	9	9	9	9	9	6	6	6	6
1	3	6	9	9	9	9	9	6	6	6	6
2	3	6	9	9	9	9	9	6	6	6	6
3	3	6	6	9	9	9	6	6	6	6	6
4	3	6	6	9	9	6	6	6	6	6	6
5	3	6	6	9	6	6	3	6	1	3	6
6	3	6	3	1	1	3	3	1	1	1	3
7	3	6	3	1	1	1	1	1	1	1	1

The results from simulating these sets can be seen in Figure 144. The average number of handovers is depicted, together with relative frequencies of having no frame erasure or less than 2 frame erasures within a SACCH multiframe.

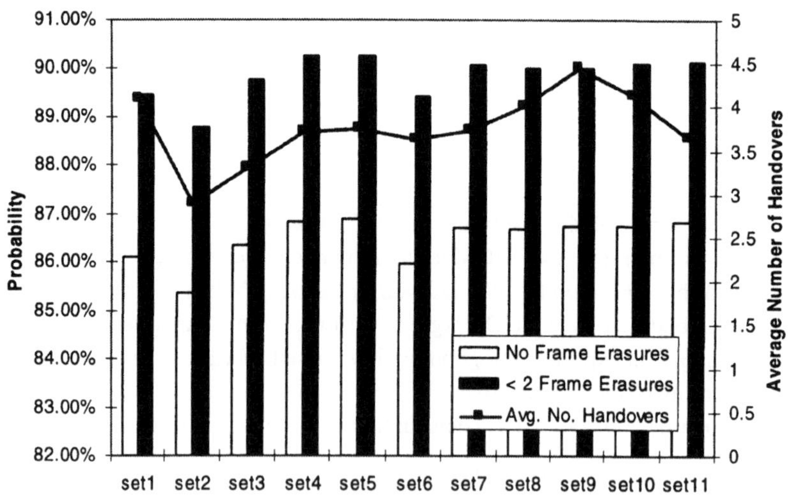

Figure 144. Average number of handovers and the relative frequency of having no frame erasures or less than 2 frame erasures within a SACCH multiframe.

Higher HOMargin values lead to less handovers, as expected. Better quality can be achieved by using the dynamic HOMargin. Sets 4 and 5 seem to give the best results. Generally it is found that:

- The HOMargin at the low RXQUAL values can be set to 9 dB. This seems to give better results than 6 dB and decreases the number of unnecessary handovers.
- The HOMargin at RXQUAL 6 and 7 should be very small, e.g. 1 dB, so that mobile stations suffering from bad quality can be saved.

It should be noted that in GSM networks there often is the possibility to assign a different HOMargin to the quality handover. In other words when a quality handover is triggered, the HOMargin can be set different, compared to other kinds of handovers. This works a bit like the dynamic HOMargin, but only with 2 levels of HOMargin, like for example set 4 in Table 31.

5. SUMMARY ON HANDOVER ALGORITHMS IN A GSM NETWORK

This chapter has dealt with handover in a frequency hopping GSM networks. In Section 1, an introduction in handover principles and an overview of the literature on handovers was given.

In Section 2 the simulation model of the handover algorithm in CAPACITY was presented and simulations showed the influence of different network parameters on the handover performance. Results from a live GSM network were used to verify the model and it was shown that the measurements and the results from the simulation tool are very alike. The influence on the handover of introducing frequency hopping in a non-hopping GSM network was shown. This showed that intra cell handover should be turned off in a frequency hopping network, since it only increases the number of handovers. In a frequency hopping network the quality is averaged over all frequencies, so making a handover to another frequency in the same cell does not improve the network quality. When going from non-hopping to frequency hopping the RXQUAL thresholds for the handover algorithm can also typically be set higher, since higher RXQUAL values can be tolerated, i.e. higher RXQUAL values give still acceptable quality. For example in a non hopping network a handover will be made at an RXQUAL equal to 4, since RXQUAL 5 is considered to be bad quality. In the same network with frequency hopping the threshold can be set at 5, since RXQUAL 5 represents still good quality. Possibly even RXQUAL 6 can be used.

Section 3 deals with theoretical models for handover algorithms. It starts with an analysis of the minimal number of handovers per cell depending on cell radius and speed. In this analysis a handover is assumed every time a mobile crosses a cell border. Also models, using Markov processes, for a single cell as well as for multiple cells are presented. One of the results in this section is that the mobility dependency has no effect on the blocking.

In Section 4 a couple of improvements to the handover algorithm are introduced and studied. An improvement is either a quality improvement in terms of FER, an improvement in trunking efficiency or a decrease in the number of handovers per mobile connection.

Channel reservation for handover

Reservation of one or a few channels in every cell exclusively for calls needing to make a handover, decreases the blocking on handovers. The prize to pay is higher blocking on the new calls, but since the blocking of handover calls is considered worse than the blocking of new calls, channel reservation is often a good idea.

Queuing

To lower the blocking on new calls, queuing can be used. The prize to pay is a small delay. It can be combined with channel reservation for handover, so that the gains from both features can be used.

Traffic Reason handover

Traffic reason handover tries to equalise the traffic over the network, so that the trunking efficiency is improved. It has been analysed theoretically and it was found that a capacity is depends on the size of the overlapping area between cells, the number of channels per cell and distribution of the traffic. When the cell overlap only covers 9% of a cell, and the neighbouring cells carrying just as much traffic as the considered cell, a capacity gain of 6 % was achieved. Higher gains were found in other situations.

Simulations of a 3/9 reused network with uniform distributed traffic show that the blocking can be decreased from 2% to about 0.75% when using traffic reason handover, while the quality of the network remains the same. It was also shown that a capacity gain of about 7% was achieved by using traffic reason handover. The simulations give a worst case result, since the traffic was uniform distributed. In a real life network with moving hotspots, the gain will be higher.

The prize to pay for the gain of traffic reason handover is that the average number of handovers per call increase.

Dynamic HOMargin

HOMargin is an important parameter is GSM. It has influence on the number of handovers and the place where the handovers are made. To decrease the number of handover, the parameter HOMargin is made dynamic.

Simulation results show a decrease in the number of handovers, while the quality is remained or even improved, if the RXQUAL dependency is set right.

Chapter 9

COMBINING REUSE PARTITIONING AND FREQUENCY HOPPING IN A GSM NETWORK

A lucrative yet complex way of optimising capacity in a cellular mobile GSM type of network is by employing frequency reuse partitioning, as briefly stated in Chapter 2. The general idea behind this technique is to increase the cell capacity by reducing the overall frequency reuse pattern. As will be described later reuse partitioning can be implemented in practice using different methods.

A proposed concept, enabling an even larger increase in capacity in specific types of cellular environments than with regular reuse partitioning, is described and analysed utilising CAPACITY. All the features developed are treated specifically for GSM, however the principle ideas can applied to any type of network based on a multi-carrier standard. Throughout the chapter a macro cellular type of network has been assumed, however if modelling the appropriate environments the functionality could be utilised in a micro or a pico cellular environment as well, making it even more interesting.

To be more precise, the proposed concept is based on the combination of two existing capacity enhancing features, frequency hopping and frequency reuse partitioning. The aim has been to achieve the combined effect of the two. What makes this proposal interesting for a network operator is that the capacity gain is achieved without additional base stations, i.e. without large investments. Because the acquisition of additional base stations in city centre environments can often be quite difficult, alternative cell types such as micro and in-building cells are utilised to compensate for the capacity short fall. However, in comparison to a macro-cell, the investment in establishing a micro- or a pico-cell is traditionally not matched by the same traffic capacity uplift. Therefore it has until now been highly desirable for cellular radio network operators to provide additional capacity using existing macro cells.

On the other hand, during the last couple of months, commercial base stations for the micro cellular environment with as much as 4 TRX's per cell have been introduced. With that many TRX's per cell, the concept has become relevant also for the micro cellular environment.

Concerning the two features, since frequency hopping has already been was dealt with in detail, only frequency reuse partitioning is treated in this chapter. The basic concept is described in Section 1 in terms of functionality as well as by potential capacity.

Then the existing reuse partitioning algorithm chosen for this chapter, the IUO algorithm, is described in Section 2 by the basic design as well as using an analytical traffic model to estimate the hard blocking capacity limit.

In Section 3 the capacity enhancement proposal itself is introduced. Some potential limitations were discovered where different solutions have been proposed and studied.

An improved version of the IUO algorithm is described in Section 5 in terms of functionality as well as by a simple analytical model.

In order to be able to simulate IUO in CAPACITY, different choices concerning the actual implementation had to be made. A description of how IUO has been implemented is given in Section 6, describing the required input configuration parameters etc.

To present an overview of the simulation results from CAPACITY, Section 7 has been devoted to an outline of all the simulations to be carried out.

Finally, the primary aim of this chapter, the CAPACITY simulation results are found in Section 8.

The concluding summary in Section 10 ends the chapter. Here the conclusions as well as some future prospects of the combination of frequency hopping and IUO are given.

1. INTRODUCTION TO FREQUENCY REUSE PARTITIONING

The basic idea of frequency reuse partitioning is to exploit the fact that some mobiles are ensured a high quality due to their geographic location in the cell. The idea is to deliberately degrade the overall *C/I* by lowering the frequency reuse, for mobiles that have more than adequate transmission quality (typically the ones close to the base stations). The goal is therefore to produce an overall network quality that while satisfying the general system quality objectives a general increase in system capacity is still achieved.

In practice frequency reuse partitioning is implemented by dividing the available spectrum in different bands and then reuse the frequencies in each band according to different frequency reuse patterns [134, p.35]. The principle is shown in Figure 145.

In the figure the *C/I* as a function of the distance to the base station is seen for two different frequency reuse patterns. The idea is to use the tightly reused frequencies while the *C/I* is still at an acceptable level.

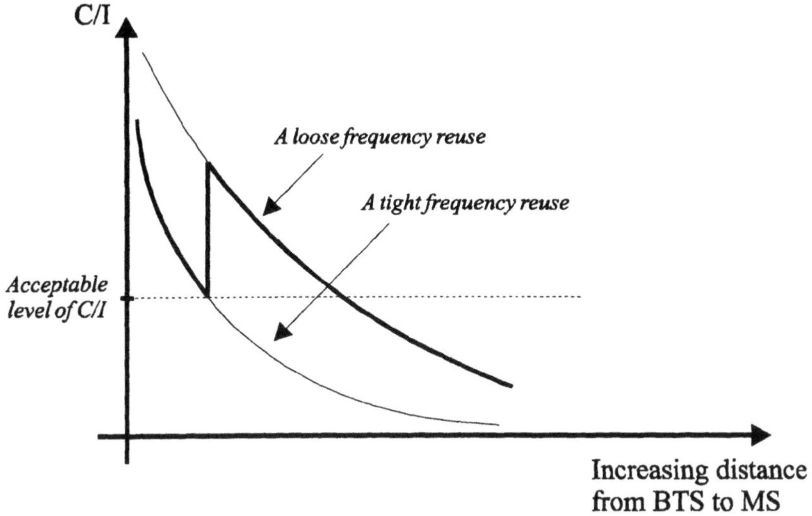

Figure 145. The basic idea of frequency reuse partitioning.

1.1 Reuse Partitioning in a Cellular Communication System like GSM

In GSM-900 a total of 124 frequencies is available. This has in Denmark resulted in 45 frequencies for each of the two national GSM-900 operators (the remaining frequencies are reserved for guard bands, NMT and cellular phones). If one of these operators would decide to use reuse partitioning it could be done as shown in the example shown in Figure 146. The available spectrum allocated to each operator is divided in two groups, where the frequencies are reused with different reuse factors. In the example, a reuse pattern of 1/3 is reserved for 9 of the frequencies and a reuse pattern of 3/9 for the remaining 36. This particular example will constitute one of the basic examples throughout the remaining part of the chapter.

The effect, in terms of capacity increase due to lowering the overall frequency reuse pattern, will in the following be illustrated for an ideal situation using analytical expressions. The idea is to use classical signal strength and propagation path loss methods to enable a simple description of the gain from reuse partitioning. Initially some traditional ideas involving frequency reuse partitioning are given.

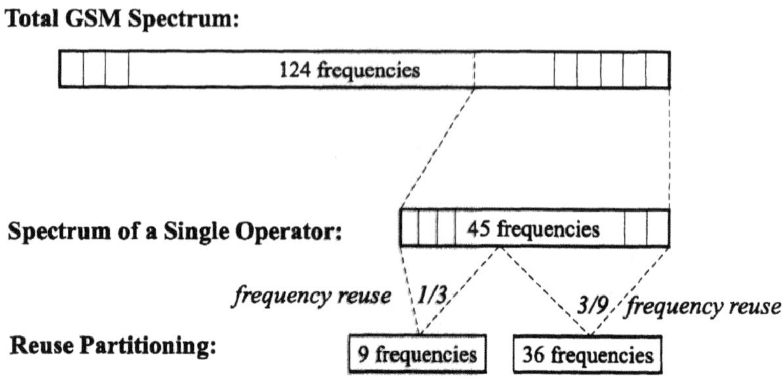

Figure 146. Splitting of the available frequencies with reuse partitioning.

1.2 Previous Frequency Reuse Partitioning Studies

Alternative ways of doing resource allocation for the radio interface has been tried out as a mean to cope with the problem of achieving an acceptable level of interference. In Chapter 2 examples such as directed retry, load sharing, queuing and soft capacity algorithms where given along with frequency reuse partitioning itself. The idea of reuse partitioning as part of the resource allocation scheme is therefore not new. In the 1980's the general functionality where studied in [87,207] and in [206] the combination with a hybrid channel assignment scheme where treated for a general analogue cellular mobile telephone system. It is described how typically a capacity increase of around 25-50% can be expected from reuse partitioning [10], depending on system parameters such as available bandwidth etc.

The idea of having one group of frequencies to provide continuous coverage, the underlay frequencies, and a second group, the overlay frequencies, which is employed to provide service to those mobiles that experience good quality, has been described for some analogue mobile systems in [122]. However, in these systems the functionality has traditionally been based on reductions in base station output power of the overlay frequencies as a mean to reduce the interference in this layer and enable the utilisation of radio channels more tightly. In these traditional studies the intra-cell handover between the layers are handled using signal level and/or distance measurements. That is, the received signal level or the physical distance from base station to the mobile was used as estimation of the quality experienced by the mobile. However, distance and received signal strength are often to simple parameters for determining the experienced quality. Determining a fixed and at the same time useful handover level thresholds is very difficult, especially in the interference limited environment. Furthermore, each cell has unique levels at which calls should be handed between the two layers, requiring extensive network optimisation and maintenance. Similar, however more sophisticated, products are available for GSM from some of the large vendors as MOTOROLA and NOKIA. The principle promoted by Motorola exploits what has been referred to as inner and outer zones and is denoted the 'concentric cell' concept [10]. NOKIA has offered a variant of the underlay/overlay principle mentioned above commercially for GSM. This concept is denoted intelligent underlay/overlay (IUO) [246]. In case of these two products no reduction in base station output power is necessary, i.e. all transceivers transmit with maximum output power.

Traditionally the motivation for studying reuse partitioning has been the need for increased cell capacity. When considering this issue the combination of reuse partitioning and various types of load sharing techniques for neighbouring sectors/cells has also been treated [32,100]. Load sharing and reuse partitioning are tightly related problems and can therefore be modelled in similar ways. Cell overlap with resulting load sharing and reuse partitioning is studied in a simple manner using steady-state models not including fading etc. in [32] and using overflow models in [232]. In [32] the capacity gain was found to be around 20 %. Furthermore, investigations concerning the network quality enhancements when using reuse partitioning have been carried out with good results [187].

Besides the similarities between reuse partitioning and load sharing, the dual-band networks arising when combining GSM-900 and GSM-1800 have very similar characteristics as the basic concept of reuse partitioning if the two networks have co-located base stations.

1.3 Idealised Frequency Reuse Partitioning Considerations

The effect of splitting the available spectrum and reusing the different frequencies using various reuse patterns is treated in the following using an analytical approach. In Figure 147 some of the network parameters used in the description are shown. In particular, the relation between the 4 different design parameters frequency reuse distance D, cell radius R_1 (loose reuse), cell radius R_2 (tight reuse) and cluster size K is shown.

In these analytical considerations, the channel allocation criteria which controls the reuse partitioning and determines whether the tight reused frequencies can be used or not, is based on an evaluation of the C/I. In other words, if the C/I is above a certain threshold, the tightly reused frequencies can be used.

Besides illustrating the principle of reuse partitioning in a cellular radio network in a simple way, the general purpose of this analytical description is to determine the overall network capacity gain or frequency reuse reduction when using reuse partitioning. It is therefore necessary to know the C/I in order to estimate the coverage area of the tight reused frequencies compared to the coverage area of the entire cell.

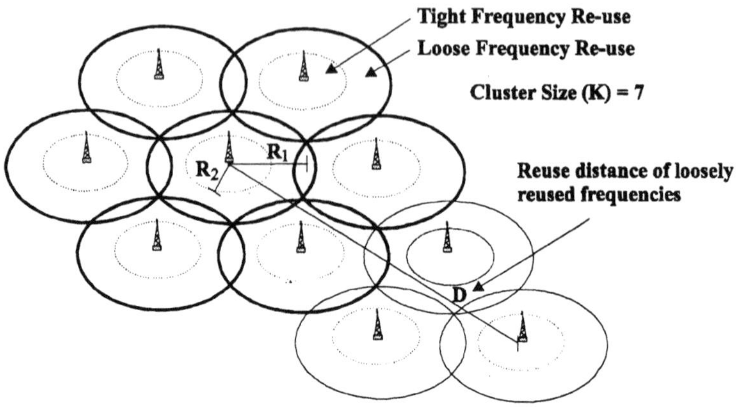

Figure 147. The network parameters used in the description of reuse partitioning.

1.3.1 Location of the Cell Edge of the Tight Reused Frequencies

The simple Equation (6.6) from Chapter 6 used to estimate the *C/I,* cannot be used to determine the *C/I* at the cell border of the tightly reused frequencies, since this border is located somewhere within the cell of the loosely reused frequencies. It is therefore necessary to know the *C/I* distribution throughout the entire cell. Another approach has therefore been taken. To make the estimation more realistic not only path loss but also log-normal fading is included in the calculations. The network structure used in the analysis is the same as before, where the reuse partitioning is made by a group of frequencies with a reuse of 1/3 and one with a reuse of 3/9. Only co-channel interference is considered.

In some parts of each cell, service is provided from both frequency groups, whereas in another parts of the cell only the loosely reused frequencies can be used, see Figure 148. Therefore the capacity in the cell is non-uniform.

The idea of this analysis is therefore to estimate the cell border of the tight reused frequencies in order to determine the size of this area compared to the total cell area. By doing that it is possible to find the effective frequency reuse factor of the tightly reused frequencies. Knowing the frequency reuse factor of each set of frequencies the capacity achieved from reuse partitioning can be determined.

Since independent shadow fading is included on both the carrier itself and on all 6 interferers in the first tier, the calculations become somewhat complex. The shadow fading is, as described in Chapter 5, in general modelled by describing the variations of the local mean signal envelope by a log-normal distribution, where the local mean (measured in dB) is a Gaussian random variable described by its probability density function. Since the probability density function of the received signal becomes a function of a random variable, a sum of random signals with log-normal probability is required to determine the total co-channel interference. The Schwartz/Yeh approach described in [211], also used in Chapter 6, has been used.

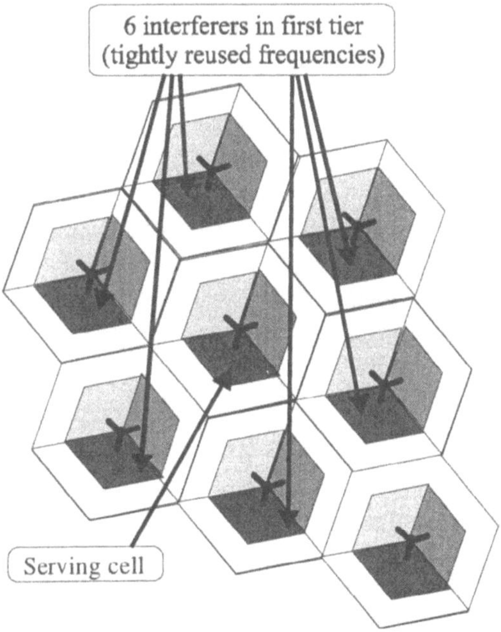

Figure 148. The scenario to be analysed of reuse partitioning between 1/3 and 3/9.

The situation to be studied is as follows: One sector of a 3 sector is analysed using a 1/3 reuse (the reuse of the tightly reused frequencies). Only the 6 closest co-channel interferers from the first tier are used, refer to Figure 148. The radiation pattern used by each of the seven antennas (one desired carrier and six interferers) corresponds to the 90° antenna type used in CAPACITY [105]. The traffic is assumed uniformly distributed throughout the network and all cells have 100 % load.

The Average Reuse Pattern with Reuse Partitioning

Using Schwartz's and Yeh's method as above, with all parameters equivalent to the previous description, except a cell radius of 3 km, since this is what is used in the CAPACITY network simulations.

The Schwartz and Yeh method has been used to determine the cell edge of the tightly reused frequencies as a function of the specified *C/I* threshold. The acceptable probability of satisfying the specified *C/I* threshold for each situation has been set to 50%. The *C/I* threshold determining the cell edge of the tight reused frequencies was varied according to the values in column one in Table 32. After having found the cell radius of the tightly reused frequencies, column two in Table 32 was made using standard geometric formulas. This column shows the coverage area of the tight reused frequencies compared to the area of the entire cell. Therefore, when a *C/I* of 14 dB is specified, the tight reused frequencies cover 33% of the whole cell (with a probability of 50%).

Table 32. The analytical calculations used to find the effective reuse pattern of the tight reused frequencies as well as the effective reuse pattern itself.

Specified C/I threshold [dB]	Probability of coverage area of the tight reused frequencies compared to entire cell area	Normalised reuse distance of tight reused frequencies (q_{tight})	Effective reuse factor of the tight reused frequencies (K_{tight})
8	0.80	3.3	**3.7**
10	0.61	3.8	**4.9**
12	0.45	4.5	**6.7**
14	0.33	5.2	**9.1**
16	0.25	6.0	**12.0**

In a network without reuse partitioning the reuse distance is normalised to the cell radius, see Equation (5) in chapter 6. In order to determine the reuse distance of the tight reused frequencies, it is correspondingly necessary to normalise the frequency reuse distance to the radius of the tight reused frequencies. This is done using Equation (5) in chapter 6. When assuming that the coverage area of the tight reused frequencies has the same form as the loose reused frequencies and this form is assumed to be a part of a circle, the effective reuse distance can quite easily be calculated. It is simply the normalised reuse distance (q) of the complete cells (in this example equal to 3), divided by the square root of the coverage area of the tight reused frequencies. This is done in column three in Table 32.

Since it is not the normalised reuse distance of the tightly reuse frequencies, but the effective reuse factor that determines the network capacity, it is also found. Again using Equation (6.5) the corresponding cluster size, i.e. the effective reuse factor can be found as a function of the normalised reuse distance. The results are shown in column four in Table 32. It is seen how the effective reuse factor increases significantly as a function of increasing *C/I* thresholds.

To calculate the overall average cell frequency reuse, i.e. the network capacity, it is necessary to know the relationship between the amount of allocated channels in each frequency group. The following 2 different situations have been studied:

1. Number of channels in the loosely reused group: 32 (four TRX's in GSM)
   Number of channels in the tightly reused group: 24 (three TRX's in GSM)

2. Number of channels in the loosely reused group: 24 (three TRX's in GSM)
   Number of channels in the tightly reused group: 48 (six TRX's in GSM)

The first configuration will from here on been denoted 4+3, while the second one will be referred to as the 3+6 configuration. Using these numbers and knowing that the loose reuse pattern is 3/9, the overall network cell capacity can be found using the effective reuse factor of the tight reused frequencies. The result is shown in Figure 149.

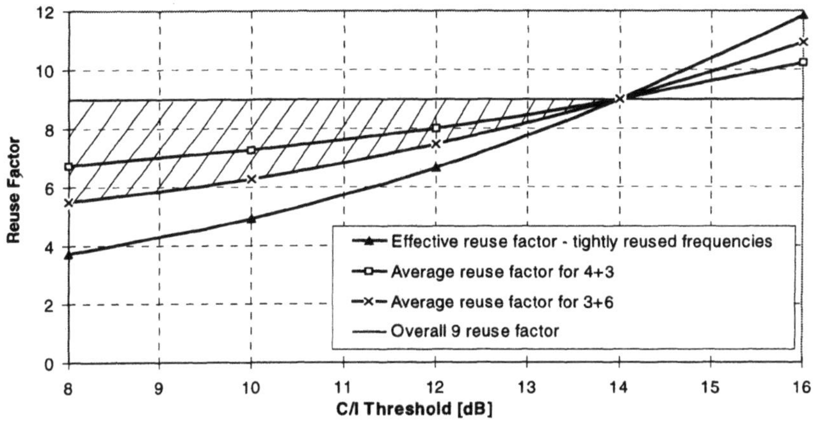

Figure 149. Examples of the average cell reuse factor as well as the effective reuse of the tightly reused frequencies themselves, as a function of the specified C/I threshold.

Using this simplified study and looking at the average network reuse factor, the capacity increase from reuse partitioning can be found directly by comparing the average reuse factor to e.g. the reuse of 3/9 used on the traffic channels in a normal single layer network exploiting FH. In Figure 149 this gain is shown for the 3+6 case as the shaded region. If assuming an available spectrum of 45 frequencies and for simplicity forgetting the BCCH frequencies, a flat network without reuse partitioning can have 5 TRX's per cell (from the 9 reuse). Using the results from above, reuse partitioning with an acceptable *C/I* threshold of 10 dB effectively corresponds to 7.3 TRX's (from the 6.2 reuse) in the 3+6 case. That is, a potential capacity gain of 43 % is achieved in this situation. The absolute figures for all the studied situations are shown in Table 33.

Throughout this study a uniform traffic distribution has been assumed. If that is not the case, the capacity gain from reuse partitioning changes accordingly. With irregular traffic distributions the *C/I* threshold can be exploited using alternative network capacity planning strategies as will be seen.

Table 33. Idealised capacity gain from reuse partitioning as a function of the *C/I* thresholds and the network configuration type used, when compared to a flat 3/9 network.

Reuse partitioning configuration	Using a C/I threshold of 8 dB	Using a C/I threshold of 10 dB	Using a C/I threshold of 12 dB	Using a C/I threshold of 14 dB	Using a C/I threshold of 16 dB
4+3	34 %	23 %	13 %	0 %	-12 %
3+6	64 %	43 %	21 %	0 %	-17 %

1.4 Practical Considerations Concerning Reuse Partitioning

In an operating live network the largest potential problem of reuse partitioning is associated directly to the mobiles that are moving or more exactly on how to design an effective handover algorithm. As described in Chapter 8 the handover algorithm has to control the allocation of the mobiles when moving between the different cells and layers. Designing a normal handover algorithm taking care of all possible situations and thereby compensating for interference, bad quality, low power budget, slow moving mobiles, fast moving mobiles etc. in any possible combination, is very difficult. Combining it with reuse partitioning does not make it easier.

The use of downlink power control in combination with reuse partitioning can influence positively as well as negatively on the capacity potentially achieved by reuse partitioning when using *C/I* as the separation criteria. Power control contributes with a reduction in interference, which of course influences on the *C/I*. In that sense proper use of downlink power control increases the potential capacity gain even further. On the other hand, since the fundamental idea of reuse partitioning is based on exploiting the surplus quality of the mobiles close to the base station, equivalent to the aim of power control, they both treat the same issue. Therefore they partly eat from the same cake.

2. THE INTELLIGENT UNDERLAY-OVERLAY ALGORITHM

As described, the reuse partitioning concept, to be investigated in terms of capacity increase, is based on exploiting an estimate of the level of interference as the handover criteria between the frequency groups. The reuse partitioning principle is in this case integrated into the handover algorithm in order to have a direct influence on the traffic management. Layers will in the remaining part of the chapter denote the different partitions (in terms of frequency groups).[24] Throughout the rest of this chapter the original layer (with the loose frequency reuse pattern) will be denoted the regular layer and the tight frequency reuse group as the super layer. The two layer reuse partitioning algorithm used here is equivalent to the intelligent underlay-overlay (IUO) handover algorithm originally introduced by NOKIA [1,2].

IUO is a feature designed to allow a tighter frequency reuse pattern for some of the available radio frequencies and hence achieve a higher network capacity in terms of handled traffic per cell. The available radio spectrum is split in two groups, a regular and a super frequency group. The notation used in describing how many TRXs that are allocated to each cell in each layer is as follows: 2+1 corresponds to 2 TRXs on the regular layer and 1 TRX on the super layer. To the existing handover algorithm based on quality, interference, level, power budget, distance etc., which is operating between the existing macro or equivalently regular cells, an intra-cell handover trigger is added to the network functionality; the *C/I* intra-cell handover trigger. Provided the *C/I* measured by the mobile is higher than a predefined 'C/I-good' threshold, the mobile is allowed to enter one of the super frequencies. In the same way when a mobile is already operating on a super frequency, it is continuously evaluated whether the *C/I* becomes worse than another predefined value, denoted the 'C/I-bad' threshold. If that is the case the mobile is handed back to one of the regular frequencies in the same cell. The intra-cell handover principle of the IUO algorithm is shown in Figure 150.

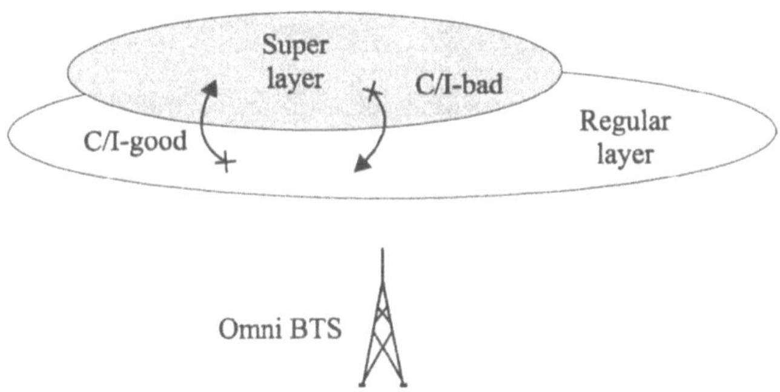

Figure 150. Intra-cell handover principle of the IUO algorithm based on interference.

If the two *C/I* thresholds are specified in an optimum way, it should be possible to carry the cell-traffic with little interference (and therefore good quality) on the super layer. This enables a tight reuse of the super frequencies. In the original algorithm [1] it is only possible to do a handover from a super frequency to a regular

[24] This is the way the partitioning was originally denoted by the inventors.

frequency in another cell by performing the imperative handover. The imperative type of handover allows a handover to a location where the quality is unknown to the mobile at the instant of handover. It is sort of a last attempt to avoid a dropped call. A typical situation where an imperative handover could be used is if all regular channels are hard blocked, but a handover from the super layer is necessary in order not to drop the call. The imperative handover is not included in this analysis, but a part from that, the handover algorithm that has been developed corresponds to the one of NOKIA. The new handover possibilities implemented due to IUO are shown in Figure 151.

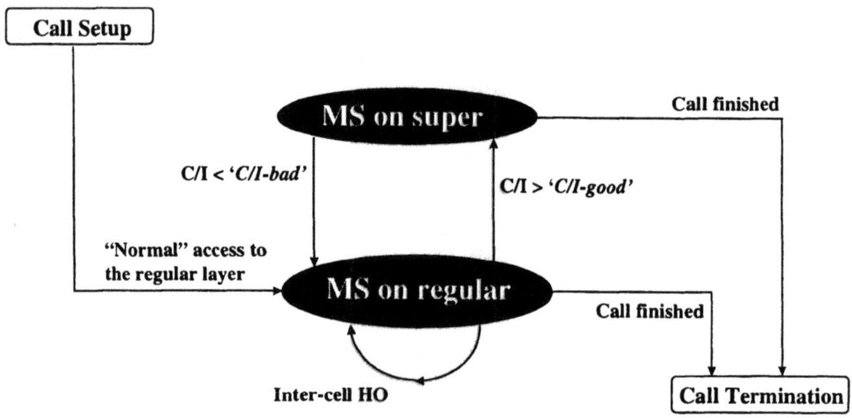

Figure 151. State diagram showing the handover possibilities of the original IUO algorithm.

In the initial analysis a new call in an IUO cell therefore always has to initiate on a regular frequency before it can be handed over to a super frequency, see Figure 151. In the same way when a mobile has to be handed off a super frequency, it always has to be done to a regular frequency in the same cell. As just described, this means that it is only possible to make inter-cell handovers between two regular frequencies. This affects the trunking efficiency quite a lot, since hard blocking on the regular layer will cause cell blocking or even dropped calls despite having free channels on the super layer. The faster the mobiles move the larger this problem becomes since the number of handovers per call will increase. This issue is treated later on since it turned out to be quite a limitation to the performance of the IUO algorithm.

2.1 Estimating *C/I* in GSM

Due to the technical limitations in GSM the implementation/design of the IUO algorithm introduces a certain level of complexity in terms of frequency planning. In GSM signal strength measurements of the neighbouring cells are related exclusively to the neighbouring BCCH carriers. Measurements of the remaining TRXs in the neighbouring cells are not received. Therefore in the case of IUO, the estimation of the *C/I* ratio looks like this:

$$\frac{C}{I}=\frac{Signal\_level\_of\_serving\_carrier}{\sum_{i=1}^{6} Signal\_level\_of\_interfering\_super\_BCCH_i} \tag{1}$$

The parameter *Signal_level_of_serving_carrier* corresponds to the desired measured received signal strength (RXLEV) as described in the GSM specifications [64,65].

The interference is estimated by summing the received signal strength from the BCCH frequencies of the cells located in the list of super frequency neighbours, referred to as the IUO reference list.[25] The parameter *Signal_level_of_interfering_super_BCCH* corresponds to the measured RXLEV of the BCCH frequencies in the cells of the IUO reference list, where *i* is the number of the interfering cell. The reason why the summation in Equation (1) is limited by 6, is because in GSM only the 6 strongest neighbours can be reported, see Chapter 3. This GSM specific limitation of course influences on the performance of the IUO algorithm, since there is no way of insuring that any of the interfering super frequencies are contained in the reported list of the strongest interferers. In the simulations using CAPACITY all 6 are always reported, unless anything else is specifically described.

In the Nokia implementation, two methods can be used by the handover algorithm to calculate the downlink *C/I*. The 'average taking method' or the 'maximum taking method' [1]. The prior method calculates an average *C/I* using the interference from all the super references contained between the 6 reported neighbours, while the latter simply takes the largest interferer and uses it as the complete interference contribution. Furthermore, if no references are available within the 6 reported values, the estimated *C/I* is assumed infinite and the mobile can be handed over to the super frequency.

The method used for the estimation of the *C/I* represents the worst case situation when considering the use of downlink power control. The interference contributions are measured on the BCCH frequencies, which are never applied power control. The carrier signal strength can on the other hand be influenced by downlink power control, but the controlling BCS, is constantly aware of the power reduction and can compensate for this.

[25] The IUO reference list is a NOKIA specific feature in which the operator should predefine the neighbouring cells, which use the same super frequencies as the serving cell.

2.2 Practical Frequency Planning Difficulties of IUO

Basically the *C/I* calculation, as described above, estimates the *C/I not* using the received co-channel interference signal strength, but the received signal strength of the corresponding BCCH frequency, i.e. an entirely different frequency. This can easily cause difficulties in the frequency planning procedure, since if this link between the neighbouring BCCH and TCH frequencies is not considered the best possible allocation of the BCCH and the super frequencies may not take place. Figure 152 contains 3 sites each consisting of 3 sectors. An example of a problematic frequency plan using IUO is shown in Table 34. Here the frequencies are allocated to TRXs in Figure 152.

Table 34. The frequency plan of a potentially problematic IUO network.

SITE NAME	TRX NAME	ALLOCATED FREQUENCY & TYPE
SITE 1	B1	100 (BCCH)
SITE 1	R1	110 (TCH)
SITE 1	S1	80 (TCH)
SITE 2	B2	101 (BCCH)
SITE 2	R2	111 (TCH)
SITE 3	R3	101 (TCH)
SITE 3	S2	81 (TCH)
SITE 3	B3	102 (BCCH)

If a mobile station is attached to TRX R1 on SITE 1 and would like to perform an evaluation of whether it is possible to make the intra-cell handover to TRX S1 planned with frequency number 80, it is necessary to estimate the *C/I* as described above. Since SITE 2 also uses super frequency number 80 the cell is in the super reference list and the signal level of the BCCH is used in the *C/I* calculation. The BCCH TRX in this cell (denoted B2) is however, seen from the mobile station, interfered by the regular frequency from TRX R3 on SITE 3. They both use frequency number 101. This means that because the BCCH B2 on SITE 2 is interfered by the regular frequency of SITE 3 the mobile allocated on SITE 1 is may not be able to decode the right BCCH and correspondingly not perform a correct *C/I* estimation. In the live network situation this could be crucial since the mobile may be allowed to enter the super layer in situations where the C/I-good criteria is not satisfied. However, seen from a practical point of view, it is not necessarily too critical since the handover algorithm typically will ensure that the mobile is handed back to the regular layer before the call has dropped.

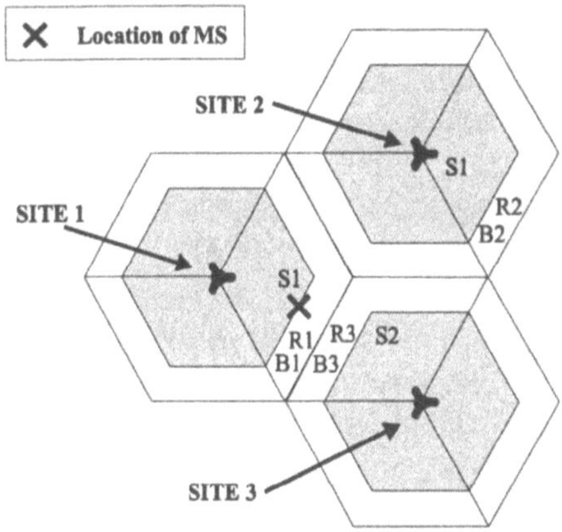

Figure 152. Example of a situation where inconsiderate planning degrades the IUO algorithm performance.

Despite this planning complexity, the IUO feature has operated well in several existing GSM networks both in Europe (by e.g. SONOFON) and Asia, Hong Kong (CSL and others). In Hong Kong an alternative planning strategy has been chosen due to the micro cellular, bad urban, propagation environment. Here the probability of having one of the super references between the reported 6 strongest receive signals is some times low. Actually often the weakest of the 6 reported neighbours is even much higher than the relevant IUO reference frequency. Therefore, instead of just using the lowest of the 6 (since the difference can be significant), the BCCH frequency of a completely different cell, typically located between the relevant IUO interferer and the serving cell, is used as super frequency reference. In that way the probability of having the desired BCCH between the 6 strongest measurement reports is increased. This method has according to CSL proven quite powerful.

2.3 Estimating the Hard Blocking Limit of an IUO Cell

As described in Chapter 6 the Erlang-B formula [109,134] is typically used to model the hard blocking limit of a cell in a one layer cellular network without capacity enhancing features like queuing, directed retry etc. However, in the case of reuse partitioning, controlled by a handover functionality like IUO, it is easily understood that the Erlang-B formula cannot be used. The most obvious argument is that the call attempts traditionally modelled statistically by a call arrival distribution, is not Poisson distributed for all frequencies. This is a fundamental assumption of the Erlang-B formula, why the formula does not work.

Actually the traffic flow of a cell exploiting IUO reminds more about an overflow system, where the traffic rejected from the super layer is the overflow traffic. A simple IUO traffic model, originally used during the introduction of IUO in the live SONOFON network in corporation with NOKIA, able to estimate the hard blocking limit was investigated by means of live network measurements [125].

2.3.1 The Original IUO Hard Blocking Traffic Model

The model is based on classic theory of overflow traffic. The merits of this traffic model does not lie in the design and development of the model since it is quite idealised when compared to a real live network situation. Also, the basic principle of the method is described in [103, p.190]. It has its strengths in the computational efficiency. Especially the performance or accuracy of the estimations made using the model has proven quite good.

The traffic model is an analytic model treating one cell at a time and only incorporates absolute amounts of traffic. That is, the actual traffic flow is treated be specifying which resources and exactly how the traffic can move to and from. It means that the model does not consider time as a parameter. Despite the simplifications, five input parameters are needed for this model to estimate the amount of traffic a cell configured with IUO can carry. They are:

1. Number of channels on the regular layer.
2. Number of channels on the super layer.
3. Desired total cell blocking probability.
4. The traffic distribution within the cell, specified by the percentage of the total cell traffic carried on the super layer.
5. A description on the time needed on a regular channel to perform the evaluation of the *C/I* compared to the entire time spent in the cell.

The actual traffic model is based on the traffic stream diagram in Figure 153. Each parameter is explained in Table 35.

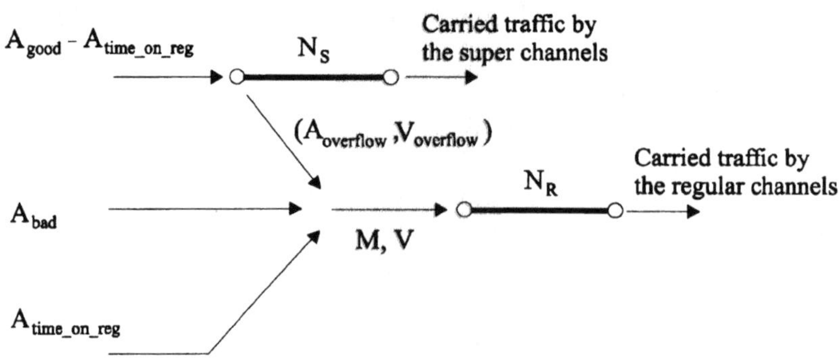

Figure 153. Traffic model of the traffic streams handled by an IUO type of cell.

Table 35. The different traffic parameters used to describe the IUO hard blocking model.

MODEL PARAMETER NAME	MODEL PARAMETER EXPLANATION
A_{bad}	The traffic generated by the users on the regular frequencies only
A_{good}	The traffic generated by the users on the super channels
$A_{time\_on\_reg}$	The traffic on the regular channels (because of the *C/I* evaluation) of the traffic that will eventually be carried by the super channels
$A_{overflow}$	The mean of the part of the A_{good} traffic that have to stay in the regular channels because of congestion in the super channels
$V_{overflow}$	The variance of the overflow traffic
M	The mean of the total traffic being offered to the regular channels
V	The variance of the total traffic being offered to the regular channels
N_S	The number of traffic channels of the super frequencies in the cell
N_R	The number of traffic channels of the regular frequencies in the cell

All the originating cell site traffics (A_{good}, A_{bad} and $A_{time\_on\_reg}$) are assumed to be Poisson distributed, but the same assumption cannot be applied to the overflow traffic. The overflow traffic is represented by the mean ($A_{overflow}$) and the variance ($V_{overflow}$). Call holding time is assumed to be exponentially distributed and a call is lost if it is blocked (Erlang-B assumption). The mean traffic offered to the regular channels consists of:

1. A_{bad} traffic, which can only use the regular channels.
2. $A_{time\_on\_reg}$ traffic, which is the traffic on the regular channels (due to the *C/I* evaluation) of the traffic that will eventually be carried by the super channels.
3. $A_{overflow}$ traffic which stays on the regular channels because of congestion in the super channels. The term 'overflow' is used for the traffic as it is the traffic being rejected to be carried by the super channels. Here, for modelling purpose, the overflow traffic is assumed to stay on the regular frequency without re-attempting the handover to the super channel and the overflow traffic is also assumed experiencing the blocking of the regular channel. These are of course *not* true in the actual system. The first assumption makes the model overestimating the blocking probability of the regular channels, where as the second assumption underestimates the blocking.

In this model, the traffic carried by the super channels is equal to (A_{good} - $A_{overflow}$ - $A_{time\_on\_reg}$). If z is defined as the ratio of the time required on a regular channel to perform the *C/I* evaluation to the mean holding time of a call, it can be approximated that

$$A_{time\_on\_reg} = z \cdot (A_{good} - A_{overflow}) \tag{2}$$

The traffic offered to the super channels is equal to $(A_{good} - A_{time\_on\_reg})$. From that,

$$A_{overflow} = (A_{good} - A_{time\_on\_reg}) \cdot E(N_S, A_{good} - A_{time\_on\_reg}) \tag{3}$$

where $E(N_S, A_{good} - A_{time\_on\_reg})$ is the Erlang B blocking probability for $(A_{good} - A_{time\_on\_reg})$ traffic being offered to N_S number of traffic channels. Here a problem arises: $A_{time\_on\_reg}$ depends on the overflow traffic, $A_{overflow}$, which in turn also depends on $A_{time\_on\_reg}$. It is expected, for most IUO system designs that $A_{time\_on\_reg}$ should be relatively small compared to the total traffic being offered to a cell. To reduce the complexity of the calculation, an approximated formula is used, which is:

$$A_{time\_on\_reg} = z \cdot (A_{good}) \tag{4}$$

A problem arises since we do not know the traffic stream distribution of the overflow traffic. Therefore it is necessary to describe it by its mean and variance. The variance of the (blocked) overflow traffic can be calculated by the following formula derived by [244]:

$$V_{overflow} = A_{overflow} (1 - A_{overflow} + \frac{(A_{good} - A_{time\_on\_reg})}{N_S - (A_{good} - A_{time\_on\_reg}) + A_{overflow} + 1}) \tag{5}$$

The combined mean traffic offered to the regular channels is:

$$M = A_{overflow} + A_{bad} + A_{time\_on\_reg} \tag{6}$$

and since the variance of Poisson distributed traffic equals the mean, the combined variance of the traffic offered to the regular channels is:

$$V = V_{overflow} + A_{bad} + A_{time\_on\_reg} \tag{7}$$

By using the Equivalent Random Theorem (ERT) [244], the mean (*M*) and variance (*V*) can be represented by an equivalent number of channels (*N'*) and equivalent amount of traffic (*A'*) offered to these *N'* channels. What is then achieved is simply a system of $(N'+N_R)$ channels offered A' traffic, which is much simpler than the initiating overflow system. I.e. if A' traffic is offered to N' channels the resulting blocked traffic will exhibit similar mean and variance as M and V. This is illustrated in Figure 154. For more information on this method see [103, p.192].

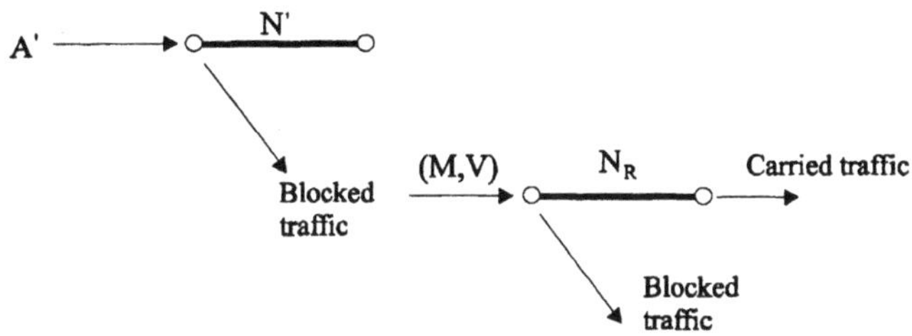

Figure 154. Equivalent Random Theorem of the combined traffic.

Y. Rapp has found an approximate solution to this problem and states the following formulas, which estimates A' and N' from V and M [103,196]:

$$A' = V + \frac{3V}{M}\left(\frac{V}{M} - 1\right) \tag{8}$$

and

$$N' = A' \cdot \frac{M + \frac{V}{M}}{M + \frac{V}{M} - 1} - M - 1 \tag{9}$$

Rapp's approximation is only valid if A' is not small [103, p.194]. After obtaining expressions for A' and N', the average blocking probability to any user in the cell should therefore be able to be calculated as:

$$\text{Average cell blocking prob.} = \frac{A' \cdot E(N'+N_R, A')}{A_{good} + A_{bad}} \tag{10}$$

However, since all the calls in the real system have to access one of the regular channels first and a mobile being blocked within the good area, will simply stay on the regular channel, all the users will experience blocking based on the number of regular channels. To find the blocking probability, a formula proposed by E. Brockmeyer [25] is therefore used instead. He shows that for a full-availability group[26] of $N'+N_R$ channels offered the traffic amount A' with channels selected in

[26] A full availability group is a group of channels where incoming calls are only blocked if *all* channels are occupied [103].

order, the proportion of time that all the last N_R channels are busy, irrespective of the state of the first N' channels, is given by:

$$B_R = E(N'+N_R, A') \cdot \frac{\sum_{v=0}^{N'} \frac{(N_R+v)!}{v!N_R!} \cdot \frac{(A')^{N'-v}}{(N'-v)!}}{\sum_{v=0}^{N'} \frac{(N_R-1+v)!}{v!(N_R-1)!} \cdot \frac{(A')^{N'-v}}{(N'-v)!}} \tag{11}$$

Since all traffic generated from the cells are Poisson distributed and since the blocking probability experienced by a random Poisson traffic is the same as the proportion of the time that all the channels are busy, the above formula gives the blocking probability to all the Poisson users accessing to the regular channels in a cell.

2.3.2 Network Measurements and Model Accuracy Estimation

Due to the simplicity of the IUO traffic model it could be expected that the accuracy of the model would be quite bad when compared to real measurements. Therefore to get a general idea of the accuracy it was investigated elaborating on measurements from the live network.

Data were collected from the operating and maintenance centre (OMC) using the available network measurement counters. From these counters, it was possible to obtain the following statistics:

1. The actual blocking probability to the regular channel - B_R^m
2. The amount of traffic carried by the super channel - A_{super}^c
3. The amount of traffic carried by the regular channel - $A_{regular}^c$
4. An estimation of the overflow traffic - $A_{overflow}^m$

In the actual system, some of the offered good traffic (A_{good}^m) can experience blocking due to lack of resources on the regular channels. However, the A_{good}^m traffic that has successful access to the regular channels will stay on the regular channels for z % of the total call holding time. Then the mobile will handover to one of the super channels or the calls will stay on the regular channels if the super channels are congested at the time of handover. This situation can be approximated by the following formula:

$$A_{super}^c = A_{good}^m \cdot (1 - B_R^m) - A_{good}^m \cdot z \cdot (1 - B_R^m) - A_{overflow}^m \tag{12}$$

Re-arranging Equation (12) yields:

$$A_{good}^{m} = \frac{A_{super}^{c} + A_{overflow}^{m}}{(1-z)(1-B_{R}^{m})} \tag{13}$$

The total traffic offered to a cell is:

$$A_{total}^{m} = \frac{A_{super}^{c} + A_{regular}^{c}}{1-B_{R}^{m}} \tag{14}$$

Finally, the percentage of the good traffic to the total traffic (*good%*) can be obtained by:

$$good\% = \frac{A_{good}^{m}}{A_{total}^{m}} = \frac{A_{super}^{c} + A_{overflow}^{m}}{(1-z)(A_{super}^{c} + A_{regular}^{c})} \tag{15}$$

The live data were collected over a period of two weeks. Different types of IUO cell configurations were included in the test, such as 2+1, 1+2, 2+2, 2+3 and 3+2. After having collected the OMC data, the various cell parameters were exposed to the IUO traffic model. Finally the two sets of data were compared to estimate the accuracy of the model. The principle of the method is shown in Figure 155.

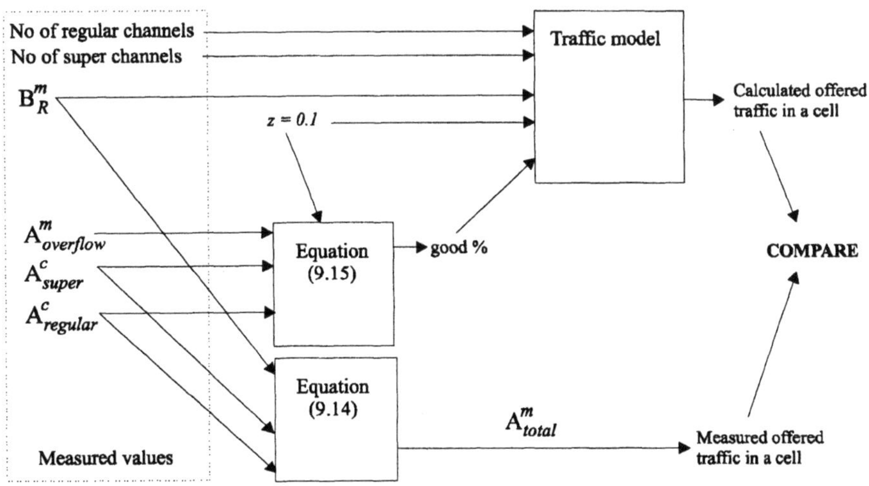

Figure 155. Comparison of the data from the IUO traffic model with the measured data from the OMC used to analyse the accuracy of the IUO traffic model.

The result from the comparison is shown in Figure 156. The absolute number of measurements as a function of the difference between the calculated and the measured value of offered traffic. The analysis was carried out for 680 different busy hour cell measurements.

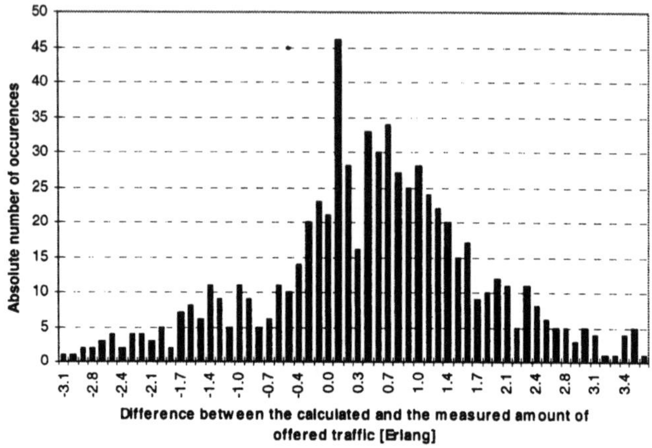

Figure 156. Distribution of the difference between the estimated offered amount of traffic in a cell configured with IOU and the actual measured traffic.

It is seen from Figure 156, that the model corresponds quite nicely to the measured data. The overall mean value of the difference is 0.4 Erlang and the sampling standard deviation of the difference is 1.3 Erlang. The overall average percentage of the differences to the measured offered traffic is found to be only 3%. The model has been tested once again in the fall of 1996 using an entirely different frequency plan. Here the comparison of real live network data measurements and the calculated model traffic also corresponded quite nicely. The overall mean value of the difference was here found to be 0.3 Erlang and the sampling standard deviation of the difference to 1.7 Erlang [178].

2.3.3 The Original IUO Hard Blocking Limit

Assuming an acceptable accuracy of the model, the theoretical hard blocking limits for different IUO configurations were made. They are shown in Figure 157. This figure is identical to the one presented in [125] except it is expanded to also describe the larger IUO configuration such as 4+3.

The number of traffic channels (TCHs) in each configuration and the percentage of good traffic (traffic carried on the super layer) used in the traffic model to produce Figure 157 is shown in Table 36. The percentages of good traffic shown to the right in the table are estimations taken directly from the OMC data measurements, except from the configurations larger than the 5+0 configuration, where the *good%* has simply been set to 60 %. It should furthermore be mentioned that the traffic model parameter z was estimated from the live network to 0.1. In other words, it was assumed that 10% of the time spend in a cell was required to perform a proper *C/I* evaluation on the regular layer. The terms 'Regular TCHs' and 'Super TCHs' in the table refers to the number of allocated traffic channels. That is, for the 1+1 IUO configuration 6 channels are available for the regular frequencies and 8 traffic channels for the super frequencies.

Table 36. Simulation input parameters used to produce the IUO hard blocking curves.

Cell Configuration	Regular TCHs	Super TCHs	*Good %*
1+1	6	8	56
2+0	13	-	-
2+1	13	8	56
3+0	21	-	-
2+2	13	16	68
4+0	29	-	-
3+2	21	16	60
5+0	37	-	-
3+3	21	24	60
4+2	29	16	60
6+0	45	-	-
4+3	29	24	60

In Figure 157 it is shown how the IUO handover limitations affects the carried traffic pr. frequency. It is indicated how the use of the IUO algorithm actually decreases the carried traffic per frequency compared to the Erlang-B case. It means that if it is possible to plan 4 frequencies pr. cell in the network without IUO, then by moving to an IUO cell configuration with 2 regular and 2 super frequencies the maximum carried cell traffic will decrease. For the individual network operator the limited available spectrum and the allowed minimum frequency reuse factor however may not allow the frequency planning of 4 frequencies in a single layer network. Therefore such a comparison would be unfair. In the case where an operator has an available spectrum of 45 frequencies and is not using frequency hopping, the minimum frequency reuse factor would be around 12. Therefore each cell can maximally contain 3-4 frequencies. In the case of IUO the same operator could use a 3+2 configuration assuming a conservative frequency reuse of 12 on the regular layer and 5 on the super layer. With 2% blocking, comparing the 4+0 and the 3+2 configurations, this corresponds to a capacity increase around 25 %, see Figure 157.

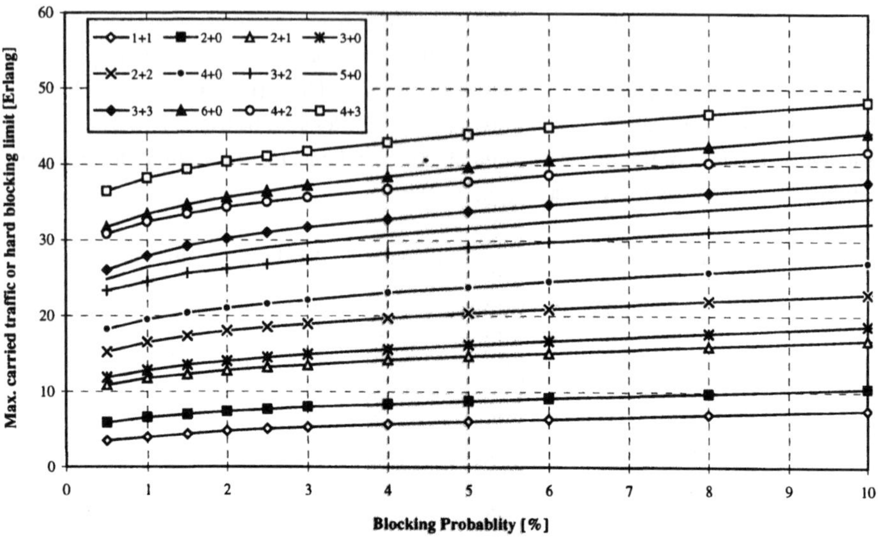

Figure 157. Calculated hard blocking limit for different IUO cell configurations.

The most important conclusion to be drawn from this simplified study of the hard blocking limit of IUO is that due to the evaluation time necessary (on the regular frequencies) to estimate the *C/I*, a non-negligible trunking loss arise when comparing to a one layer network. This trunking loss is of course highly undesirable.

2.4 Remarks on the IUO Algorithm

Based on the description above, the IUO feature is found to create a frequency reuse partitioning system with several advantages. It is a concept that can be applied to an existing network configuration with little additional cost. For large bandwidths where several super frequencies can be allocated per cell a high capacity gain may be possible. If the trunking problem described above could be minimised, an even larger gain could be achieved.

In general it is believed to be a good idea to use interference as an intra-cell handover criteria between the regular and super frequencies, since a dynamic flexibility of adjusting to the actual network conditions is accomplished. If a mobile is moving e.g. from the outdoor environment on the sidewalk to an indoor, the received signal level might easily be reduced by 10 - 15 dB instantaneously. However, the estimated *C/I* will not necessarily be affected that much since the interferers as well as the serving carried experiences correlated fading.

The are some obvious problems with the feature. Of course the decrease in trunking efficiency is a problem, but also when looking at the frequency planning of cells using IUO the procedure becomes more complex, see Figure 152.

Concerning the actual *C/I* estimation, the mobile only reports the six strongest signals that it receives back to the network. This means that at least one of the 6 reported signals should arise from one of the super frequency neighbours in the reference list for the *C/I* estimate to be useful. If this is not the case, the problem can be dealt with in several different ways:

1. The *C/I* can be assumed to be good enough and the mobile station is handed over to a super frequency. This seems reasonable since none of the interfering super frequencies are measured between the 6 strongest interferers. Furthermore, if there is too much interference on the super frequency anyway, the mobile is ideally handed back to the regular due to bad quality. This is quite an aggressive approach, with a certain risk of introducing dropped calls.
2. The *C/I* is assumed *not* to be good enough to be sure that a mobile, which is handed over to the super layer, has a satisfactory quality. This is a quite conservative solution.
3. The weakest of the six reported interferers (*not* one of the super cell neighbours) is used as an estimate of the interference. This gives some kind of a worst case interference level indication, since the interference experienced from one of the strongest of the six super cell neighbours is always lower than the weakest of the six reported signal levels. In case of the micro-cellular CSL network this solution has not been possible, since the level of the weakest of the 6 reported measurements is a lot higher than the strongest IUO interferer.

The importance of having at least one of the super references between the 6 reported values is studied later, where results are shown in Section 8.

In general, since the handover algorithm (with or without IUO) is very robust, the IUO will work even if the relative complex parameter settings are not set in the optimum way. Maybe a mobile, which should stay on the regular layer, is handed over to the super layer, but then it will return to the regular layer due to bad quality or one of the other intra-cell handover possibilities. The mobiles are not necessarily dropped.

3. THE CAPACITY ENHANCEMENT PROPOSAL

As indicated, both FH and IUO reduce the overall frequency reuse pattern and thereby enable a capacity gain in a cellular mobile system. IUO by allowing a tighter frequency reuse of a selected number of the frequencies, and FH by averaging the existing network interference out on all the cell frequencies, thereby allowing a higher minimum acceptable overall interference level in each cell. With FH also an increased coding gain is achieved (fractional loading) as well as the frequency diversity gain from averaging the fast fading out. The proposed idea is to combine the two features and thereby get the combined effect from both [175,166]. In Figure 158 an example of the principle is shown using baseband hopping and a configuration of 4 regular frequencies and 3 super frequencies.

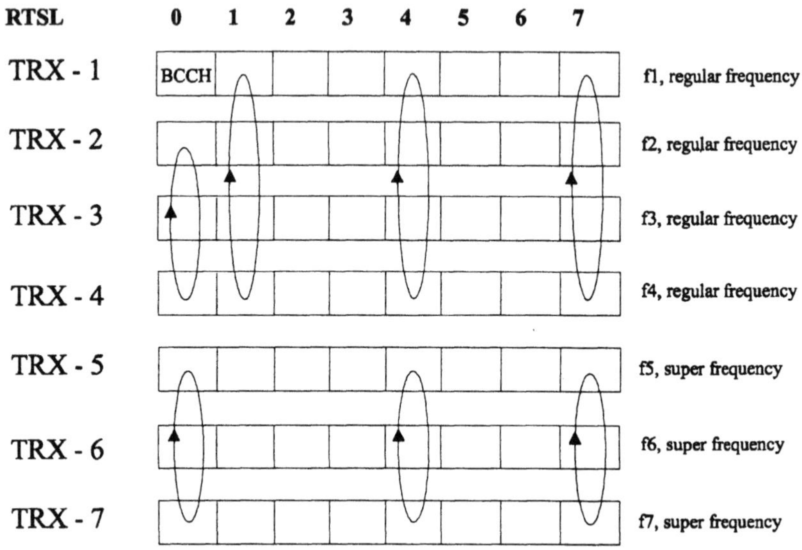

Figure 158. Basic principle of combining baseband FH and IUO for 4+3 configuration.

In the example above, FH between the individual timeslots can be carried out in the following way:

- Timeslot 0 of TRX 2, 3 and 4 hop over f2, f3 and f4.
- Timeslot 1-7 of TRX 1, 2, 3 and 4 hop over f1, f2, f3 and f4.
- Timeslot 0-7 of TRX 5, 6 and 7 hop over f5, f6 and f7.

Some precautions have to be taken before the two features can be combined. Firstly, in order not to violate the functionality of IUO it should not be allowed to hop between the super and the regular frequencies. This means the MA list, specifying the frequencies to hop between in the GSM standard [62], either consists only of the cell frequencies in the regular layer or of the frequencies in the super layer. Secondly, due to this spectrum splitting quite a lot of frequencies has to be allocated in each cell. To get a significant gain from FH the MA has to contain at least 3 frequencies. Therefore cell structures containing as much as up to 7 frequencies, 4 on the regular layer and 3 on the super layer, should be considered. This is quite a lot of frequencies per cell, however e.g. a 3/9 frequency reuse on the regular layer and a 1/3 frequency reuse on the super layer demands a total of 45 frequencies (if letting the BCCH have a frequency reuse of 9). A spectrum of that size is available for network operators such as each of the two Danish GSM-900 network operators SONOFON and TeleDanmark, where other European operators have spectrums as large as 75 frequencies.

In practice when using FH in cells with IUO, the super frequencies have to be planned group-wise. If not, it would be very difficult to estimate a useful *C/I*. This, as a consequence, means that the interference diversity gain from FH is decreased for the super frequencies. It should be noted that when allocated to a super frequency the mobile is (from the nature of IUO) in most cases ensured a low level of interference. Therefore the interference diversity gain will always be limited.

When combining the different baseband FH and IUO, 4 different combinations are possible. They are shown in Table 37 along with the most fundamental characteristics. In case of no hopping, the intra-cell functionality can be used instead, however this functionality is primarily powerful in situations of low load.

Besides the combinations shown in Table 37 RF FH can be used in various ways, however this has not been treated in here. Other studies have shown that potential gain can be expected if the network planning is done carefully, e.g. using MAIO-management as described in Chapter 6.

Table 37. Characteristics of the possible combinations of baseband FH and IUO.

	Regular Layer: NO HOPPING	**Regular Layer: BASEBAND HOPPING**
Super Layer: NO HOPPING	Normal IUO network, not combined with FH.	Requires at least 3 TRX's on the regular layer to get the FH gain. BCCH is hopping. No FH interference diversity gain on super layer. Minimum reuse factor of 9 on the regular TCH layer. Minimum reuse factor of 4 to 6 on the super layer.
Super Layer: BASEBAND HOPPING	Requires only 1 TRX on the regular layer if intra-cell handovers are not used. No FH gain on regular layer. Minimum reuse factor of 12 on the regular layer. Minimum reuse factor of 3 on the super layer.	Requires at least 3 TRX's on the regular layer to get the FH gain. BCCH is hopping. Requires at least 3 TRX's on the super layer to get FH diversity gain. Minimum reuse factor of 9 on the regular TCH layer. Minimum reuse factor of 3 on the super layer.

4. PRELIMINARY SIMULATION STUDIES OF IUO WITH FREQUENCY HOPPING

During the development process some preliminary simulation results describing the capacity enhancement proposal was made. They were presented in [166]. Here limitations of the functionality were discovered and some solutions were proposed for further work. These limitations and proposed solutions are described in the following.

4.1 Problems Discovered with the Original IUO Algorithm and FH

The studies revealed that the IUO algorithm did not work as well as it could. Whenever situations of high load where simulated, deadlock situations sometimes started to occur. It was not necessarily a problem when all channels on the super layer were occupied, because the mobile station that desired a handover to the super layer can stay on the regular layer. However, when all regular channels were occupied, a mobile allocated on the super layer could not get away from its channel. A deadlock situation resulting in dropped calls had therefore occurred.

Another problem, that seemed to occur if the regular layer was congested, while free channels were available on the super frequencies, was that a new mobile requesting a free channel was rejected. This was independent on the experienced quality, since it had to access the regular layer in order to evaluate the *C/I*. This causes the trunking loss also found from the traffic model in Section 2.3.

There were two different reasons to why these problems were found from the simulations. The primary reason was probably the fact that in the, at that time, newest version of CAPACITY (version 4.2), only one out of the 8 timeslots per frequency were simulated. It meant that in case of IUO, there were not L regular channel candidates (corresponding to $L/8$ regular frequencies in GSM), but only $L/8$. This limitation in the implementation effected the simulated number of dropped calls, since when a mobile was not able to make a handover to a regular frequency, it could not move away from its super frequency and eventually it would be dropped. Furthermore, the basic design of the original IUO algorithm of always requiring an evaluation of the *C/I*, while stationed on the on the regular layer, introduces a trunking loss as described. If implementing all 8 timeslots per frequency, the fact that the only way from a super frequency goes via the regular frequency of the same cell will cause the same type of problems.

4.2 Improvements to Enhance the IUO Algorithm

Proposals for some modifications to the IUO algorithm on how to minimise the problems described above were made [166]. Such modifications will inevitably increase the complexity of the IUO algorithm, but if the capacity gain is significantly high the complexity is believed acceptable. Two potential improvements were proposed:

1. **Allow direct access to the super layer.** The idea is to add one more handover criteria to the IUO handover algorithm, namely a criterion that allows a direct call access to the super layer. Provided the received signal strength (RXLEV) is above a certain pre-defined level, the quality is assumed to be good enough and the mobile station should be allowed direct access to the super layer. Since this criteria is an estimation of the quality, it is not known for sure whether the quality is good or not, and therefore this direct-access-to-super-layer-criteria should probably only be used when it is absolutely necessary. In other words, the criteria should only be activated when congestion occurs (or is on its way) on the regular layer. A possible way of controlling this could be by specifying a percentage, e.g. 70%, as a threshold to how much of the available regular channel capacity should be occupied before the direct access functionality is enabled. In general this improvement treats the issue of getting the mobiles onto the super layer.
2. **Timeslot-reservation for intra-cell handovers**. When the *C/I* estimated by a mobile allocated on a super frequency becomes small, the mobile should be able to make a handover to a regular frequency, otherwise the call will eventually be dropped. Therefore some free resources always have to be available on the regular layer. The idea is to dedicate one or more specific timeslot(s) to be used only for this intra-cell handover (from super to regular layer). After having performed the handover from the super layer to a dedicated timeslot, the call should be handed further on to another timeslot on the same frequency or another frequency in the same cell or maybe even in another cell. By using this principle a queuing type of system is introduced. In general this improvement treats the issue of getting the mobiles off the super layer.

In order to treat the problems discovered, the simulator as well as the IUO algorithm itself was improved. CAPACITY was enhanced to model the GSM network better by simulating 8 timeslots per frequency. Initially the improvements implemented in the IUO algorithm, which are actually more extensive than proposed above, are described.

5. THE IMPROVED IUO ALGORITHM

As described, the primary limitation is the reduction in trunking efficiency due to the fact that every access to and from the super layer goes through the regular layer in the same cell. This can be summarised in the following two situations.

1. New calls in a cell and inter-cell handovers to the cell can be hard blocked on the regular layer while free channels are available on the super layer. The size of the problem depends of the traffic distribution and is especially large if the traffic is located around the cell border.
2. Moving away for the coverage area of the super frequency can be denied if hard blocking occurs on the channels on the regular layer.

Both of the two proposals from above [166], allowing a direct access to the super layer and reserving timeslots for handovers, have been implemented in CAPACITY. The direct-access-to-the-super-layer-feature is treated in detail in the following, along with some new handover possibilities have been introduced to enhance the IUO algorithm performance. The second proposal has been studied for the case of the 4+3 IUO configuration using random baseband FH with mobiles moving at 3 km/h and an offered network load of 83 %. For this scenario, with as little as one channel reserved for handovers, an increase in blocking was discovered. As will be shown in the result section, the hard blocking of this scenario arises primarily based on blocking on the regular layer. The reason for should be found in the increased number of handover possibilities introduced in the improved IUO algorithm, which is described in the following. They degrade the problem of being able to get of the super frequencies and thereby treat the same problem as the timeslot reservation feature. Further studies of allowing timeslots to be reserved for handovers in relation to IUO has therefore not been carried out.

5.1 Improved Handover Characteristics with IUO

The handover possibilities of the IUO algorithm to and from a particular IUO cell of a 3-sector site are depicted in Figure 159. The dotted lines indicate the improved handover possibilities made to the IUO algorithm, while the remaining ones illustrates the handover possibilities of the original algorithm.

From Figure 159 it is seen how 12 different types of handovers can be carried out with the improved algorithm, while with the original only 6 types are possible. A detailed description of the design of the improved IUO algorithm as it is implemented, is described in Section 6 on page 239, where the software implementation is described.

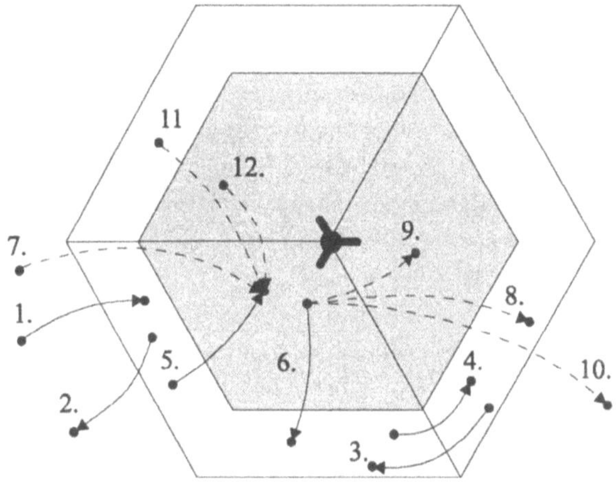

Figure 159. All handover possibilities to and from an IUO cell of the improved IUO algorithm. The shaded region is the super layer for 3 cells of a 3-sector site.

5.2 Hard Blocking Traffic Model of the Improved IUO

The traffic model, used in Section 2.3.1 to estimate the hard blocking limit of an IUO cell, has to be changed in order to be able to model the improved IUO algorithm. This is done in the following, where the direct-access-to-the-super-layer handover possibility is compensated for.

The basic idea was to design a model using the classic tele-traffic theory, as described in [25,196,244], including the direct-access-to-the-super-layer feature. This change can be visualised as a third layer covering part of the super layer with the coverage area depending on the specified signal strength threshold.

5.2.1 The Traffic Model

To simplify the calculations it is assumed that the direct-access-to-the-super-layer signal level threshold is specified so that it covers exactly the area covered by the super frequencies. In reality this is probably difficult and therefore describes the optimum situation. The coverage area of the 3 different layers are shown in Figure 160, where the area covered by the direct access level threshold, for better illustration, only covers part of the entire super layer.

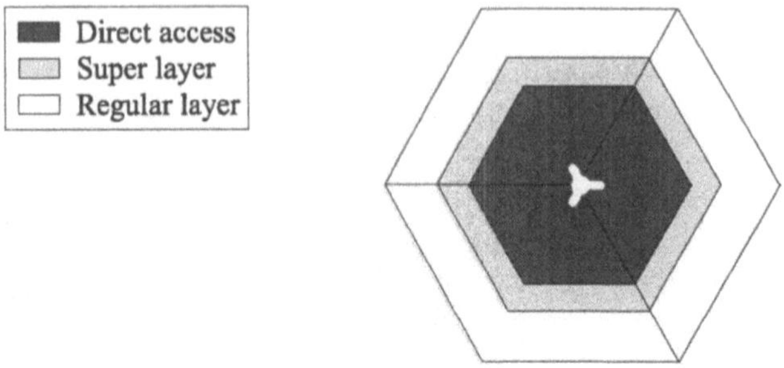

Figure 160. Coverage areas of the IUO cell with direct-access-to-the-super-layer.

The model of the traffic streams of an IUO cell using the improved algorithm is simpler than for the original IUO cell. It is shown in Figure 161. The only difference is that the time required to perform a proper *C/I* evaluation, resulting in the traffic named $A_{time\_on\_reg}$ in the model of the original IUO algorithm, is completely removed.

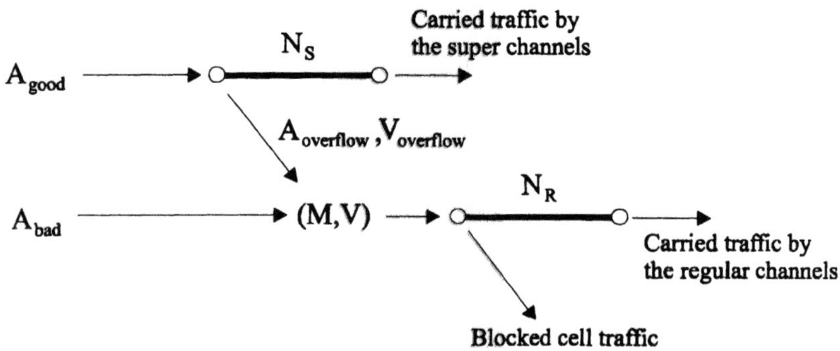

Figure 161. Traffic streams of a cell using the improved IUO algorithm.

The parameters in Figure 161 are equivalent to the parameters described in Table 35.

The total amount of offered cell traffic (A_{total}) corresponds to the traffic offered to the area covered by the super frequencies (A_{good}) and the traffic offered to the area covered by the regular frequencies only (A_{bad}).

$$A_{Total} = A_{good} + A_{bad} \tag{16}$$

Again the originating cell traffics (A_{good} and A_{bad}) are assumed to be Poisson distributed, while the same assumption cannot be applied to the overflow traffic, since traffic rejected due to congestion in general cannot be assumed to have a Poisson distributed arrival process [103, p.192]. Its statistical mean ($A_{overflow}$) and variance ($V_{overflow}$) therefore instead represent the overflow traffic, as previously. The call holding time is assumed negatively exponentially distributed and a call is lost if it is blocked (BCC assumption). Handovers due to movement of the mobiles during a call are not considered.

The traffic offered to the super channels, equal to A_{good}, generates the overflow traffic from the super layer. The mean overflow traffic can therefore simply be calculated as shown in Equation (17) using the Erlang-B formula.

$$A_{overflow} = A_{good} \cdot E(N_S, A_{good}), \tag{17}$$

The variance of the overflow traffic is also required and can be calculated by the following formula [103,200,244]:

$$V_{overflow} = A_{overflow} \cdot \left(1 - A_{overflow} + \frac{A_{good}}{N_S - A_{good} + A_{overflow} + 1}\right) \tag{18}$$

The combined mean traffic offered to the regular channels is therefore:

$$M = A_{overflow} + A_{bad} \tag{19}$$

and since the variance of Poisson distributed traffic equals the mean, the combined variance is:

$$V = V_{overflow} + A_{bad} \tag{20}$$

The total traffic offered to the regular channels, characterised by (M, V), is then set equivalent to the traffic rejected from a full availability set of channels with same mean (M) and variance (V). This is done using the Equivalent Random Theorem. With this method it is possible to convert any overflow system containing both overflow traffic and new Poisson traffic offered to a second group of channels into a simple overflow system. The new model is shown in Figure 162.

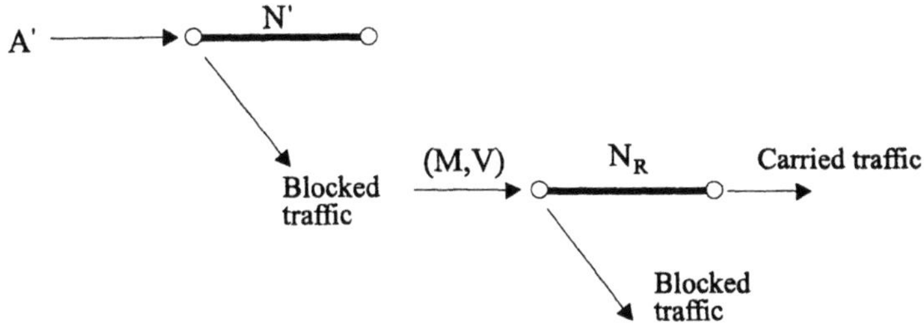

Figure 162. Equivalent Random Traffic of the combined input traffic to the regular layer.

Using Rapp's approximated solutions again it is possible to calculate the traffic (A') and number of channels (N') in a fictive group delivering exactly the overflow traffic with a mean value of M and variance V. I.e. if A' traffic is offered to N' channels of full availability the resulting amount of blocked traffic will exhibit similar mean and variance as (M,V). The approximation states [103,196]:

$$A' = V + \frac{3V}{M} \cdot \left(\frac{V}{M} - 1 \right) \tag{21}$$

and

$$N' = \frac{A'}{1 - \dfrac{1}{M + \dfrac{V}{M}}} - M - 1 \tag{22}$$

After obtaining the fictive values A' and N', according to Figure 162, the average blocking probability to any user in the cell can be calculated as:

$$\text{Average cell blocking prob.} = \frac{A' \cdot E(N' + N_R, A')}{A_{good} + A_{bad}} \tag{23}$$

Equation (23) then describes the blocking probability of a cell using the improved IUO algorithm.

5.2.2 Performance of the Improved IUO Algorithm

The hard blocking traffic limits of the improved IUO algorithm is studied for the same configuration types as used previously. They are:

Table 38. IUO configurations studied for the hard blocking model of the improved IUO.

Configuration Type	Regular Channels	Super Channels	Good Traffic Percentage
1+1	6	8	56
2+1	13	8	56
2+2	13	16	68
3+2	21	16	60
3+3	21	24	60
4+2	29	16	60
4+3	29	24	60

To investigate the hard blocking of the improved IUO, compared to a flat network not using IUO, Figure 163 has been made.

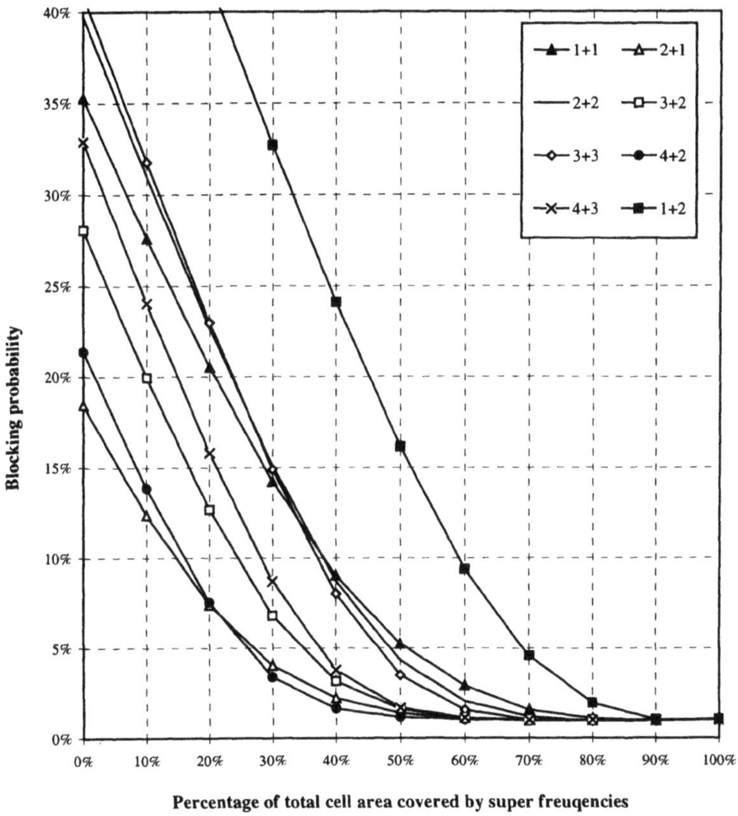

Figure 163. Probability of blocking of the improved IUO algorithm.

In Figure 19 the probability of blocking of the improved IUO algorithm when assuming an amount of offered traffic corresponding to the 1 % Erlang-B traffic of a flat network with the equivalent number of TRX's, as a function of the size of the coverage area of the super frequencies, is shown. This corresponds to varying the C/I-good and C/I-bad thresholds. In all situations the traffic amount used, is equivalent to the amount of offered traffic when using the 1% Erlang-B blocking probability of a flat network using the same number of available resources.

With the sizes of super-area-percentages considered here (above 50%), the average cell blocking is always below 5 % for all investigated configurations (except the hypothetical 1+2 configuration), when modelling the amount of offered traffic equal to the 1 % Erlang-B blocking limit of a flat network. It is furthermore seen how the amount of super frequencies in the cell compared to the total number of cell frequencies is very important. The more super frequencies, the large a super coverage area percentage is required to keep the blocking percentage down. This seems reasonable. The question of whether more super frequencies than regular frequencies can be used with the modified IUO algorithm has frequently been raised. Therefore one hypothetical example, the 1+2 configuration, is included in Figure 163. It is clearly seen how a very large coverage area has to be ensured by the super frequencies in order not to increase the blocking significantly.

From Figure 163 it can furthermore be observed that with 100 % coverage of the super frequencies, the blocking probability is 1%, i.e. equivalent to the Erlang B probability of the flat network. Also with 0 % super coverage area the blocking probability is equal to the Erlang B blocking probability of the regular channels.

The hard blocking limits of the different IUO configurations are shown in Figure 164 ranging from a blocking percentage of 0 % to 8 % and using the good traffic percentages from Table 36. The Erlang-B probabilities are included, for comparison.

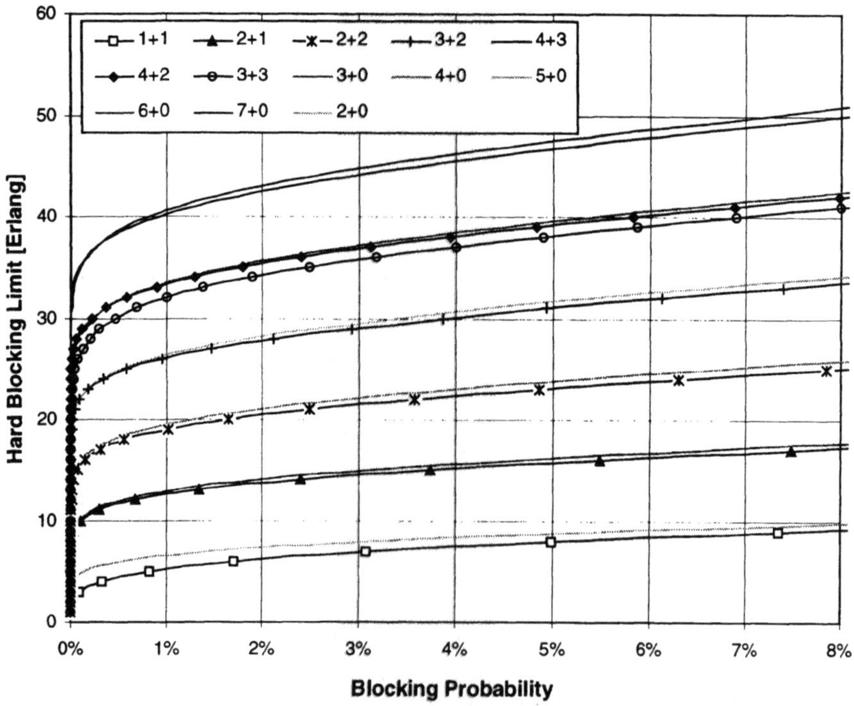

Figure 164. Hard blocking of the improved IUO algorithm using configurations in Table 36.

It is seen how the improved IUO algorithm, with the good traffic percentages (*good%*) in Table 36, has a trunking loss almost equivalent to what is found from the

Erlang-B formula except for 2 configurations (1+1 and 3+3). Here the ratio between the number of channels available on the regular layer and on the super layer of 1 influences on the trunking loss.

5.2.3 Comparison of the Two Hard Blocking Limitation Models for IUO

Examples of the blocking, when offering the amount of traffic corresponding the 1 % Erlang-B, for the flat network, the improved IUO including direct-access-to-the-super-layer and the original IUO, are shown for two different numbers of TRX's per cell, namely 2 and 6. These configurations clearly show the difference in trunking loss for the various configurations. Note that the ordinate for the 6 TRX configurations does not start at 0 Erlang, but at 29 Erlang.

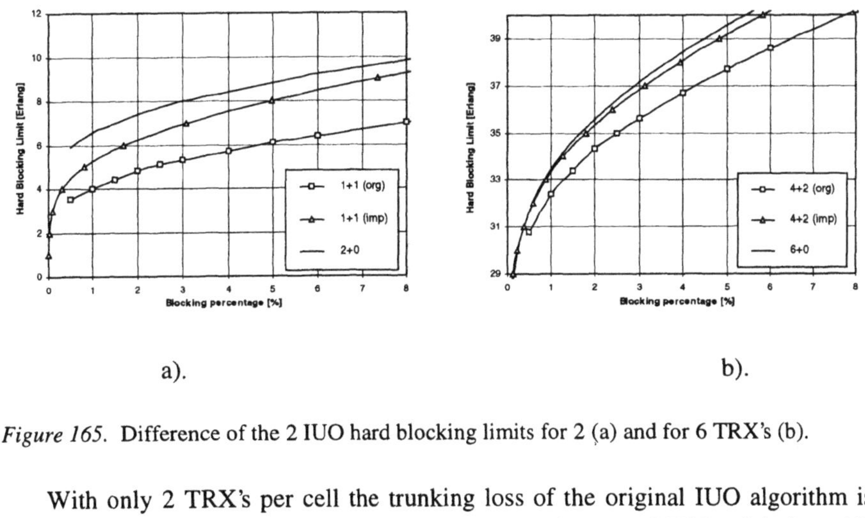

a). b).

Figure 165. Difference of the 2 IUO hard blocking limits for 2 (a) and for 6 TRX's (b).

With only 2 TRX's per cell the trunking loss of the original IUO algorithm is quite large compared to the Erlang-B formula. The improved IUO minimises this loss but there is still a difference compared to the flat network. In case of 6 TRX's the improved IUO has almost removed the trunking loss when compared to the flat network. It should be remembered that these comparisons are made with the specific cell traffic distribution assumptions in Table 36.

6. IMPLEMENTATION OF IUO IN CAPACITY

The most important aspects of the implementation of IUO in CAPACITY are described in this section. The feature has been implemented as close as possible to the algorithm of NOKIA with the improvements described in the previous sections. That is, it becomes possible to perform a thorough study of the IUO functionality as well as the improvements. Section 6.1 specifies the input parameter list required to configure IUO in CAPACITY, while Section 6.2 describes the implementation of the handover algorithm.

6.1 The IUO Input Parameter List

Some input configuration parameters have been implemented exclusively for the IUO algorithm in CAPACITY. These parameters are listed in Table 39.

Table 39. IUO configuration input parameters required in CAPACITY.

THE IUO PARAMETERS
SuperReuseGoodCiThreshold
SuperReuseBadCiThreshold
HOAveragingIUO
QualityHandover
LevelHandover
Cir_goodHandover
Cir_badHandover
Dir_acc_th

SuperReuseGoodCiThreshold and SuperReuseBadCiThreshold are the two thresholds defining the C/I-good and C/I-bad thresholds. HOAveragingIUO in the number of values used for averaging before each *C/I* comparison. A value of e.g. 5 means that for every TDMA frame the last 5 *C/I* values are averaged, before being compared to the C/I-good or C/I-bad threshold. QualityHandover, LevelHandover, Cir_goodHandover and Cir_badHandover are all enable parameters for the IUO algorithm implemented in CAPACITY. Dir_acc_th specifies the received signal strength threshold used for direct-access-to-the-super-layer (specified in dBm). If the received signal strength of the individual mobile is higher than this threshold the direct access is allowed.

6.2 Implementation of the Handover Algorithm

As described in Section 2 the IUO algorithm can been seen simply as an enhancement to the existing handover algorithm. The different states and handover possibilities a mobile can have with the improved IUO implementation is shown in Figure 166.

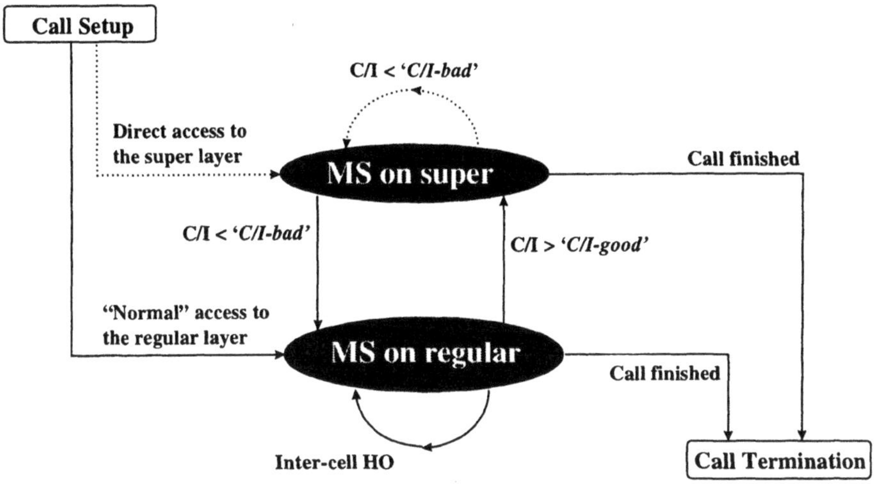

Figure 166. State diagram with handover transitions for the improved IUO algorithm.

In the following the different implemented types of handovers are described along with an explanation of why each particular type is allowed and what situation the handover is supposed to handle. Basically only 4 different types exists.

1. **Regular inter-cell handovers**. These are handovers from a regular frequency in the home cell (the name used in Figure 167) to a regular frequency in a neighbouring cell. They correspond to handover type number 1, 2, 3 and 4 in Figure 159. In CAPACITY the regular inter-cell type of handover can be carried out from the home cell to all cells depicted in Figure 167, corresponding to the cells in all neighbouring sites. The best one will be chosen based on the different handover criteria's. Compared to the real GSM network this is a sufficient number of candidates, since usually no more than 10-15 handover candidates can be measured.
2. **Super inter-cell handovers**. These are handovers from a super frequency in one cell to a super frequency in one of the neighbouring cells. They correspond to handover type number 9 and 12 in Figure 159. This type of handover can be activated due to a low estimated *C/I* (below C/I-bad threshold), bad quality or a low signal level. It can be made from the super frequencies of the home cell to one of the neighbouring sectors on the same site. The reason for this type of handover is to allow the mobiles close to the base station to handover directly from one sector to another sector of the same base station. When a mobile station attempts this kind of handover the received signal strength on the BCCH channel from the new antenna has to be higher than the predefined direct-access-to-the-super-layer threshold Dir_acc_th. If this estimated received signal strength does not satisfy this requirement, a handover to a regular frequency is attempted.

3. **Super -> regular handovers.** These are the handovers from a super frequency to any possible regular frequency. They correspond to handover type number 6, 8 and 10 in Figure 159. This kind of handover only occurs when it is not possible to make a handover to a super frequency in one of the neighbouring sectors, i.e. when the received signal strength of all neighbouring sectors is lower than the specified Dir_acc_th. The handover candidates are frequencies in all super layer cells in Figure 167.
4. **Regular -> super handover.** These are the handovers from a regular frequency to a super frequency. They can occur when the estimated *C/I* of a mobile station is above the C/I-good threshold and correspond to handover type number 5, 7 and 11 in Figure 159. When the occurring handovers are based on the C/I-good handover cause, the mobile is always handed over to the super frequency in the same cell. The regular to super type of handover can furthermore occur using the Dir_acc_th, provided the received signal strength is high enough. If using this threshold the handover can be carried out from a regular frequency of a neighbouring site, as well as from a regular frequency of a neighbouring sector of the home site.

In CAPACITY as many as 6 handover attempts can be made at each time instant. In other words, if a handover attempt is blocked the second best candidate is tried out and so on. If no available candidates can be found, the mobile simply stays on the channel it is already allocated to. Then a handover will be attempted again after a certain (predefined) time-period.

Furthermore, to emulate a real time system even more accurately, to compensate for the sequential execution in the simulator, handover prioritising has been implemented as well. By handover prioritising is meant that handovers are prioritised higher than new (incoming) calls. When a new call attempt occurs in a certain cell at the same time as a handover attempt occur to the same cell, the handover is carried out first. The handovers are also prioritised so that the mobile that has been most eager to perform the handover is the first in line. Then the second most eager is taken care of etc. This functionality is implemented by weighing the mobiles according to how many handover attempts has been made. Of course it is possible that the mobile station may not want to perform the handover anymore, if it has moved to another physical location during the time interval between the two handover attempts, where it is not necessary. When this is the case the mobile is simply removed from the queue.

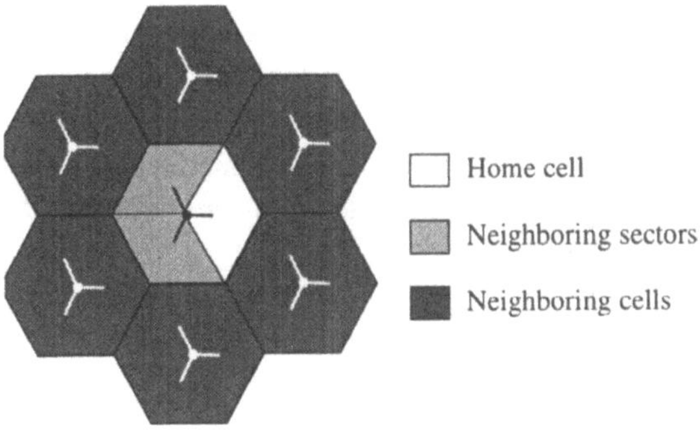

Figure 167. Simulated cell layout with home-cell, neighbouring sectors and cells.

7. OUTLINE FOR CAPACITY SIMULATIONS CONCERNING IUO

Determining exactly what simulations to carry out has not been a simple task. The primary aim of this IUO study is to investigate if the network capacity of FH combined with IUO. Can the network capacity be increased without degrading the network quality and if so how much? To answer these questions a wide variety of IUO configured networks can be imagined relevant. Examples are IUO networks with and without baseband FH on one or on both layers. Furthermore, network parameters as the reuse factors of the regular as well as the super layer and the available spectrum have a significant influence on the performance and are therefore also essential.

First of all, the IUO parameters have to be configured according to the existing handover and power control parameters. To find the maximum capacity of a network using IUO, the best settings of the IUO parameters, as e.g. the C/I-good and C/I-bad thresholds, of course also have to be determined. As described several parameters are essential to configure the algorithm properly and since the optimum specification of these parameters depends heavily on the network configuration characteristics, this is quite a complex task. Network characteristics are the parameters describing the type of network that is dealt with. Examples are; the number of super and regular frequencies available per cell. The total spectrum available. Is frequency hopping used or not? Is power control activated? Is the network configured with omni directional antennas or antennas to be used for 3-sector sites? Secondly the actual capacity and corresponding quality has to be determined.

The type of simulations to be carried out therefore can be divided into two groups. First of all the different IUO algorithm parameter settings are treated. The idea of this initial part is also to study the functionality of the IUO algorithm in a GSM network. Also, the proposed improvements, as the direct-access-to-the-super-layer, are investigated. They are all studied using a 4+3 configuration with baseband FH, power control, DTX, a mobile speed of 3 km/h, a frequency reuse of 3/9 for the regular and 1/3 for the super frequencies, a network load of 70 % and an available spectrum of 45 frequencies. The BCCH frequencies are not simulated. Secondly categories of different network characteristics are studied. Several network characteristics are treated.

7.1 IUO Parameter Settings

As described, a complete optimisation of the IUO algorithm for each treated situation is an impossible task. Therefore, the aim has here been to find a set of parameters, default IUO parameters, which are close to optimum and makes the IUO algorithm perform well for the scenario's dealt with. The chosen IUO parameters optimised upon are shown in Table 40. The rest of the IUO parameters have been fixed. Concerning the remaining handover and power control parameters the default parameters found in respectively Chapter 7 and 8 have been used.

Table 40. The parameters studied concerning the functionality of IUO.

IUO PARAMETERS SIMULATED IN CAPACITY
C/I-good and C/I-bad thresholds
Direct-access-to-the-super-layer signal level threshold
The *C/I* evaluation time required on the regular layer
The impact of only reporting the 6 strongest neighbours

7.2 Network Parameter Settings

The general CAPACITY network configuration used for the IUO simulations is shown in Table 41.

The combination of IUO and FH is compared to different types of flat networks of respectively a tight 1/3 reuse (with RF FH) and a looser 3/9 reuse (with baseband FH). All the simulations have been carried out for mobiles moving at 3 km/h as well as 50 km/h, since the general network as well as the number of handovers per call depends on the user mobility. In all cases random FH has been used, primarily due to the frequency planning, which has always been with grouped frequencies.

Four different scenarios have been simulated. They are 2 different IUO cell configurations: 4+3 and 3+6, of which 4+3 is the primary configuration. Furthermore, two sets of reference simulations have been carried out. They are flat baseband FH networks with a frequency reuse of 3/9 and 1/3. In the 3/9 case 5

TRX's are allocated per cell, while with the 1/3 reuse 7 TRX's are used per cell of which the BCCH is non-hopping.

Table 41. Network system simulation parameters identical for the CAPACITY simulations concerning IUO.

PARAMETER	VALUE
Path loss model	$L_p = 35 \log(\text{distance})$
Shadow fading standard deviation	6 dB
Shadow fading de-correlation criteria	$1/e \approx 0.37$
Shadow fading correlation distance	110 m
Call mean hold time	80 s (negative exponential distribution)
Time slots used	8
Cell radius	3 km
Antennas	90 ° sectorized
Type of frequency hopping	Random
Handover algorithm	Based on interference, quality, level, power budget and IUO
Power control	Exploiting both level and quality
DTX factor	0.5
Dropped call threshold	19
Dropped call RXQUAL threshold	5
Cell edge	Wrap around grid
Inter-timeslot handovers	Enabled
Network topology	6 * 8 sectorized sites (144 cells)
Control channels	With RF hopping, BCCH is simulated

The specifications of the simulations initially carried out, are shown in the four tables below. It is seen how various network loads have been treated. In the tables '*BB*' means baseband FH, '*RF*' means synthesised FH and '*No-FH*' means non-hopping.

The five available performance parameters RXQUAL, *C/I*, FER, dropped calls and hard blocking are all found for every simulation. Another feature, found relevant to include in the network along with the combination of IUO and FH, is load sharing.[27] One of the drawbacks of the IUO algorithm, the sensitivity to the traffic distribution, can be compensated for by sharing the resources between cells. The overlap between the regular cells can be used to reduce the trunking problem and,

[27] The implementation of the feature made in CAPACITY is described in Chapter 8, using dynamic handover margins.

depending on the situation, reduce the hard blocking. The concept is shown in Figure 168. One such simulation has been run, simply to see the effect with IUO.

The 4+3 IUO Configuration:

Simulation name:	IUO (4+3)	
Network type	IUO baseband FH	
MS speed:	3/50 km/h	
Network load:	77%,80%,83%,86%	
	REGULAR	SUPER
TRX's:	4	3
Reuse:	3/9	1/3
Frequencies required:	36	9
Features:	BB	BB

The 3+6 IUO Configuration:

Simulation name:	IUO (3+6)	
Network type	IUO baseband FH	
MS speed:	3/50 km/h	
Network load:	50%,55%,60%	
	REGULAR	SUPER
TRX's:	3	6
Reuse:	3/9	1/3
Frequencies required:	27	18
Features:	BB	BB

Flat Network with 3/9 reuse and BB:

Simulation name:	Flat network (5+0)	
Network type	flat network with BB FH	
MS speed:	3/50 km/h	
Network load:	78%	
	REGULAR	SUPER
TRX's:	5	
Reuse:	3/9	
Frequencies required:	45	
Features:	BB	

Flat Network with 1/3 reuse and RF:

Simulation name:	Flat network (7+0)		
Network type	flat network with RF FH		
MS speed:	3/50 km/h		
Network load:	78%		
	BCCH	REGULAR	SUPER
TRX's:	1	6	
Reuse:	4/12	1/3	
Frequencies required:	12	33	
Features:	No-FH	RF	

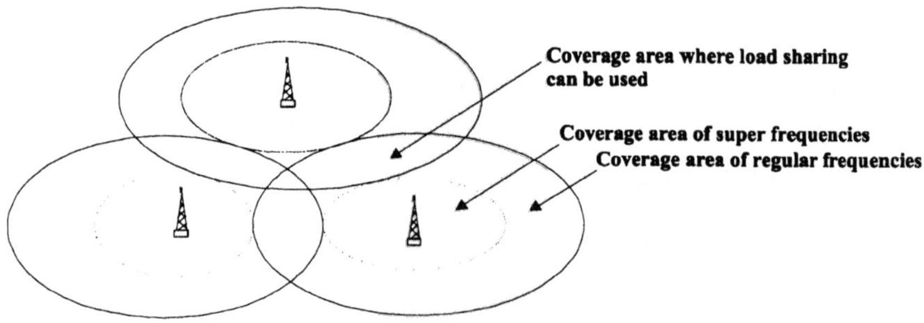

Figure 168. Combining IUO, FH and load sharing to maximise the cell capacity even further.

8. CAPACITY SIMULATION RESULTS

A large part of the results presented here have been devoted to the 4+3 IUO configuration using baseband FH, since this is the scenario that the proposal originally was intended for.

The section is divided in two parts, where the selected simulations are either concerned with the functionality of the IUO parameters or the performance of IUO combined with baseband FH.

In general the FER is believed to be the quality parameter with the highest correlation to the quality experienced by the subscriber. Despite the FER is not available in a live network, the BER, *C/I* and dropped call statistics are also shown whenever soft blocking is treated. Initially, when considering the functionality in Section 8.1, the aim is not necessarily to compare the soft blocking parameters of different alternative network features, but rather to describe the performance of the IUO algorithm. Therefore other performance parameters, as e.g. the different handover causes, have been used.

8.1 Simulations of the Functionality of IUO and FH

Concerning the functionality of IUO, including configuring the IUO algorithm parameters properly, the following 4 subjects have been studied: The C/I-good and C/I-bad thresholds, the evaluation time necessary on the regular layer to perform a "proper" *C/I* estimation, the direct-access-to-the-super-layer signal strength threshold and finally the impact of only reporting the 6 strongest neighbours.

The relevance of each of these parameters has already been dealt with previously in the chapter and will therefore not be treated further. The network characteristics used for this functionality study have previously been given in Section 7.

8.1.1 The C/I-good and C/I-bad Thresholds

In Figure 169 the effect of varying the fundamental parameters of the IUO algorithm, the *C/I* thresholds, is shown. The abscissa holds the value of the C/I-bad threshold. In all situations the C/I-good threshold has been specified 2 dB above C/I-bad.

When the thresholds are lowered the number of handovers to the super layer is increased, but also the handovers back to the regular layer. It is also seen how the reason for mobiles to return to the regular layer, is more and more frequently caused by bad signal quality or low received signal strength. Furthermore, what could be seen from these simulations was, that it would, for this simulated scenario, probably not be desirable to have lower thresholds than 9 and 11 dB. In that case the number of handovers per call would start to increase significantly. The C/I-good threshold, to be used in the performance simulations later on, is in the simulations specified to 11 dB, while the C/I-bad threshold is 9 dB. It is also seen how the number of inter-cell handovers from a regular to regular frequency is almost not influenced by the *C/I* thresholds.

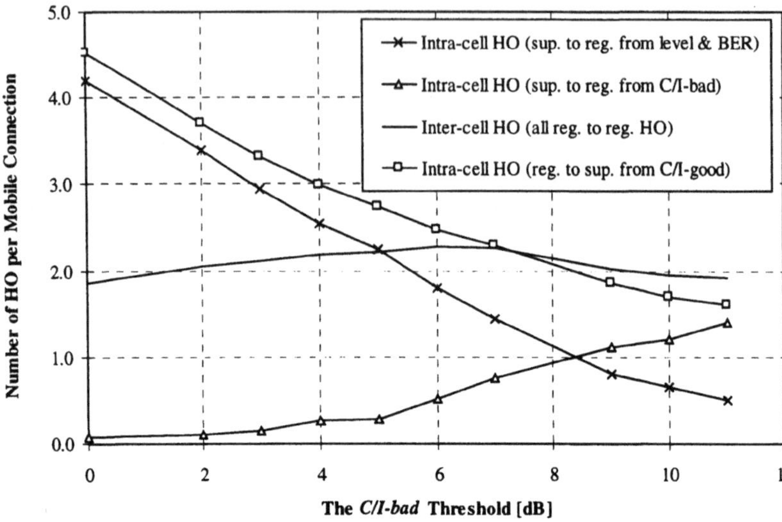

Figure 169. Influence of the C/I-good and C/I-bad thresholds on the number of handovers per call connection. Only the relevant handover causes are shown.

It has been discussed whether the algorithm should be configured so that the hand-off from the super layer should be based on the C/I-bad threshold (the original idea), or maybe a step further could be taken to let most of the handovers occur due to bad quality. Since the two parameters both describe the level of interference, the reason for proposing to use the quality, has been that the accuracy of the *C/I* estimation will probably be by far the worst of the two.

In a live network situation this type of figure, indicating the different handover causes, can be used to find the two proper *C/I* thresholds.

8.1.2 The Influence of Direct-Access-to-the-Super-Layer

The improvement to the original IUO algorithm of allowing access to the super layer without accessing the regular layer is studied here for different signal level thresholds ranging from –70 dBm to –105 dBm. The relevant parameters, the dropped call rate, the types of handovers, the hard blocking and the FER have been treated for various thresholds of direct access.

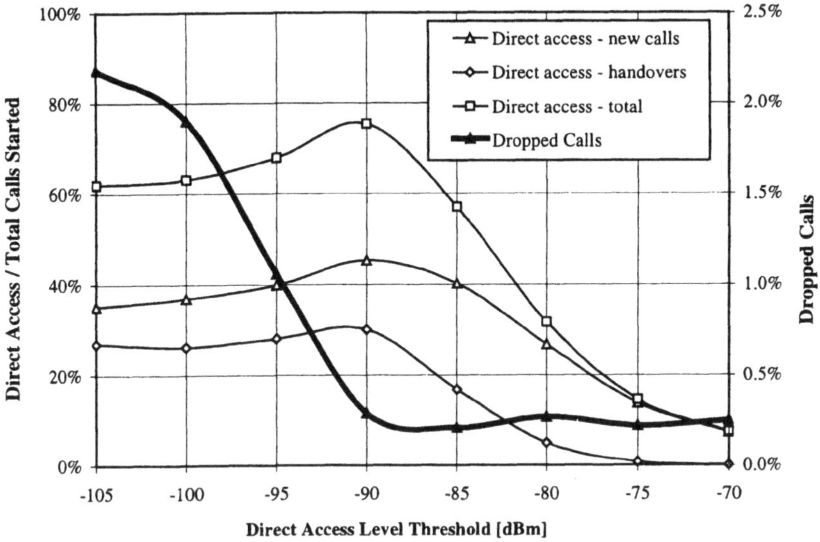

Figure 170. Influence on the dropped call rate of the direct-access-to-the-super-layer signal strength threshold. Also the effect of the feature as a function of handover utilisation is shown.

The most important thing is to see that by setting the direct-access-to-the-super-layer signal level threshold to aggressively will introduce dropped calls. Some mobiles are allowed to enter the super layer without have a satisfactorily good quality. If the regular layer is blocked, the mobile will eventually drop. What is also seen in the figure is the distribution of the two different types of direct access, from a new call or from a handover. Since uniformly distributed traffic has been used, it could be expected that primarily new calls would utilise the direct access. However, since handovers are prioritised ahead of new calls, as described previously, many handovers are also carried out using this feature.

Handover statistics are presented in Figure 171. Initially the number of handover causes, related to handing off the super layer, for different direct-access-to-the-super-layer thresholds, is given. The number of handover to the regular layer is unaffected, however at a threshold lower than –80 dBm the number of attempts starts to increase quite significantly. This is undesired, why a level threshold of –80 dBm has been chosen for the further studies.

Figure 171 (bottom) illustrates where the blocking occur for the simulated 4+3 network. Almost all of the hard blocking occurs from the regular layer. This has been the reason why the timeslot reservation proposal, described in Section 4.2, only degraded the performance. The mobiles does not seem to have any problem with getting of the super layer, simply due to the newly implemented handover possibilities (see Section 5) and the handover prioritising.

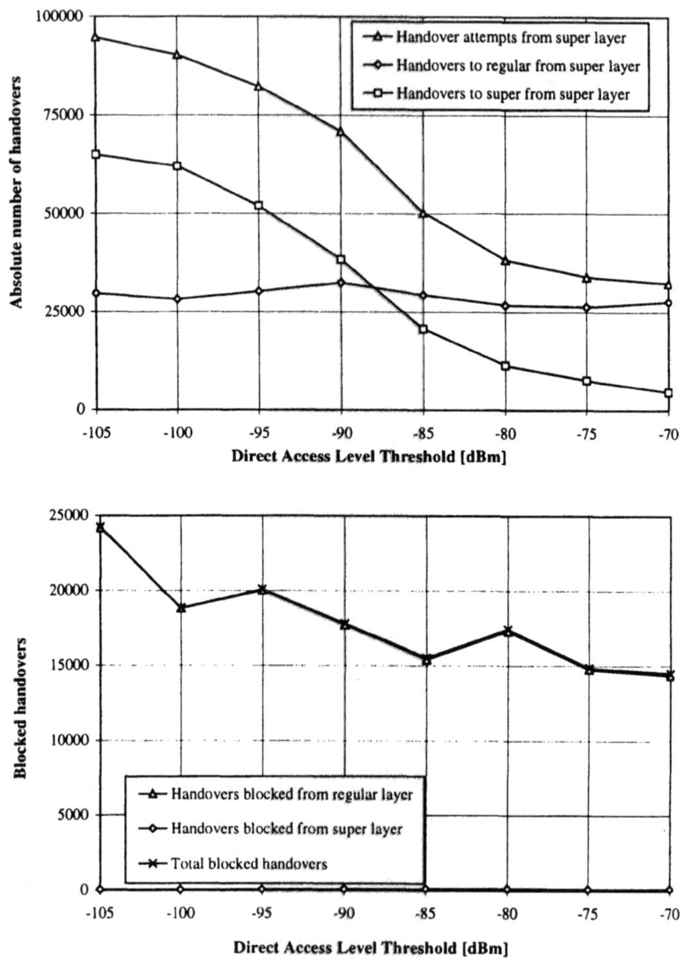

Figure 171. Handover statistics concerning the number of attempts from the super layer (top) and concerning the reason for hard blocking (bottom).

The soft blocking performance has also been studied for both layers using various direct-access-to-the-super-layer thresholds. The FER is also shown in Figure 172.

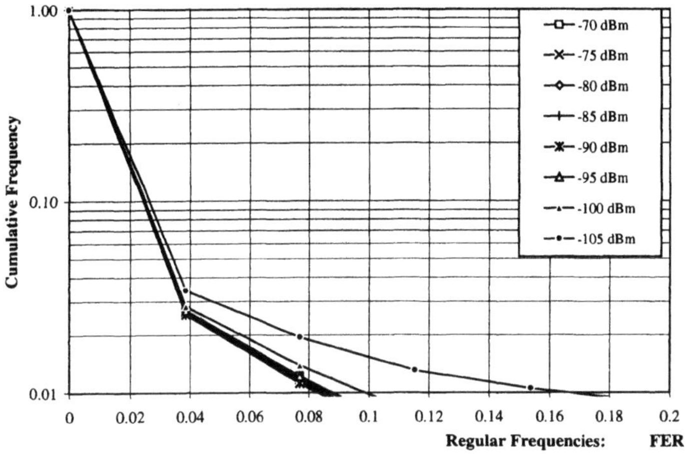

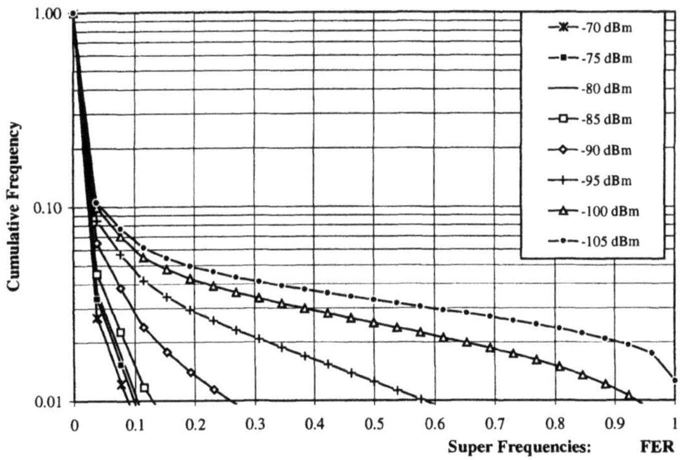

Figure 172. FER of regular (top) and super layer (bottom) for various direct-access-to-the-super-layer values.

The network quality on the regular layer is almost unaffected by changing the thresholds where as for the super layer a clear degradation is seen when lowering the threshold to more than –80 dBm. This is as expected.

8.1.3 The Evaluation Time Necessary to Perform the C/I-Estimation

Without the direct access feature, all mobile stations have to stay on the regular layer for a certain period. By adding the direct access feature to the IUO algorithm, less mobile stations have to go through the regular layer. In Figure 173 the influence of the time spend on the regular layer can be seen for a 4+3 network with mobiles moving at a speed of 3 km/h. The carried amount of traffic per cell, with 2 % blocking, is depicted.

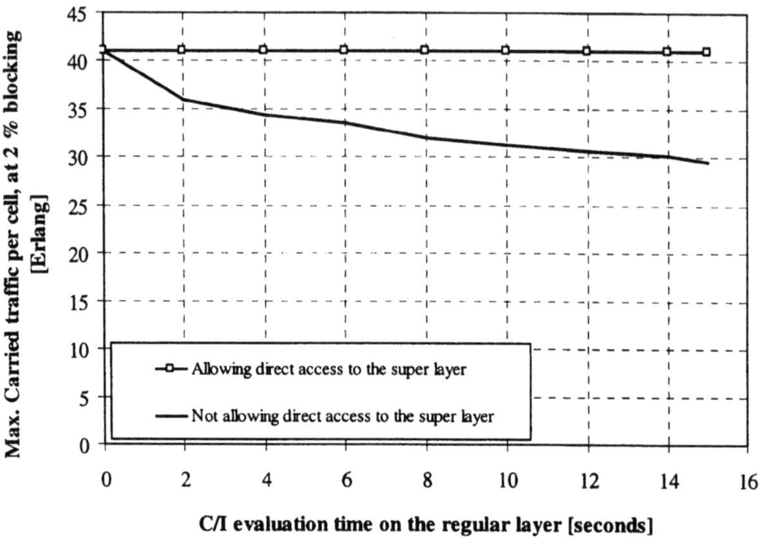

Figure 173. Influence of allowed evaluation time to carry out the C/I-evaluation. The figure reflects the trunking loss both with and without direct-access-to-the-super-layer.

It can be seen that with the direct access feature the evaluation time on the regular layer has become less critical. IUO without direct access can in worst case situations loose about 25 % of the maximum amount of carried traffic when the *C/I* evaluation time is increased. On the other hand the amount of carried traffic is kept unaffected for IUO with direct access.

8.1.4 The Impact of Only Reporting the 6 Strongest Neighbours

In GSM only the 6 strongest neighbours are reported from the mobile station, despite the fact that often many more are received. This limitation has an impact on the implementation of the IUO feature, as was described in Section 2.1. The interference estimate is in reality not based on the six super cell neighbours, but on the worst one. This means that the signal level of at least one of the super cell neighbours has to be between the six reported values. Otherwise no estimate of the interference of the super frequency exists. To determine the impact of this limitation, the available number of super references between the 6 strongest neighbours has been studied. To do this, two different frequency plans were made. The result using the first one, denoted best-case, is shown in Figure 174. The reason for denoting it the best-case is because a completely regular optimum frequency plan is used.

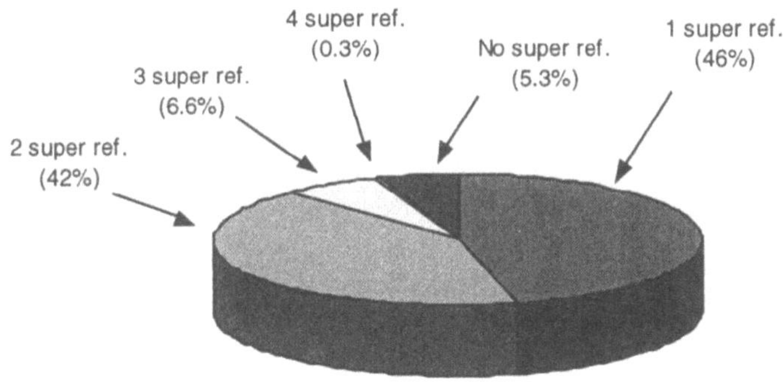

Figure 174. Simulation results of the amount of super references existing among the 6 strongest neighbours using the base-case frequency plan of the super frequencies.

It can be seen that in 46 % of the cases only one of the super neighbour signal levels is reported, while one in 5.3 % of the cases there are no super neighbour among the six reported signal levels.

The result using the second one, denoted worst-case, is shown in Figure 175. The reason for calling this scenario worst-case is because serious co-channel interference is deliberately introduced between the super frequencies. The frequency plan is made so that the interference from one super frequency to the neighbouring super frequencies is as high as possible. Thereby a higher probability of having the measurements of the interfering super neighbours between the 6 strongest neighbours is achieved, i.e. they are reported back to the network. The trade-off is that a higher level of co-channel interference is introduced (a tighter reuse).

It can be seen how here approx. 68 % of the time the evaluation is carried out using 2 of the super references. Furthermore only approx. 2 % of the evaluations carried out does not contain any of the super references between the 6 strongest neighbours. Therefore, for the simulated scenarios, with regular grid, uniform traffic distribution etc., it rarely happens that no super neighbours can be measured between the six strongest. A *C/I* estimate based on the strongest super reference interferer is possible in at least 95 % of the cases (and 98 % in case of a worst case frequency plan). As described in Section 2.2, this situation can change quite a lot for irregular inhomogeneous networks (as in e.g. Hong Kong), where alternative methods have to be used.

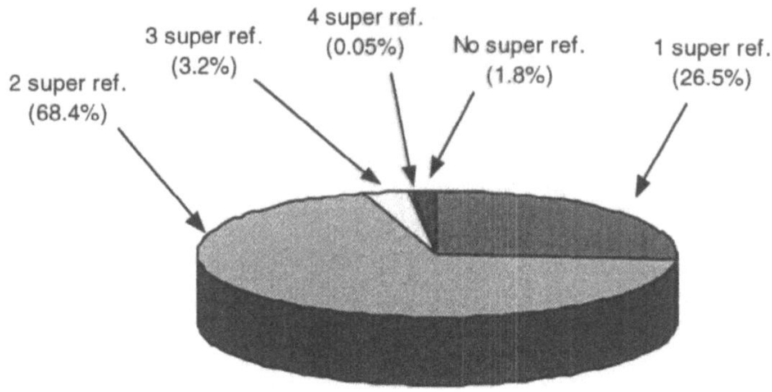

Figure 175. Simulation results of the amount of super references existing among the 6 strongest neighbours using a worst-case frequency plan of the super frequencies.

When one or more of the super references can be measured, a reasonable *C/I* can be maintained. The issue left is therefore to deal with the situations when none of these are known. This problem has been thought of in the design of the original IUO algorithm, were several solutions have been described. The more complex solution is as follows. The received signal strength (RXLEV) of the carrier is always known and can be used. This signal level is compared with a threshold (exactly as with the proposed direct-access-to-the-super-layer feature). When the signal level on the carrier is higher than this level, the mobile is allowed to access the super layer and otherwise rejected.

8.2 CAPACITY Simulations of IUO and Baseband FH

Concerning the combination of baseband FH and IUO, 4 different scenarios have been simulated. The different configurations are described in Section 7.

Concerning the performance investigations of the 4+3 and 3+6 IUO configurations as a function of the network load, extensive simulations have been run. To limit the number of simulations here it has been chosen only to shown the curves describing the hard blocking and dropped call rates. The primary conclusion from the other statistics, which in all situations are highly correlated, is that combination of IUO and baseband FH is not very sensitive to the network load.

8.2.1 Hard blocking and Dropped Call Statistics as a Function of the Load

For the 4+3 IUO configuration with slowly moving mobiles at a speed of 3km/h the results in Figure 176 was found. As well as the simulated blocking and dropped call rates, also the corresponding limit of Erlang-B using (56 channels) is included along with the maximum limit according to the IUO hard blocking traffic model of the improved IUO algorithm.

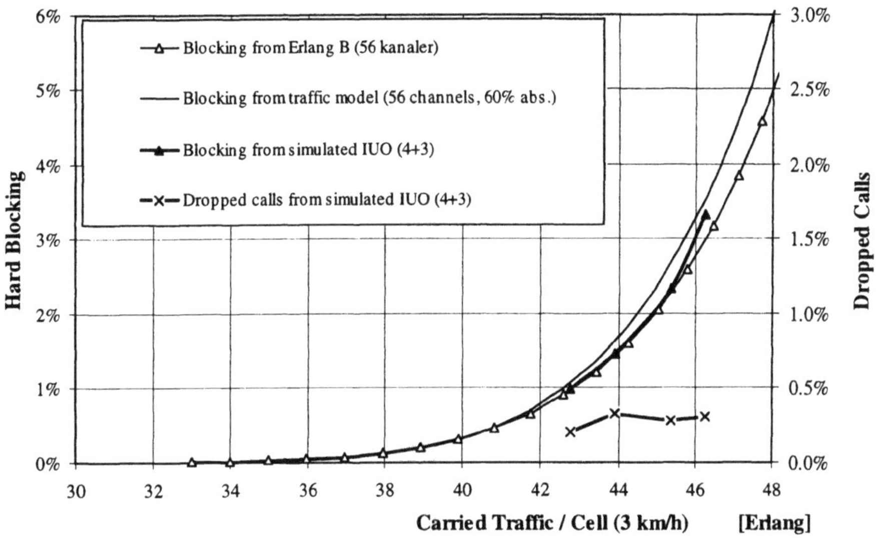

Figure 176. Hard blocking and dropped call statistics as well as a comparison to both the Erlang-B and the IUO traffic mode for the 4+3 configuration at 3 km/h.

It is seen how the simulated available capacity is almost equivalent to the Erlang-B case. It is better than found using the hard blocking analytical traffic model. The 2 % hard blocking case is reached with a carried out of traffic per cell of 45 Erlang (having 56 channels). It is also seen how the dropped call rate is somewhat independent on the network load.

For the faster moving mobiles of 50 km/h the same type of results are shown in Figure 177. Now the simulated IUO capacity has decreased, so it is a bit worse than the analytical traffic model. The 2 % hard blocking is reached at 44 Erlang per cell. Furthermore, the dropped call rate has almost doubled compared to before. Here a correlation between load and dropped calls is seen, however still very few dropped calls occur.

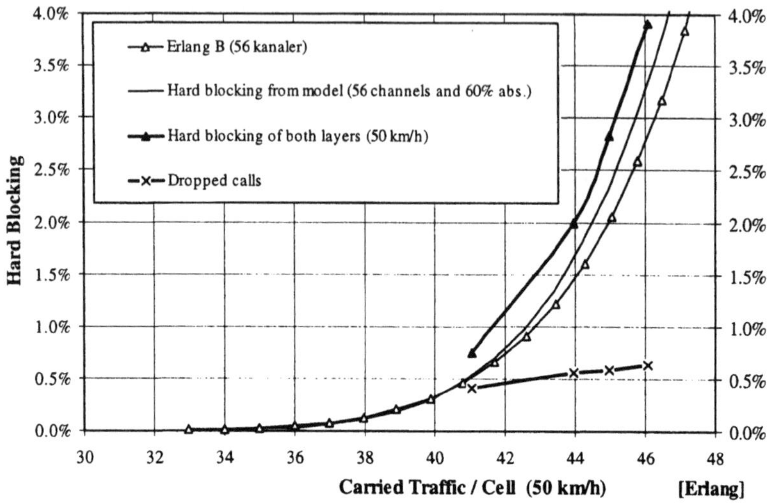

Figure 177. Hard blocking and dropped call statistics for the 4+3 configuration at 50 km/h.

The situation of having 3 regular and 6 super frequencies per cell (3+6) have also been studied. It should be remembered that the IUO parameters are specified according to the 4+3 configuration. This will degrade the performance of the 3+6 configuration, especially with uniformly distributed traffic. The results of the mobiles moving at 3 km/h are shown in Figure 178, while for 50 km/h the results in Figure 179 were found.

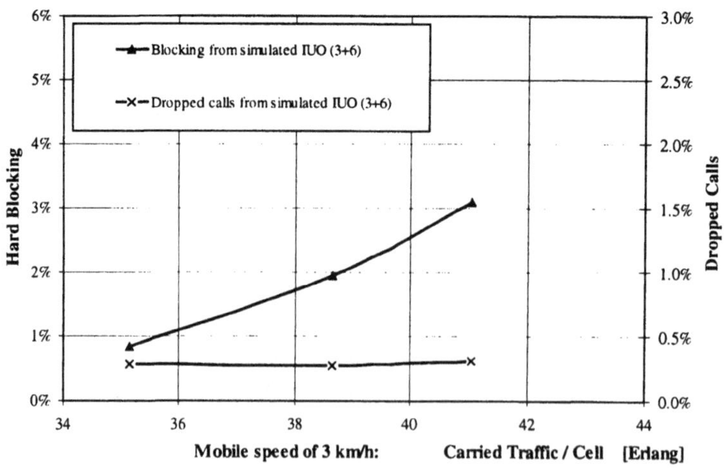

Figure 178. Hard blocking and dropped call statistics for the 3+6 configuration at 3 km/h.

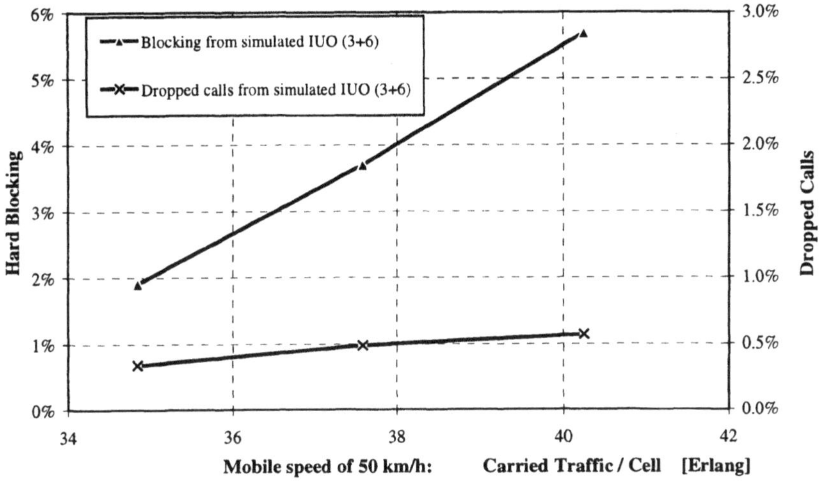

Figure 179. Hard blocking and dropped call statistics for the 3+6 configuration at 50 km/h.

In both cases the available cell capacity has decreased when compared to the 4+3 case. For the slow an available capacity of 38.8 Erlang could be achieved while and only 35 Erlang per cell for the faster. Also, it seems like the dropped call rates are quite similar for the simulated 4+3 and 3+6 configurations.

8.2.2 Comparison of the 4 Network Scenarios

To find the available network capacity when considering the combination of base FH and IUO, the performance of the four networks (1/3, 3/9, 4+3 and 3+6) has been compared. The amount of offered traffic has in each situation been determined based on the 2 % hard blocking limit. Therefore different amounts of traffic is carried. The 4 configurations to be compared are:

IUO (4+3), 80 % offered load	= 44.8 Erlang/cell
IUO (3+6), 55 % offered load	= 39.6 Erlang/cell
3/9 Flat network with 5 TRX/cell, 78 % offered load	= 31.2 Erlang/cell
1/3 Flat network with 7 RTX/cell, 78 % offered load	= 43.7 Erlang/cell

The two flat networks (the 1/3 reused and the 3/9 reused) are only simulated with what corresponds to the 2 % hard blocking limit if using the Erlang-B formula (with respectively 56 and 40 channels). For these flat networks the capacity can in general be determined from this formula, as described in Chapter 6.

It is seen how the 4+3 IUO configuration allows most traffic when considering the hard blocking. The question is then, what is the network quality at these different amounts of traffic. The statistics of the *C/I*, the BER and the FER are shown in Figure 180, Figure 181 and Figure 182 for the case of mobiles moving at 3 km/h.

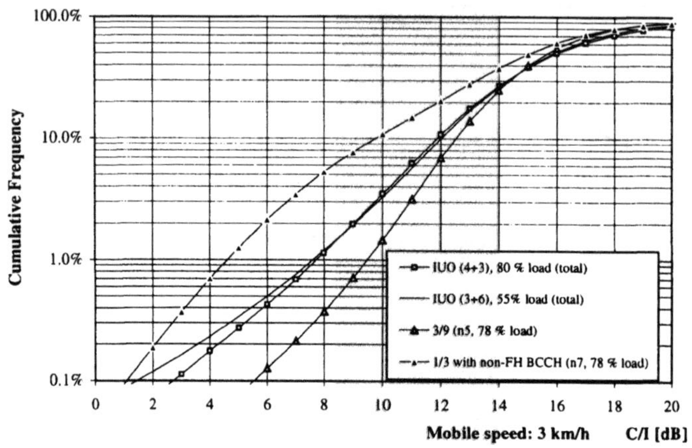

Figure 180. Cumulative statistics of the *C/I* for the 4 different network scenarios at 3 km/h.

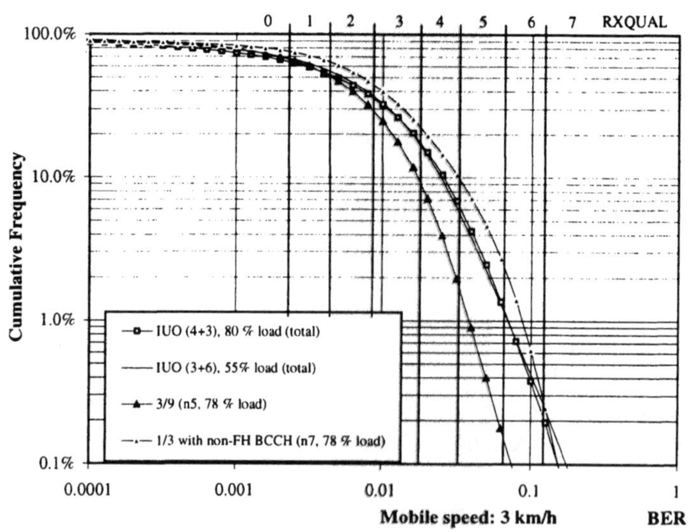

Figure 181. Cumulative statistics of the BER for the 4 different network scenarios at 3 km/h.

Since the 3+6 IUO configured and especially the 3/9 reused network a seriously limited by hard blocking and thereby offer a lot less traffic to be carried, the real contest is between the 4+3 IUO configuration and the tightly 1/3 reuse network. They can both handle almost identical amounts of traffic per cell, when only considering the hard blocking limit.

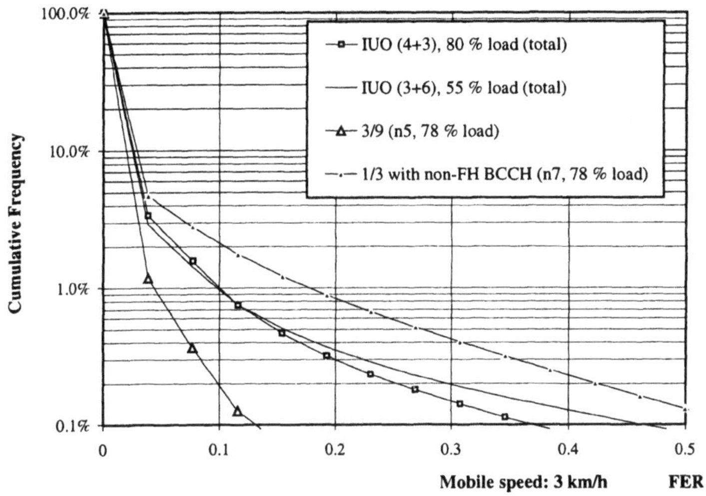

Figure 182. Cumulative statistics of the FER for the 4 different network scenarios at 3 km/h.

In all 3 set of statistics the 1/3 reused network is the worse. The two IUO configurations are close to identical, while the 3/9 reused network is by far the best. Remembering that because of the hard blocking a lot less traffic is carried in the 3/9 case when comparing to the 1/3 and IUO cases. Therefore the real contest is between the 4+3 IUO configuration and the flat network with frequency reuse of 1/3. To conclude this 3 km/h comparison, all parameters except the FER are summed in Table 42.

Table 42. Simulations of IUO and FH compared to the reference networks at 3 km/h.

Name of simulation	Hard blocking	Dropped calls	BER > 0.064 (RXQUAL 6 & 7)	CIR (10 % outage)
IUO (4+3), 80 % load (total)	1.5%	0.3%	1.3%	11.8 dB
IUO (3+6), 55 % load (total)	1.9%	0.3%	1.3%	12.0 dB
3/9 (n5, 78% load)	2.8%	0.2%	0.2%	12.7 dB
1/3 win non-FH BCCH (n7, 78 % load)	1.5%	2.4%	2.7%	9.8 dB

It is seen how the network quality of the 1/3 reused network is significantly worse than that of the 4+3 IUO configuration. Whether looking at dropped calls,

BER, *C/I* or FER a clear difference is seen. For e.g. the *C/I* statistics a gain around 2 dB is seen, where as in case of the BER more than twice as much bad quality is seen for the 1/3 case when compared to the 4+3 IUO configuration. In case of hard blocking they have similar values.

An identical number of simulations were run for the mobiles moving at 50 km/h. The most important values are summed in Figure 183 and Table 43.

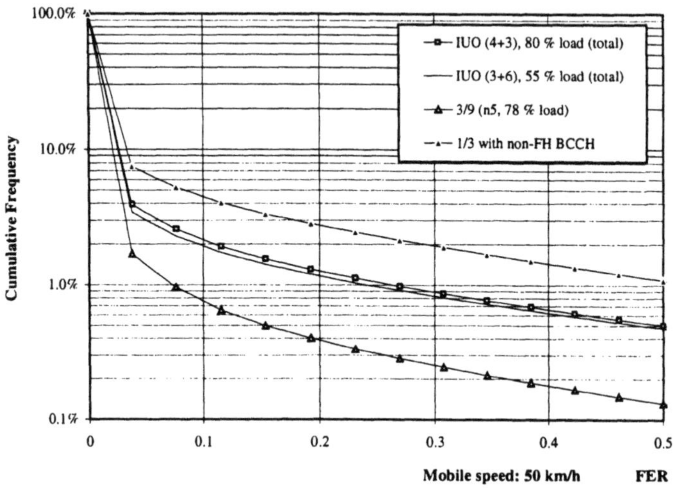

Figure 183. Cumulative FER statistics for the 4 different network scenarios at 50 km/h.

Again the FER statistics show almost identical performance of the two IUO configured networks, with the 1/3 much worse and the 3/9 much better. The summed results are shown in Table 43.

Table 43. Simulation results of IUO and FH compared to the reference networks at 50 km/h.

Name of simulation	Hard blocking	Dropped calls	BER > 0.064 (RXQUAL 6 & 7)	CIR (10 % outage)
IUO (4+3), 80 % load (total)	2.0%	0.6%	5.8%	10.0 dB
IUO (3+6), 55 % load (total)	3.7%	0.5%	5.0%	10.4 dB
3/9 (n5, 78% load)	2.3%	0.2%	1.9%	11.2 dB
1/3 win non-FH BCCH (n7, 78 % load)	1.6%	3.7%	8.2%	7.5 dB

When considering the remaining parameters in Table 43 the tendency is the same as with the slow mobiles except when looking at the hard blocking. Here it is shown that with IUO an increase in blocking can be seen when the mobiles start to move faster. This was also found when studying the impact from load variations in Section 8.2.1. Furthermore a general network quality degradation can be observed for all soft blocking statistics of all 4 simulations equivalent to what previous CAPACITY simulations of flat networks have shown in Chapter 6.

In general the conclusion is that a significant capacity increase can be achieved with IUO combined with baseband FH for the situation studied, when compared to a RF FH network with reuse 1/3. For slow mobiles almost no trunking loss arise when compared to Erlang-B, but with increasing speed the hard blocking starts to increase slightly. Another advantage found from the study was that the two layer principle is not very load sensitive. An example, varying the load from 74 % to 86 % for a 4+3 IUO configuration with mobiles moving at 3 km/h hardly changes the network quality when looking at the BER, *C/I* or FER statistics.

8.2.3 Combining Baseband FH and IUO with Load Sharing

The idea of combining baseband FH and IUO with load sharing has also been thought of and treated using the same method as the one described in Section 7. The idea is to compensate for the natural sensitivity towards the traffic distribution introduced with IUO by allowing load sharing on the regular layer, see Figure 168. The configuration parameters used, are the ones referred to as 'set 3' in the traffic reason handover study in Chapter 8. Only one simulation, simply to show the effect, has been carried out. Again the 4+3 configuration with a mobile speed of 3 km/h has been used. From the results the general tendency was a small degradation in all the soft blocking parameters, while the hard blocking was reduced from 2.3 % to 2.0 % for the simulated case of 83 % load. Since the load sharing parameter configuration used was optimised for a flat network, better settings could probably be found thereby reducing the soft blocking degradation. At the same time the hard blocking could perhaps correspondingly be decreased even further.

9. LIVE NETWORK TRIALS RELATED TO THE COMBINATION OF IUO AND FH

Largely based on this study the combination of FH and IUO has actually been implemented by NOKIA as a network feature, denoted intelligent frequency hopping (IFH). Also the improved IUO algorithm functionality's, of the direct-access-to-the-super-layer as well as the direct super-to-super handover possibility, have been implemented enhancing the IUO algorithm. Several different live network trials have already been carried out using IFH, of which the most extensive trial has been conducted in China. Here IFH (using RF FH and the 2+2 IUO configuration) was compared to other capacity enhancement features like the 1/1 and 1/3 frequency reuse of a flat RF FH network. The results were very positive, with IFH being superior to the others when considering measurements of quality, dropped calls and call success rate [137]. By looking at different bandwidths (from 6.2 to 10.4 MHz), a clear capacity gain were found when comparing to the flat FH network as well as a network using IUO by itself. One of the reasons for these very good results should be found in the fact that downlink power control and DTX where not used, favouring the IUO algorithm.

Also CSL from Hong Kong, who originally proposed IUO, uses the combination of IUO and FH. In their network only the regular layer utilises FH, where the planning of the frequencies are carried out using the MRP principle [162]. Correspondingly SONOFON has carried out an initial functionality test of the feature to be used for future hot spot cases.

10. CONCLUDING COMMENTS ON THE COMBINATION OF IUO AND FH FOR GSM

A proposal on the combination of frequency hopping and the intelligent underlay and overlay reuse partitioning scheme, has in this chapter been studied for the GSM network. Different methods have been used to analyse this concept, where the main results have been found by the CAND approach using CAPACITY.

10.1 Analytical Calculations and Network Simulations

Using a simplified analytical one-cell approach the functionality of reuse partitioning, as well as the capacity gain, has been modelled. For the idealised reference case (4 regular and 3 super frequencies per cell) a potential gain as high as 34 % was shown, when comparing to a flat baseband FH network of reuse 3/9.

An existing reuse partitioning algorithm, the IUO algorithm, was chosen and utilised for the proposed idea of combining FH and reuse partitioning. Based on another analytical study of the functionality of this algorithm, as well as preliminary simulation studies using CAPACITY some immediate problems associated to this original IUO algorithm were discovered. A model reflecting the hard blocking limit of a cell utilising IUO was treated. Based on this model an extra trunking loss (added to the one arising from Erlang-B) was discovered due to the necessary evaluation time required on the regular layer before allowing any mobile to enter the super layer. Initially two proposals were made on how to minimise this problem. The idea of the first one was that by measuring the received signal strength of the desired new frequency before actually entering the cell, direct access to the super layer should be allowed, provided a certain predefined signal strength. The second proposal, arising from preliminary CAPACITY simulations, of reserving specific channels for handovers from super to regular was also implemented and tried out for the 4+3 configuration. Due to some later IUO algorithm improvements no gain were found from this feature. These later improvements were primarily based on two additional handover possibilities. The possibility of doing an intra-site handover, from one super frequency to another of a neighbouring cell, were initially implemented. Furthermore, the possibility of performing a handover from a super frequency to a regular frequency other than the one of the same cell was enabled. These modifications, combined with a general decision on prioritising handovers ahead of new calls, solved the general problem of having the regular layer congested while attempting a handover from the super layer.

An improved model of the IUO algorithm, including the functionality's described above, was implemented in CAPACITY as well as analytically. From the simplified analytical studies, the desired reduction in trunking efficiency was observed. The CAPACITY implementation has been treated in more detail with various objectives. Initially some of the limitations as well as the basic functionality have been studied. Then an estimate of the potential capacity increase has been found using IUO with baseband FH and by comparing to the flat reference types of networks of reuse 1/3 and 3/9.

Characteristics of the Improved IUO

By simulating a GSM network with an available bandwidth of 45 frequencies, a mobile speed of 3 km/h, including power control and DTX, IUO cells configured as 4+3 with reuse 3/9 and 1/3 and using baseband FH, the essential issues of the improved IUO algorithm have been studied.

First of all the configuration of the basic IUO parameters, the *C/I* thresholds, has been dealt with. Here the aim is to avoid an increase in number of handovers by ping-pong effects between the layers arising thresholds set to low. On the other hand if specifying the thresholds to high the offered capacity is reduced. Thresholds of 9 and 11 dB were chosen.

The improvement of allowing access to the super layer without accessing the regular layer has been studied for different signal level thresholds ranging from –70 dBm to –105 dBm. A threshold of –80 dBm has been shown to be a good choice. With this threshold the dropped call rate, the number of handover attempts as well as the *C/I*, BER and FER statistics are kept unaffected. In general the network quality on the regular layer is almost independent on the direct-access-to-the-super-layer feature, while it is almost exclusively on the super layer the quality degradation occurs if the threshold is specified to low.

Without the direct-access-to-the-super-layer feature, all mobile stations have to stay on the regular layer for a certain period of time to evaluate a proper *C/I*. By adding the direct access feature to the improved IUO algorithm, less mobiles have to stay on the regular layer. This has in CAPACITY been shown to remove the trunking problems for the studied scenario. In has been shown how the IUO algorithm, without the direct-access-to-the-super-layer feature, can in some situations cause a trunking loss of as much as 25 % if the *C/I* evaluation time becomes as high as 10 sec.

The improved IUO algorithm has some practical difficulties caused by the method used for the *C/I* estimate. In GSM only the six strongest signal levels measured by the mobile are reported back to the base station. It means that there is no guarantee that any of the super reference neighbours are measured among these. This potential problem was investigated using two different frequency plans. If using a completely regular frequency plan it was shown how 5.3 % of the times no super references could be measured. For the other plan, with deliberate co-channel interference between neighbouring sites, this number was lowered to 1.8 %. Different other methods can be used to deal with this problem either in the planning process of the super references or simply by using the absolute received signal level as an indicator. Simulations have further shown that if only one of the super references is available for the *C/I* evaluation no influence on the overall network capacity can be seen.

In general, having studied the functionality of the improved IUO, running several hundreds of simulations, it seems like the algorithm is very robust. Since the IUO algorithm is integrated into the existing handover algorithm, which is very robust, the same robustness is found with IUO. If a mobile, by a mistake, is handed to the super layer, quite often it is simply handed back to the regular layer with no dropped calls.

Network Capacity when Combining Baseband FH and IUO

Four different scenarios have been simulated in CAPACITY to be able to determine the gain by comparing IUO and baseband FH to other network features. They are two different IUO configurations: 4+3 and 3+6. Furthermore, two types of reference simulations have been carried out. They are both flat FH networks with a frequency reuse of respectively 3/9 and 1/3. In the 3/9 case 5 TRX's were allocated per cell, while with the 1/3 reuse 7 TRX's were used of which the BCCH is non-hopping. Again the available bandwidth has been 45 frequencies, where also power control and DTX has been included.

In general the network quality, measured by dropped calls, BER, *C/I* or FER, is quite insensitive to the offered network load. For the 4+3 case, when varying it from 74% to 86%, literally no quality degradation is seen. The 2% hard blocking limit is reached with an offered network load around 82% for the slow mobiles and at 80% for the faster mobiles. In other words, with increasing speed the blocking increases as well.

Having treated the blocking and dropped call rates as a function of the load, the statistics of 4 networks (2 IUO configurations, 4+3 and 3+6 and 2 flat FH networks with reuse 1/3 and 3/9) have been compared, using the 2 % hard blocking limit of each network. Since the hard blocking limit of the 3+6 IUO configuration and the flat baseband FH network with a frequency reuse of 3/9 is a lot lower than for the tow others, the essential comparison was between the 4+3 IUO network and the 1/3 reused RF FH network. Therefore if only equipment able to utilise baseband FH equipment is available, the combination of FH and IUO enhances the available capacity substantially. Looking at the same amount of offered traffic (around 80 %) the quality of the 1/3 reused FH network and the 4+3 IUO network have been compared. Actually the 4+3 reused network could have been compared using an offered amount of traffic of 86 %, since the network quality is quite load independent. With 80 % offered traffic the network quality (*C/I*, BER, dropped calls as well as FER) is quite a lot better than for the flat FH network for slow as well as for the faster moving mobiles.

Finally the combination of baseband FH and IUO has been studied with load sharing. Using load sharing a hard blocking reduction, for the 4+3 IUO configuration with an offered network load of 83 % and simulating with slow mobiles, from 2.3 % to 2.0 % were found at the price of a slight degradation in network quality. By optimising the parameters of the load sharing algorithm a further improvement can be expected offering even more network capacity.

10.2 Ideas for Future Improvements of IFH

Probably the largest problem to overcome with most of the radio related network performance enhancement features is the practical problem of how to handle the fixed parameters that have to be specified. The most problematic of these are the ones that have to be specified individually for each cell, like e.g. the *C/I* thresholds of the IUO algorithm. The same problem occurs with the power control algorithm, where the fixed power regulation parameters have to be specified. Furthermore, one of the general trends is to combine more and more network features to maximise the performance. This increases the complexity of the network and makes configuration of fixed parameters even more problematic. A potential way of enhancing the features would be by allowing the fixed parameters to adjust themselves according to the network load. This is also believed relevant for the IUO algorithm, where the C/I-good and C/I-bad thresholds could be made dynamically adjustable according to the network load. That would create a dynamic absorption to the super layer. With a low network load there is no reason to use the super layer and the *C/I* thresholds should be increased to lower the absorption. Then when the network load increases the *C/I* thresholds should dynamically decrease to increase the absorption and compensate for the required capacity.

To compensate for the IFH performance dependency on the speed of the mobiles, estimation of the speed of the mobiles (if this becomes possible) could be used to allow or reject access to the super layer. It is desirable to lower the number of handovers per call and therefore very fast mobiles should not be allowed to enter the super layer.

Chapter 10

FREQUENCY PLANNING OF FREQUENCY HOPPING GSM NETWORKS

Frequency planning is currently one of the most challenging and time consuming tasks in designing a GSM type of network. Effective usage of the frequency spectrum, being one of the scarcest resources for any operator, can lead to both better network quality and increased capacity. The key lies in solving the most fundamental problem in cell planning - allocating the frequencies in the network in a way, so that the interference is minimised. Presently, planning tools that can help the operator during this process exists. However a large part of the planning is done manually due to the inaccurate modelling used in the planning tools.

Typically current planning tools are only able to plan a non-hopping type of network, not incorporating the frequency and interference diversity gain from frequency hopping. Therefore it is not optimal to do frequency planning for a network, which is using frequency hopping. This chapter presents a tool developed specifically for planning of a frequency hopping network, taking both the frequency and interference diversity gain from frequency hopping into account, enabling a frequency plan with both high capacity and high quality.

1. INTRODUCTION

As shown in [238,29] the quality and capacity of a network is determined by its reuse. In those studies a regular network was assumed, i.e. the size and form of all cells is the same for all cells. The traffic is assumed uniformly distributed and the environment is the same in all cells, i.e. the same fading patterns and statistics were used. However, in reality the traffic is not uniformly distributed. If looking at the traffic distribution for different European countries, it appears that for most countries 50% of all traffic is generated in a geographical area of less than 5% of the whole country [66], as can be seen Table 44. The spatial statistics of cellular traffic can be described by a log-normal distribution [78].

Table 44. Approximate national geographic area generating 50% of the traffic [66].

BELGIUM	15.0 %	**SPAIN**	3.0 %
Finland	0.1%	**Sweden**	1.5%
France	10.0 %	**United Kingdom**	3.0 %
Germany	2.0 %	**Hungary**	0.6 %
Ireland	1.5 %	**Lithuania**	5.0 %
Portugal	3.0 %	**Romania**	<5.0 %

Neither the environment is the same for all cells. Typically there is spoken of urban, suburban, rural and hilly terrain environment, for which there are defined channel profiles in the GSM standard [64]. Not only the channel profile is different, but also the shadow fading changes for the different environments. The value of the standard deviation lies generally between 5 and 10 dB, with a tendency to larger values in heavy built areas [135], due to high-rise buildings. In [248] the standard deviation for the log-normal fading in urban areas is stated to be 6.5 dB for a 900 MHz radio signal.

The fact that the environment and the traffic distribution are not the same for every square km leads to non-uniform cells. Omni-directional sites are used aside of sectorized sites; cells have different sizes, different antennas, use different tilting, are placed on different heights and so on. In the centre of cities small macro cells with a radius down to a few 100 meters or micro cells are used, while in rural areas macro cells with a radius up to 30 km, are more likely to occur.

All this means that there cannot be spoken of a regular network, so there is no such thing as a regular reuse. This makes frequency planning one of the most challenging and time consuming tasks in designing a GSM type of network. Effective usage of the frequency spectrum, being one of the scarcest resources for any operator, can lead to both better network quality and increased capacity. The key lies in solving the most fundamental problem in cell planning - allocating the frequencies in the network in a way, so that the interference is minimised.

This chapter deals with frequency planning. At the end of this introduction an outline of the whole document is given, but first the problem of frequency planning is clarified and an overview of the existing tools and techniques is given.

1.1 The Frequency Planning Problem

The main tasks of a planning department of a mobile phone operator are coverage prediction and frequency planning. In this document we concentrate on frequency planning. The goal of frequency planning is to increase the quality of the network with a given capacity within certain costs.

The costs of a network can be split into initial investments and annual costs. The initial costs are the costs that are needed to implement base stations and base station controllers. This includes site acquisition, purchasing of radio equipment, network connections and construction. These costs depend on site configurations and site types. In Table 45 the costs distribution of a number of operators can be seen. It was concluded in [66] that increasing the number of base stations is approximately 30-35% more expensive than increasing the capacity per base station. [28]

Table 45. Distribution of costs in a cellular network [66].

	SITE ACQUISITION	RADIO EQUIPMENT	NETWORK STRUCT.	BUILDING CONSTRUCT.	OTHERS
Operator 1	6%	47%	14%	29%	4%
Operator 2	3%	47%	1.5%	47%	1.5%
Operator 3	-	69%	4%	27%	-
Operator 4	8%	77%	6%	2%	7%
Operator 5	-	67%	4%	27%	3%
Operator 6	-	41%	7%	45%	7%
Operator 7	-	75%	13%	3%	9%
Operator 8	5%	50%	20%	15%	10%
Average	**5.5%**	**59.1%**	**8.1%**	**24.4%**	**5.2%**

After the initial costs, there are the annual costs. The annual costs represent a remarkable portion of the total costs. There are three main categories in annual costs. These categories are transmission costs, rents and O&M costs. O&M costs are most likely the largest part of annual costs, but at the same time, also, the most difficult to determine [66].

The costs of planning fall under the O&M costs. It consists of the costs for placement of new sites, for changing the frequencies and/or parameters on base stations plus the salaries of the employees in the planning department.

If we exclude the situation where the network is built, then we know quite accurate how much traffic there is per cell. So we know how much capacity there should be per km^2. In practice the busy hour traffic is used and a certain safety margin in build in, so that the traffic can increase without the network having to be changed.

As mentioned above, the goal is to create a frequency plan, which maximises the quality. This is done by minimising the network interference. Predictions are used to predict how much interference there is coming from one cell to another cell. By

[28] It was commented in [69] that the costs of implementing new base stations mainly depending on the equipment used. Therefore, the 30-35% difference mentioned above can be substantially lower or higher.

combining these predictions with a frequency plan, the total amount of interference can be calculated. In simple terms, making a frequency plan is trying to avoid the strongest interferers for each cell.

Typically only predictions for the downlink are used, even though the uplink has shown to be the limiting link in [29]. But if a method of antenna diversity is used at the base station the downlink is the limiting link [29,231]. Another advantage of using the downlink is that for predictions in the downlink direction the traffic distribution does not have to be known, where it has to be known for uplink predictions, since the mobile stations are the sources of interference.

Network planning tools often are used to get the predictions, which are used to make a frequency plan. Techniques based on Uniform Theory of Diffraction (UTD) have been studied extensively [12,27] and it is shown that these models can deliver quite accurate propagation predictions. A requirement for getting accurate predictions is a high-resolution clutter database. For micro cells a clutter resolution in the order of 1-2 meter is required [245]. Obtaining such detailed clutter information can be difficult for some cities and, generally, the cost associated with a high-resolution database is relatively high. The resolution for clutter databases is normally in the order of 50 meters. Thus information on building dimensions is not provided with sufficient level of detail which leads to high inaccuracies in signal level predictions. As a consequence, errors are expected in the estimate of the quality of service. The quality of service is the parameter which is optimised and which has a minimum demanded value. Conservative estimates would motivate the system planner to improve the quality of service by increasing complexity, resulting in higher costs, but the quality of service is guaranteed. Optimistic results on the other hand would result in the system failing to meet the specified minimum quality of service and cause user dissatisfaction. In real live the predictions often are verified with measurements. Lots of network operators have test mobiles driving around in the network and some have the intention of using measurements instead of predictions to avoid inaccuracy.

The inaccuracy is highest for the small macro cells and micro cells in the centres of cities, since there the propagation is influenced by buildings and so on, while it is not so harmful for large cells in urban area, which have a more uniform environment.

A block diagram of the frequency planning process can be seen in Figure 184. Predictions, expected traffic and the allowed costs are inputs to a black box that produces a frequency plan as output. The black box is then equivalent to one of the automatic frequency planning tools described in the next section.

When the traffic at a certain location has increased dramatically, a new site has to be placed in the network. This can also be seen as a part of the frequency planning process, since this new cell has to be assigned some frequencies and this cell interferes with the other existing cells. Also the place of the new base station is of importance, since this has influence on the predictions and thus the interference conditions.

It is not always possible to get the exact location, which is wanted for a new base station, since landlords are not always willing to agree to site installation on their premises or they simply ask a too high price. This can easily lead to a shift from a few 100 meters up to km's, which is not always so problematic for a macro cell with a radius of 30 km, but it sure is for a micro cell of a couple of 100 meters. In practice the base stations are seldom placed on the optimum place, which makes the task of the frequency planners not easier.

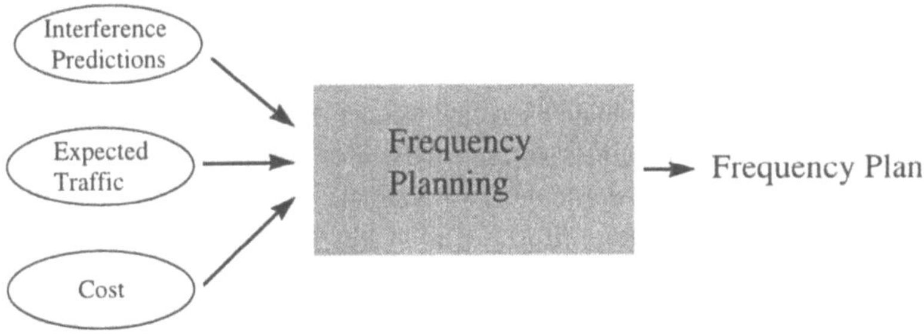

Figure 184. Principle behind the traditional frequency planning procedure.

Not only the frequencies have to be planned, also all kind of settings for each base station have to be made, like the handover, power control, frequency hopping settings and the neighbouring cells lists. The latter can be quite important when using for example IUO, as was described in Chapter 9. A concept of dynamic neighbour list planning, which should make planning of the neighbour lists unnecessary was presented in [185]. Section 4 of this chapter deals with planning of some of the other parameters.

1.2 Existing Techniques

Presently, planning tools, that can help a network operator making frequency plans, exists. They represent the black box of Figure 184. There are different methods and concepts. In this section a small overview.

The channel assignment problem is concerned with the allocation of channels to base stations such that each base station is assigned a prescribed number of channels and co-channel and adjacent channel interference is minimised. This optimisation problem can be shown to be a NP-hard problem [222], which is commonly believed to imply that it is theoretically impossible to find optimum solution within reasonable time.

In the earliest studies of how to do frequency planning the task was to find a frequency plan fulfilling some separation requirements and minimise the bandwidth, see e.g. [23]. Those separation requirements set demands for the distance in frequencies used in different cells. For example a separation requirement of *2* between *cell 1* and *cell 5* means that when frequency *34* is used in *cell 5*, then the frequencies *33, 34* and *35* are not allowed to be used in *cell 1*. The separation requirements should be chosen in a way that interference is minimised, so on places where interference is high, high separation distances should be placed. This concept of fulfilling separation requirements and minimising bandwidth however, suffers from the following drawbacks [127]: definite hard-decisions given by separation requirements are questionable, the trade-off between different requirements is impossible and the bandwidth is not the right thing to minimise since it is always fixed.

Using exhaustive search methods to find the optimal solution are not feasible for a practical planning tool, as the solution time scales exponentially with the size of the problem [126]. Therefore a number of heuristic algorithms have recently been developed to achieve a best effort solution. In those algorithms the task is formulated as a cost-function minimisation problem.

The simplest algorithm is perhaps the greedy algorithm, where beginning from some initial solution one jumps recursively to the best solution (in the sense of the cost function) found in the neighbourhood of the current solution [222]. The Hill Climb algorithm is a bit more advanced: the solutions in the neighbourhood of the current solution are tested in some order and the first one having lower cost-function value is accepted [218]. The problem with these algorithms is that there are a lot of local minima in the frequency allocation problem and the solution these algorithms come up with often will be a local minimum. Some more advance algorithms have been studied to overcome this problem.

Simulated annealing is one of those algorithms. Simulated annealing is a well documented technique for finding nearly optimal solutions to hard combinatorial optimisation problems [3]. With this method an initial usually random solution to the problem is generated and a search for improvement is performed by iteratively making changes to this solution. Changes that decrease the energy value, i.e. the cost function, are always accepted. In order to escape from local minima, simulated annealing accepts energy increasing changes with a certain probability, which decreases during the run. This probability is characterised by a parameter, called temperature. When the temperature is high, the probability for accepting changes leading to a higher energy is high, while a low temperature is associated with a low probability.

Simulated jumping was recently introduced in [7]. It is similar to simulated annealing in that it applies the same search technique. The difference between the two methods is the behaviour of the temperature. In simulated annealing the temperature has a high value in the start and is gradually decreased, as explained above. In simulated jumping the temperature starts at a low value and is rapidly increased and decreased. When the temperature is high more random jumps are made, while when the temperature cools down, the random jumps become rarer.

In [126] channel assignment with simulated annealing and simulated jumping are compared to each other and an other technique, called subspace approach [127]. It was found that simulated annealing and simulated jumping perform similar, while they both outperform subspace approach.

Also neural networks have been tried out [117], but the greediness of these methods cause difficulties since the cost-functions in channels assignment may have enormous of local minimum's [118].

There are lots of algorithms, which are a mix of different techniques, like for example the algorithms described in [74,77]. These algorithms have some identical elements: there is a number of constraints, requirements and a cost function. The form of the cost function might be different and the search method differs, but these algorithms often come up quite realistic frequency plans, since they are very practical oriented.

Multiple Reuse Pattern (MRP) is proposed by Ericsson [114]. MRP is a method, which describes the division of the frequency spectrum over different layers. In Figure 185 two cells with respectively 2 layers and 4 layers can be seen. MRP tells how many frequencies to use on each layer, not how to plan them within the layer. So MRP splits the problem in smaller problems, but a final solution it does not offer.

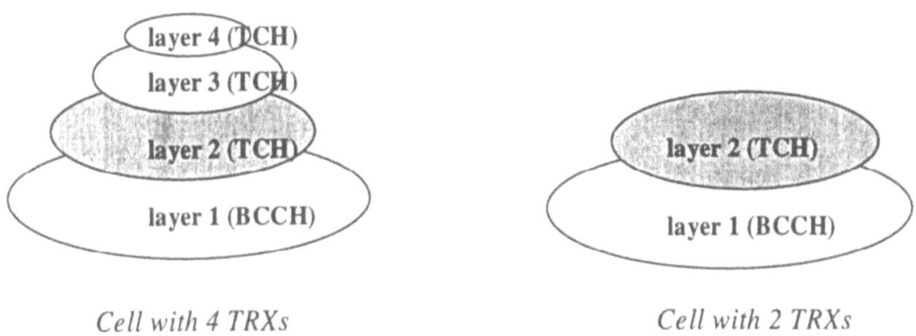

Figure 185. Example of a cell with 4 layers (left) and 2 layers (right).

All above frequency planning methods use fixed channel allocation (FCA) techniques, opposite to dynamic channel allocation techniques (DCA) [104]. In DCA all channels are available in every cell. Frequencies are assigned at the start of each call in such a way as to avoid interference with frequencies used in other cells. DCA performs better than FCA for low traffic, but not necessary for heavy traffic [199] in terms of hard blocking. Hybrid allocation (HCA) is a compromise between FCA and DCA, in which a subset of the available channels are assigned dynamically. There are many HCA schemes, like for example the ones described in [41,106,199]. They perform better than DCA in heavy loaded networks and are better than FCA in term of hard blocking. There is one great lack in the studies about DCA: they concentrate on the hard blocking issue and forget to treat the issue of the quality of the frequency plan. In addition it is nearly never possible to implement them in a GSM network, without loosing some of the performance, due to some practical issues. This is the reason why the scope of this chapter is limited to FCA.

One dynamic scheme [35] however is of interest. In this algorithm the channels are only updated after long time measuring. An optimal plan is found after some time. One by one the base stations are allowed to try to change their worst frequency to another one. It does do that by scanning all frequencies and selecting the one with the lowest received power, i.e. the lowest downlink interference. This way the network is constantly updated and it is shown that this scheme converges to a solution close to the theoretical optimum.

Current planning tools have one major limitation: they are only able to plan a non-hopping type of network, not incorporating the frequency and interference diversity gain from frequency hopping. Therefore it is not possible to do optimal frequency planning for a network, which is using frequency hopping. This chapter presents a tool developed specifically for planning of a frequency hopping network, taking both the frequency and interference diversity gain from frequency hopping into account.

1.3 Chapter Outline

In this document a frequency planning method is proposed, which includes the gain from frequency diversity and interference diversity in the case of frequency hopping networks. In Section 2 the algorithm principles and working is explained. Also the method for including the gains from frequency hopping can be found in this section. Section 3 deals with the performance of the frequency planning tool in different scenarios. The tool is among others compared to commercial available frequency planning tools. Also live results of a frequency hopping network with a frequency plan, made with the frequency planning algorithm, can be found in this section. Section 4 deals with the planning of some other parameters, while Section 5 closes this document and contains the conclusions of the frequency planning work.

2. THE FREQUENCY ALLOCATION PRINCIPLE

In channel assignment the task is to assign the available frequencies/channels to TRX's in a way, that the interference is minimised. In this section the principles behind and the functionality of the frequency planning tool *JETTPlan* are described. JETTPlan falls under the category of heuristic algorithms.

The outline of the section is as follows: first Section 2.1 deals with the propagation input to the planning tool. The subject of Section 2.2 is frequency planning in frequency hopping networks, while Section 2.3 deals with a characteristic property of JETTPlan, the splitting of bands for TCH and BCCH use. The tool itself is described in Section 2.4.

2.1 Propagation Prediction Input

A frequency planning algorithm needs to have predictions among others as input, like shown in Figure 184. The other essential input parameters are the cost and the expected traffic. These last two input parameters come from decisions, based on the strategy of an operator and the existing network, while the predictions usually come from a prediction tool.

JETTPlan does not generate these predictions, but can deal with predictions from different prediction tools. In this project, it has not been tried to improve these predictions, but to use them in the best possible way in the planning tool. If the predictions become better, then the output of the planning tool automatically will become better. The challenge lies in finding a good frequency plan, while the input is not prefect!

2.2 Frequency Planning in FH Networks

In the case of frequency hopping it is not satisfactory only to look at the worst case *C/I* for each individual frequency of each individual pixel, as is usually done in the case of a non-hopping network. With frequency hopping it is necessary to evaluate the combined effect of all the frequencies used in the hopping sequence, since the quality of each mobile depends on all frequencies in the serving cell. Therefore the quality predicted in each pixel should be a function combining the effect of all frequencies in the serving cell. This is shown in Figure 186, where the resulting effect (quality per pixel) is described by the corresponding FER per pixel.

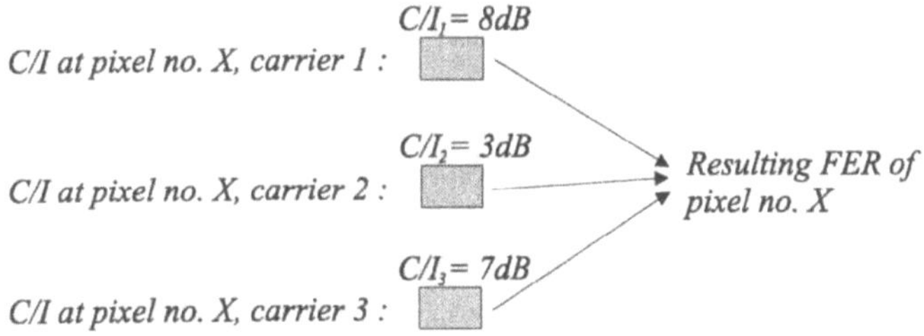

Figure 186. The quality in a pixel should represent the combined effect from all frequencies when using frequency hopping.

Various methods have been proposed to do frequency planning of frequency hopping networks [114], but a real frequency optimisation tool for frequency hopping network does not exist. The currently proposed frequency hopping planning methods are based on one of two different principles. The first is denoted frequency reuse partitioning, where Ericsson has proposed the MRP [43] procedure. The idea is that the total spectrum is divided into groups where the frequencies in each group are reused with different frequency reuse factors. Then all frequencies are planned in each band using a normal frequency allocation algorithm. An example of this is shown in Figure 187, where the total spectrum contains 36 frequencies. The first TRX is reused with the factor of 12 (the BCCH), the second with 9, the third with 7 etc.

While the first reuse factors (21 and 9) provides a good quality, the small reuses result in bad quality. The idea is that this bad quality is compensated by the frequency hopping gain. But when the reuse becomes lower than 3, the interference is getting so strong [66], that just as well no burst could be transmitted. But in practice the small reuses have greater effective reuses. Say that there is a network using MRP with reuses 12, 9, 6 and 3. Not all cells have 4 TRX's, so the reuse 3 is punctured, i.e. the actual reuse is greater. This is the main reason for the MRP schedule to work. It however also means that when a very regular network has to be planned, i.e. all cells have the same amount of TRX's, the concept breaks down. Besides that, the method does not give a solution to the planning within one of the layers.

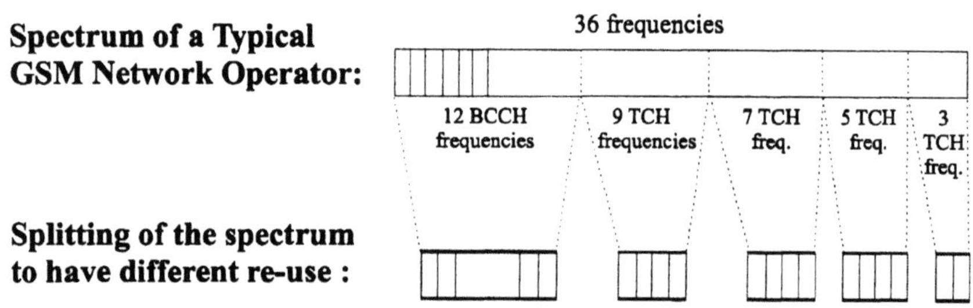

Figure 187. Example of MRP with an available frequency band of 36 frequencies.

An even simpler method, from a planning point of view, is to use a very low frequency reuse pattern like 1/3 along with fractional loading [11,238]. This latter method has originally been proposed as a network capacity enhancement method, but the method is actually also a potential frequency planning method. By having e.g. a 4/12 frequency re-use pattern on the BCCH carriers and a 1/3 frequency re-use pattern on the remaining frequencies the planning becomes very simple since all the frequencies are simply divided into three groups and allocated group wise to each sector (for 3-sector sites). The load in this 1/3 reused network has to be kept low. This can be done by installing fewer TRXs than frequencies (synthesised hopping is used) or by using a load limitation technique. Several field trials have shown promising results. However always the comparison was made with a reference plan, which not necessarily is a good frequency plan.

It should be noted that by introducing hopping in a network, the gain of intra cell handover [177] disappears. This gain comes from being able to avoid a frequency with a lot of interference or from lack of frequency dependent coverage. However, when using frequency hopping, all frequencies are used by each mobile. So every frequency is being 'heard'. This also means that without hopping a bad planned frequency in a cell may not lead to bad quality in this cell, since the mobiles can make a handover away from this frequency. Turning on hopping in this cell will make this bad frequency appear. Therefore with hopping the frequency planning has become harder since all bad planned frequencies can be seen!

2.3 Broadcast Channels versus Traffic Channels

In general, splitting a frequency band into different groups decreases the number of possible combinations to choose from. In other words this results in a loss in trunking efficiency, i.e. a loss in the number of possibilities. This problem is partly described in [77]. It is believed to be best to avoid splitting the frequencies. In the case of the MRP technique [114] this is an obvious problem since the spectrum is divided in as many groups as the largest number of TRXs in an individual cell in the network.

However the BCCH channel is fundamentally different from the TCH channel. It carries information for identifying cells for access evaluation, paging and measurements on neighbouring cells for location evaluation, etc. [151]. Because of its importance, great care should be taken to protect the BCCH from interference. It is not possible to use power control and DTX on the BCCH, since it is used for measuring by the mobile stations. Furthermore, in a system with frequency hopping, the traffic channels are hopping, while the BCCH channel is not hopping. Hence, the BCCH frequency cannot be planned according to a tight reuse pattern, like TCH channels.

An average minimum frequency reuse of 12 is recommended for the BCCH frequencies in a macro cellular network [114]. An example of this is shown in Figure 188, where the available spectrum of 45 frequencies, corresponding to the spectrum size of a typical Danish GSM operator, is shown. In this Figure the spectrum is split. It should be said that the BCCH frequency reuse of 12 is an absolute minimum, i.e. in the case of an overall higher frequency reuse, the BCCH frequency reuse should be correspondingly higher.

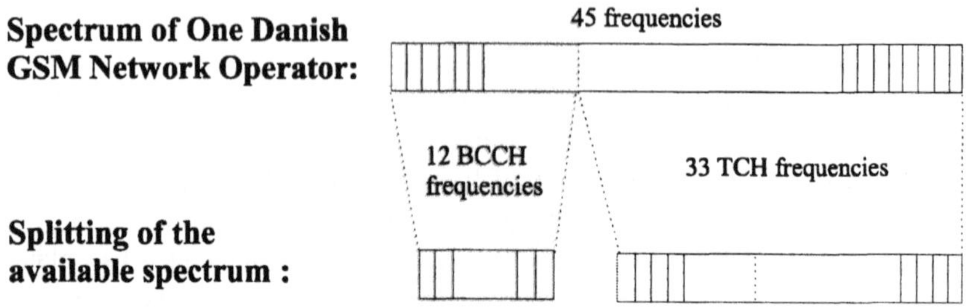

Figure 188. Splitting of the frequency band between BCCH and TCH frequencies.

Another important argument for band splitting, from a practical point of view, is that the interference rising from the TCH carriers is completely load dependent. If the spectrum is split, the interference on the BCCH frequencies is stationary, so the quality of the BCCH frequencies is load independent.

Furthermore, other secondary more practical arguments exist for having a separate BCCH band. One example is that in GSM all the dynamic information about the neighbours is measured on the BCCH frequencies of the neighbours. Since a mobile is only capable of measuring a limited number of frequencies, it is an advantage to have a good BCCH layer in a GSM network. By having a separate BCCH layer, the MA list becomes smaller, which has positive impact on the handover performance [185].

It is easier to put extra TRXs in a network, that uses a separate BCCH layer, since the BCCH frequency plan, does not have to be changed. Only a new TCH frequency for the new TRX has to be found and maybe some minor changes have to be made in the TCH plan.

For these reasons it has been chosen to use band splitting with a separate band for the BCCH and for the TCH frequencies in JETTPlan. The spectrum of M frequencies is split into two parts as can be seen in Figure 189.

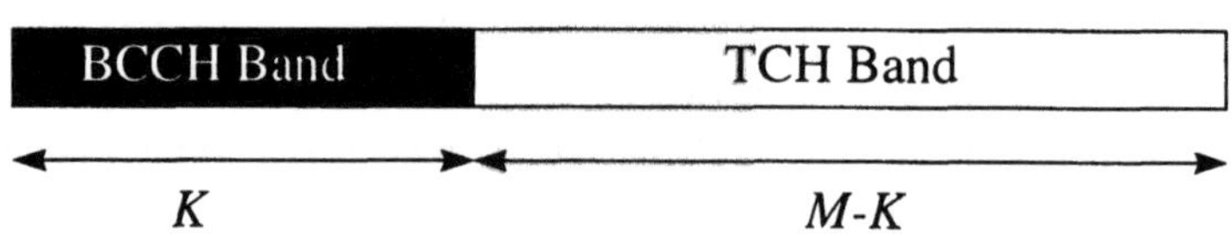

Figure 189. Splitting a frequency band of M frequencies into the BCCH and the TCH.

Another way of splitting the band can be seen in Figure 190, where interleaving is used. The BCCH frequencies are interleaved with the TCH frequencies. The advantage is that there is no adjacent interference between the BCCH frequencies. The distance between the individual TCH frequencies also increases, which is beneficial for the hopping, if a small reuse is used, since the distance between the frequencies increase.

Figure 190. Interleaving the BCCH frequencies between the TCH frequencies.

The disadvantage is that the interference on the BCCH frequencies becomes partly load dependent, since adjacent interference arises from the TCH frequencies. Furthermore, the complete planning process becomes more complex, which is the primary reason why this issue has not has been treated further.

2.4 The Frequency Planning Method

In this section JETTPlan is described. In Section 2.4.1 the overall structure is described, followed by a description of the input to the tool in Section 0. Section 2.4.3 deals with the actual search algorithm, while in Section 2.4.4 the method of implementing the gain of frequency hopping is described.

2.4.1 The Overall Structure of JETTPlan

The structure of the planning tool is shown in Figure 191. Initially all the frequencies are planned in the iterative optimisation process. The proposed frequency plan is implemented in the network and interference predictions and the mapping from *C/I* per serving frequency per pixel to FER per pixel, is made. It is evaluated whether the plan is satisfactorily good. If it is, the planning procedure stops and the plan is ready for use. If not, two types of problems can exist; a (or some) local or a global problem. If it is a local problem a manual or a limited tool update can be made in a new iterative process. If the problem is global, the optimisation process starts over by either reducing the requirements or by letting the algorithm run for a longer time period to achieve a better solution.

However, due to the limited amount of time available it has been decided not to implement the entire functionality as shown in Figure 191. Instead of designing software to graphically visualise the quality (FER) from scratch, it has been chosen to use a commercially available tool for that purpose. This means that the whole process is split into two parts. A part where the evaluation is done with the help of an external commercially available tool and a part where the frequency plan itself is made.

In the tool it is possible only to plan a small part of the network or even only one cell at a time, while still taking all interfering cells into account. While all interfering cells are taken into account all the frequencies of the cells that are not to be planned are kept fixed and cannot be changed.

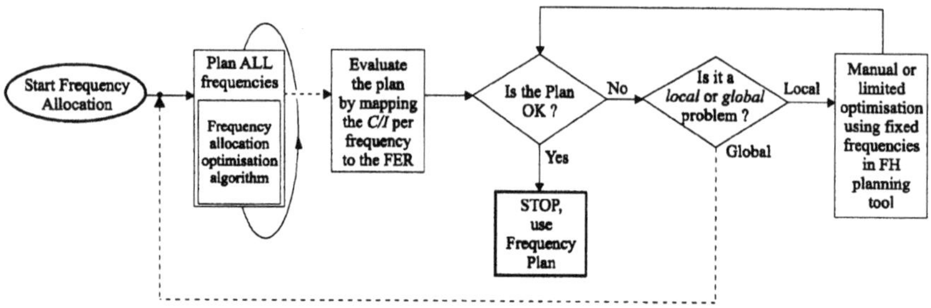

Figure 191. Flow diagram of the frequency planning tool.

The program has two different modes. One where the whole network or specified area is planned at once, the network mode, and a mode where one cell at a time is planned, the cell mode. The network mode can be seen in Figure 192. As can be seen the algorithm consists of two identical parts, the BCCH and the TCH allocation. At the start first the BCCH frequencies are planned, followed by the planning of the TCH frequencies. The two allocation parts are identical, except for the thresholds used. Basically it can be said that the BCCH frequencies are weighted more, since they are more important and they do not use DTX or power control and are always 100 % loaded.

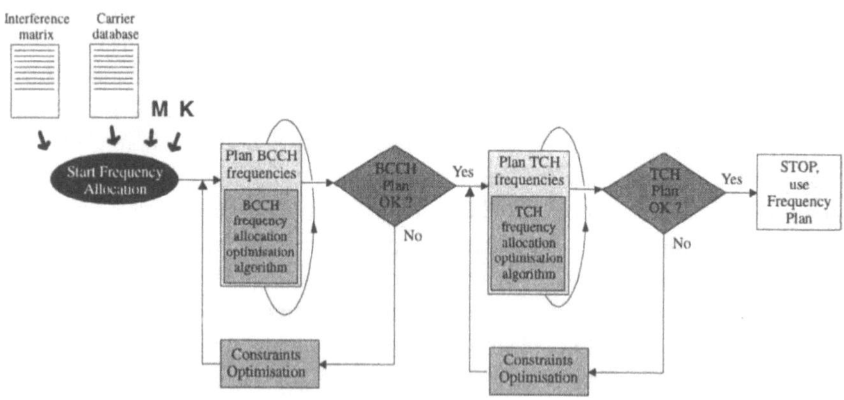

Figure 192. Structure of the frequency allocation algorithm in network mode.

It can be seen that the input consists of 4 elements: the interference matrix, the carrier database, the total spectrum M and the size of the spectrum to be used for the BCCH, K. The interference matrix and carrier database will be dealt with in detail later. In Figure 193 the structure of the frequency allocation algorithm is shown for the cell mode. The blocks basically do the same, but compared to before only one cell is planned at the same time.

In both modes there is a possibility of fixing cells, i.e. locking the frequencies of some cells, so that they cannot be changed. The actual search algorithm (the box, with 'Plan TCH/BCCH frequencies' in the Figures) is described in Section 2.4.3, as well as the constraint optimisation.

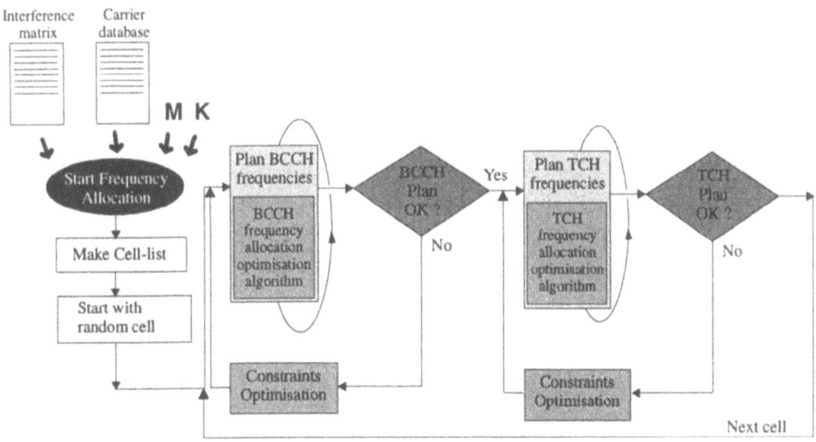

Figure 193. Structure of the frequency allocation algorithm in cell mode.

2.4.2 Inputs to the Planning Program

As described, the two most important input files are the interference matrix and the carrier database. The idea of the interference matrix is that it should reflect the frequency independent levels of interference between all cells in the network. A traditional example of the format of one cell in the interference matrix file, is shown here:

CELL	SITE1	1	4.96	200	5
INT	SITE1	2	0.194617	10	
INT	SITE1	3	1.1027	40	
INT	SITE2	1	0.04413	2	
INT	SITE2	2	0.000113471	0.1	
INT	SITE2	3	0.574312	25	

The first row describes the serving cell, which here starts with the word `CELL`. The next parameter in the row is the serving cell site identity followed by the sector number. In this case the serving cell is `SITE1` sector `1`. The next parameter is a value corresponding to the coverage area of the cell. The fifth parameter reflects the traffic to be carried by the cell. The last parameter the number of interfering cells to this particular serving cell. In this example five interferers exists of which each are described in the next five lines in the interference matrix. Each interferer is described by seven parameters. The first is always INT. Then the site identity and sector number follows, as was the case for the serving cell. The fourth parameter is a value describing the area of the serving cell potentially affected by interference from this cell. The fifth parameter reflects the traffic of the serving cell potentially affected by interference from this cell. Besides treating interference from geographical area as well as from the traffic distribution some planning tool furthermore provides information of co-channel as well as adjacent-channel interference.

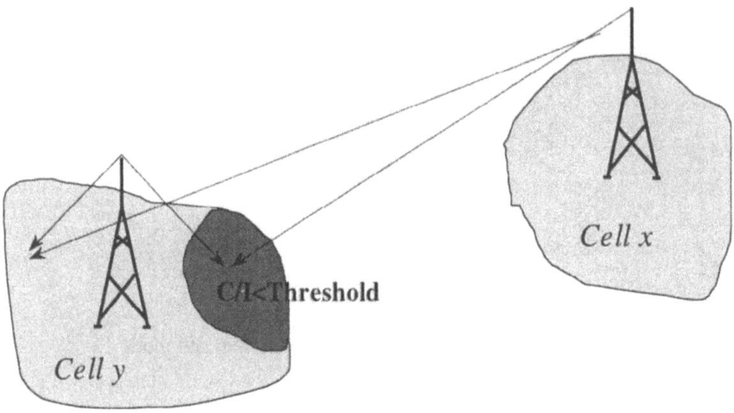

Figure 194. Principle used to calculate the level of interference for the interference matrix.

Making predictions for every pixel in a cell generates the interference matrix. The index (x, y) in the interference matrix, indicating how much interference is coming from cell x to y, is calculated as is shown in Figure 194. For each pixel in cell y the interference coming from cell x is calculated. Also the carrier power is calculated. Then the resulting *C/I* is compared to a threshold (usually 9 dB is being used). If the *C/I* is below this threshold the pixel is 'affected by interference'. By dividing the number of pixels in cell y which are effected by interference and the total number of pixels in cell y, the index in the interference matrix between these two cells is found. This could be done for traffic as well as for geographical area.

The following limitations apply to this calculation:

- Since just a single threshold is being used, no absolute level information is taken into account. In reality there are only two kinds of *C/I* values; those above and those below the threshold. If a threshold of 9 dB is used, then it does not matter if the *C/I* in a group of pixels is 8 dB or -9 dB as would be the case in reality !

- The procedure described above is carried out between all cells which means that it can very easily happen that a certain area is affected by interference from two different cells. Then this area is taken into account twice, which might not be correct since there is an overlap. The interference will probably be worse in this overlapping area and therefore by taking it into account twice may not necessarily cause a big error. The problem is shown in Figure 195, where the black area in cell y is taken into account twice.
- As mentioned above no real traffic distribution has been taken into account. The traffic per cell is uniformly spread over the cell. The use of the real locations of the traffic would give more realistic results.

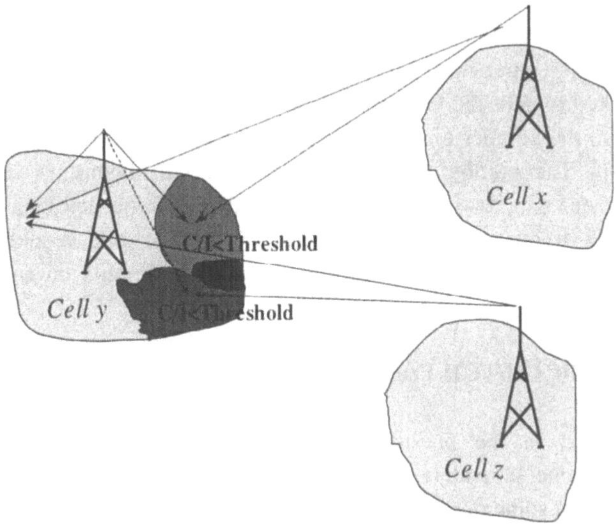

Figure 195. Problem with overlapping affected area.

The carrier database contains the requirements in terms of frequencies per cell. It also specifies the site identity names and sector numbers. Furthermore the carrier database contains the actual frequency plan (if available), which can be used to specify the initial constraints as will be explained in the next section. An example of the format of the input file carrier database, is shown here:

```
SITE1  3  2  96 108
SITE2  1  3  96 98 111
SITE2  2  4  108 96 100 110
SITE10 1  1  108
SITE10 2  4  99 98 110 119
    ...
```

Each line describes a cell. The first parameters are the site identity and the sector number. The third parameter is the number of TRXs in that specific cell. The following parameters are the actual frequencies. In the example above `SITE1` sector `3` has two frequencies, namely channel `96` and `108`.

In JETTPlan there is also a possibility to specify a RF hopping file. This should be done if RF hopping is used. This file has to contain for each cell the number of TRXs and the number of frequencies. The planning tool will then plan the right number of frequencies and use the gain coming from RF hopping (due to the lower load) to calculate the cost function (this is described in Section 2.4.3). It should be noted that the tool does not calculate the optimum MA list length (this is equal to the number of frequencies), but it does use the MA list length as specified, i.e. the tool uses the number of frequencies specified in the RF hopping file.

Beside of these input files, some other parameters are used as input. First of all the user is asked to give the total number of frequencies available, i.e. the available spectrum. Also the number of frequencies to be used for the BCCH part is an input of the program. This number of BCCH frequencies is held constant over the whole network, so if one part has to be planned with x BCCH frequencies and another part with y BCCH frequencies, the two parts of the network should be planned separately. The program will take the influence of both parts on each other into account.

2.4.3 The BCCH/TCH Frequency Allocation Optimisation Algorithms

As described in the previous section planning of the BCCH and TCH frequencies is done separately. The allocation algorithm for the two networks is identical, but with some design thresholds being different. In the BCCH part only a frequency diversity gain from frequency hopping can be taken into account (if using baseband hopping), while in the TCH besides this frequency diversity gain, the fractional loading gain from frequency hopping is taken into account.

The algorithm is kind of a search tree of which an example can be seen in Figure 196, where the frequencies within 8 cells are planned. One cell, cell number 1, has 2 frequencies, while all the other cells only have one. All cells are omni directional cells, except for cell 7 and cell 8, which are sectors of one base station.

The algorithm randomly chooses the frequencies for each cell at first instance. It starts with the cell in the top of the list. In the example frequency f3 is chosen as frequency number one for cell 2. Then the algorithm finds a frequency for the next cell in the list. This is again done randomly from all possible candidates. Not always all frequencies can be chosen from, due to constraints, like for example in cell 8, frequency number one and in cell 1, frequency number two. More about constraints can be found further on in this section.

When an acceptable solution is found for the complete area, the cost function is calculated by summing the cost of each frequency. In the example in Figure 196 the cost of each frequency is shown at the right. If the total cost function is better than the existing plan and all previously found solutions, then this solution is kept as best solution.

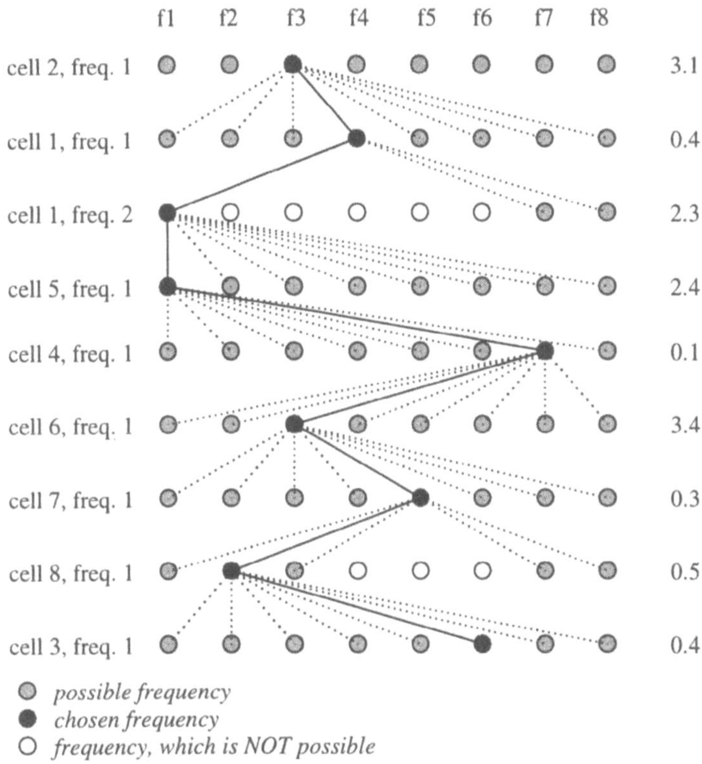

Figure 196. Example of the search tree algorithm.

After finding a solution, the program tries to optimise the solution by changing the frequencies with the highest cost functions. This stops when either a timer is finished or the frequency planner wants to stop the algorithm. After each loop, the constraint optimisation process is run, as can be seen in Figure 192 and Figure 193. Then the next loop starts at scratch (except for the fact that the best solution is kept). Using this method the tree is randomly searched for the best solution. The program stops, if the user wants to stop it, if the maximum number of loops has been finished or if the cost function becomes zero.

Setting the Constraints

There are two types of constraints, the ones to avoid situations that should never occur and the ones that make the search algorithm faster and more efficient. The first type occurs in two different situations: the distance between 2 frequencies within one cell should be at least 3, due to some hardware restrictions and adjacent channel interference. This can be seen in cell 1 of the example (Figure 196). The second situation is when the distance between 2 frequencies used in sectors of the same site should be at least 2. This is also to avoid adjacent channel interference. This can be seen in cell 7 and 8 in the example.

The second type of constraints is much more complex. By preliminary simulations it was found that the best solutions were found when settings many constraints, provided that they are set on the right places, so that the greatest interferers are avoided. It becomes harder to find a solution, but the solutions found are of better quality. Therefore the problem is: how many constraints should be set? When setting too many, no solution will be found, but by setting too few, a good solution is less likely to be found.

If the program is run in network-mode, the search algorithm becomes very complex, since the number of combinations increases exponentially with the number of cells. Therefore the number of combinations should be limited, which is done with the constraints of the second type.

In JETTPlan an automatic constraint optimisation is being used, which will is explained with an example. In Figure 197 a frequency plan diagram is shown. In a frequency plan diagram each dot represents a frequency. On the y-axis the carried traffic on that frequency is shown, while on the x-axis the affected traffic over carried traffic ratio can be found. This affected traffic over carried traffic ratio is the ratio between affected traffic in the considered cell by interference from all cells using the same frequency and the total traffic in the considered cell. Therefore this ratio can maximally become one, meaning that all traffic in the cell is affected by the interference from other cells.

The following points apply to frequency plan diagrams:

- A good frequency plan has its points on the left side of in the frequency plan diagram. The left side means that the affected traffic over carried traffic ratio is low, i.e. not much of the total traffic in the cell is affected by the interference.
- It is preferable to have points in the bottom having a high affected traffic over carried traffic ratio rather than having points lying in the top having a high affected traffic over carried traffic ratio. The points in the top are the frequencies carrying a lot of traffic, so if a lot of that traffic is interfered, then the absolute interfered traffic is higher than when it would be a point in the bottom.
- Frequency plan diagrams can be made for adjacent and for co channel interference.

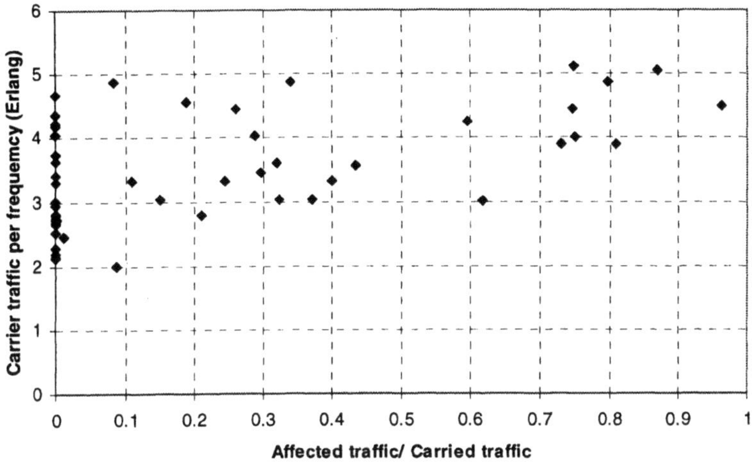

Figure 197. Example of a frequency plan diagram.

The constraint setting is done as follows. The input frequency plan is used to find a start guess. Initially the point with the highest affected traffic over carried traffic ratio of the cell with most traffic is found of the old plan. Then a line is drawn from that point to the point (0,1). Now constraints are set to the right of this line, i.e. it is forbidden to plan frequencies, so that points occur on the right side of the line. This can be seen in the left top Figure 198.

If a solution is found in the loop starting after this constraint setting, then the point at the x-axis is shift to the left (the top right of Figure 198). Note that now it becomes harder to find a frequency plan. As long as a frequency plan is found the algorithm shifts the line to the left. If no solution is found the line is moved back to the right, but never more than where the last solution was found. When a solution found again, the line is shift to the left again. This way, it is always attempted to move towards the left.

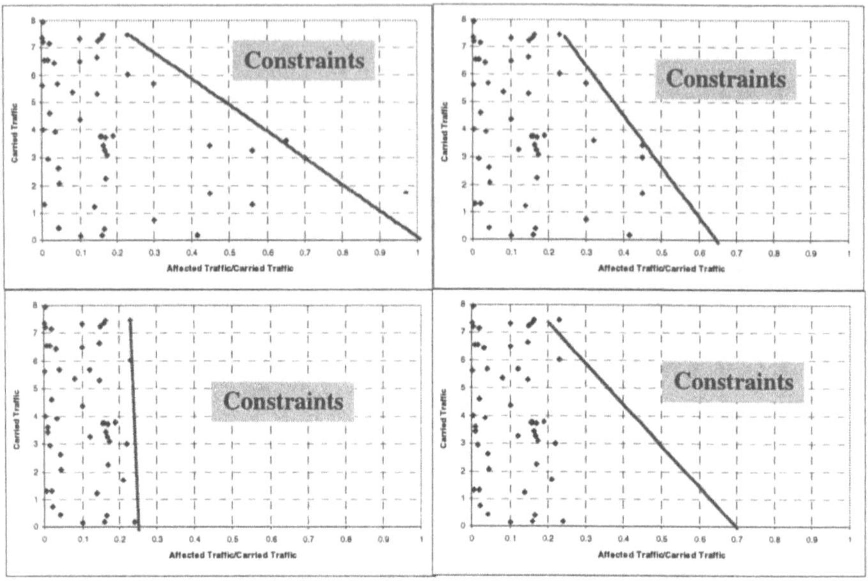

Figure 198. The process of automatic constraint setting.

When the line is almost vertical (the bottom left of Figure 198), the point at the top is shifted to the left (the bottom right of Figure 198). The point at the bottom is again set at (1,0) and the process starts over again. This way the algorithm keeps setting more constraints as long as a solution is found, while it loosens the constraints, when no solution is found. This process is carried out for co-channel interference. For adjacent-channel interference the initial line is found just as with the co-channel interference. However this line is held constant for adjacent interference, i.e. no optimisation with the help of constraints is done for adjacent interference.

If the program is run in cell-mode, then there is a very limited amount of different combinations, which should be searched through, so no constraint optimisation is done. The constraints are set so that a minimum quality is guaranteed.

Ordering at the Start of Each Loop

At the start of each loop, the cells are ordered. For this ordering two principles are used. First of all the cells depending on each other should be kept together. If two cells depends on each other their cost function contributions correspondingly have a great influence on each other. By clustering them together, the proper solution is found faster.

The second principle is that there should be some randomness in the ordering, so that every time a new loop is started a new situation occurs. This is achieved by mixing the groups of cells, which are depending on each other, randomly. The whole ordering at the start of each loop can be seen in Figure 199.

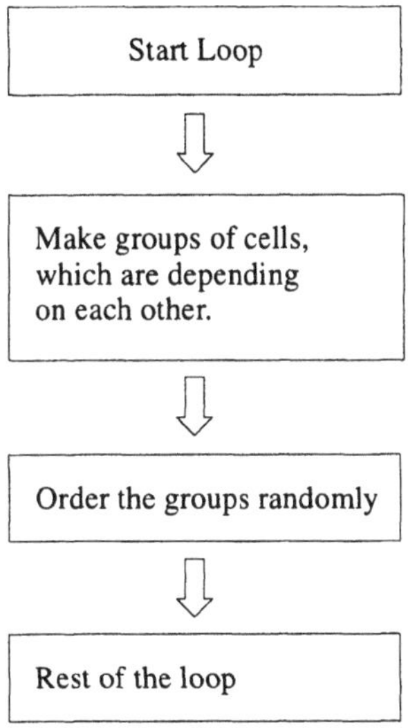

Figure 199. The ordering algorithm used at the start of each loop.

Reordering After Having Found an Acceptable Solution

After having gone through the tree and found a solution, like in the example in Figure 196, an iterative optimisation routine is started. First of all the cells are prioritised so that the cell with the highest cost function contribution (= sum of cost functions of the frequencies in that cell) is placed at the bottom of the tree. The frequency from the cell in the bottom, having the greatest contribution to the cost is removed and a new search for another frequency is started. In Figure 200 this situation is shown. The cells are ordered according to their cost. Note that the cost of the entire cell has been taken into account, i.e. cell 1, frequency 1 and cell 1, frequency 2, are summed. If it is not possible to find a new frequency for this level the algorithm moves up in the tree and starts over.

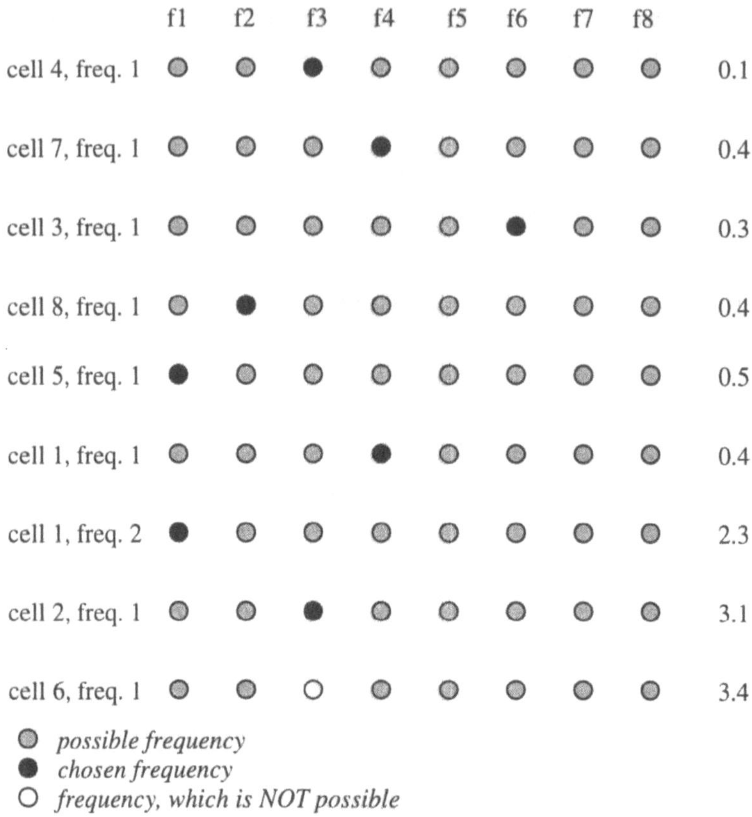

Figure 200. The situation that will occur after the situation described in Figure 196.

Calculating the Cost Function

The goal of the algorithm is to minimise the total amount of affected traffic or area by interference. To achieve this a cost function has been used. The cost function is an expression, which indicates how good or bad a frequency is planned from an interference point of view. The cost function of frequency *i* in cell *a* is the sum of the cost of the co-channel interference plus the weighted adjacent interference:

$$Cost\_level(i,a) = \sum_{k=cells\_with\_a\_co\_channel} cost\_co\_channels + adj\_factor \cdot \sum_{k=cells\_with\_a\_adj\_channel} cost\_adj\_channels \quad (1)$$

As *adj_factor* 1,5 % has been used, which corresponds to having filtered the adjacent interference by about 18 dB. The cost function can in more details be written as:

$$\begin{aligned} Cost\_level(i,a) = & \sum_{k=cells\_with\_a\_co\_channel} (Loadgain(k) \cdot J_{k,a} \cdot Hopgain(a) + \\ & \qquad Loadgain(a) \cdot J_{a,k} \cdot Hopgain(k)) + \\ adj\_factor \cdot & \sum_{k=cells\_with\_a\_adj\_channel} (Loadgain(k) \cdot J_{k,a} \cdot Hopgain(a) + \\ & \qquad Loadgain(a) \cdot J_{a,k} \cdot Hopgain(k)) \end{aligned} \tag{2}$$

where *Hopgain(x)* is the hopping gain in cell x, *Loadgain(x)* is the load gain from cell x and $J_{x,y}$ is the absolute amount of affected traffic in cell y by interference from cell x (which can be found in the interference matrix). The gains *Hopgain(x)* and *Loadgain(x)* give an interference reduction, so they are always between 0 and 1. The factors in the cost function and their relation can be seen in Figure 201. The cost function of a cell is a measure of how much traffic in the cell is affected by interference from other cells plus the amount of traffic which is affected by interference in other cells by the cell.

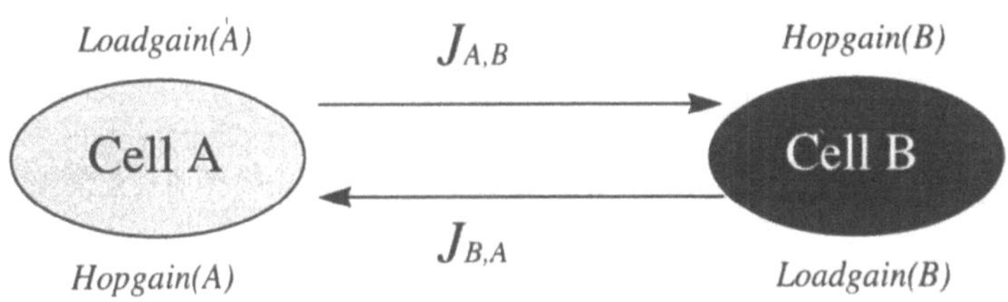

Figure 201. The factors included in the cost function of JETTPlan.

The two factors 'Loadgain' and 'Hopgain' are explained in the next section.

2.4.4 Including Frequency and Interference Diversity Gain in JETTPlan

This section deals with the loading and frequency diversity gains introduced in the last section. As explained in Chapter 3 there are 3 frequency hopping modes:

- *No hopping.* There is no gain from frequency diversity and no loading gain. The latter may sound a bit strange, since there still might be a low load in some cells, so why does that not give any gain? This is due to the fact that this gain is not spread out over all frequencies, so a worst case scenario should be considered.

- *Sequential Hopping*. Obviously there is a gain from frequency diversity, but there is no gain from interference diversity, i.e. there is no Loadgain.[29]
- *Random Hopping*. Since there is a gain from both frequency and interference diversity, there is a loading gain and a frequency hopping gain.

Below the interference and frequency diversity gain are treated.

Including the Frequency Diversity Gain

Frequency diversity is described in Chapter 3. Figure 202 the FER can be seen as function of the *C/I* for non-hopping and random hopping with 2 to 12 frequencies. The typical urban channel profile has been used and the speed of the mobiles is 3 km/h. Results for a mobile speed of 50 km/h can be seen in Figure 203.

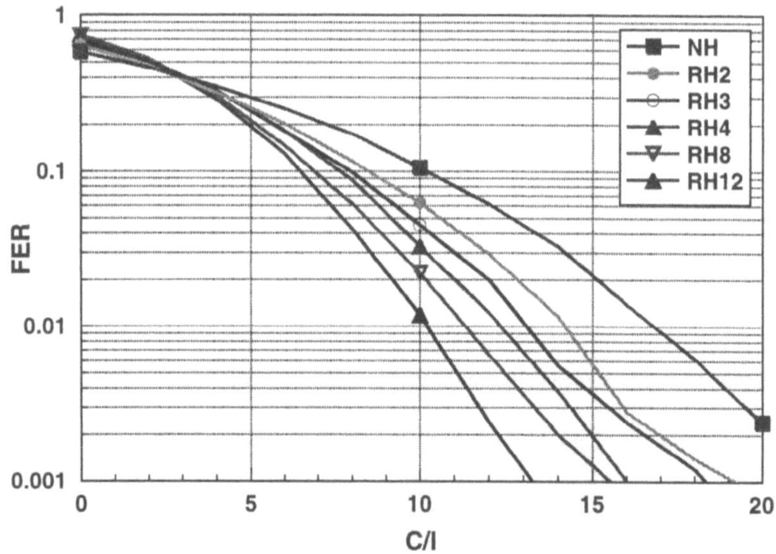

Figure 202. FER as a function of the *C/I*, using a TU3 channel profile with random hopping.

[29] Unless non-grouped frequency planning is used.

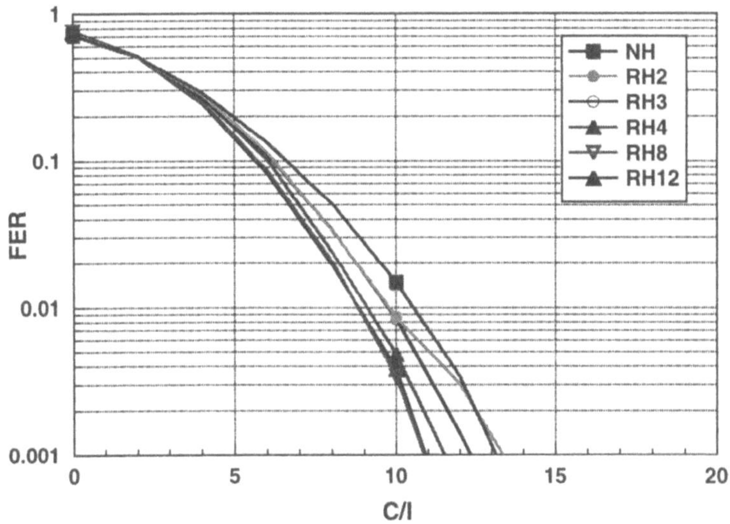

Figure 203. FER as a function of the *C/I*, using a TU50 channel profile and random hopping.

By looking at a FER of 2 % a gain can be found. This gain (in *C/I*) can be seen in Table 46. The first thing, which can be seen, is that the gain is very speed dependent. It's also clear that with 3 or 4 frequencies most of the gain is achieved.

Table 46. Gain of frequency diversity for different numbers of frequencies.

NUMBER OF FREQUENCIES	GAIN (TU3)	GAIN (TU50)
1	0 dB	0 dB
2	2.2 dB	0.8 dB
3	3.2 dB	1.2 dB
4	4.2 dB	1.3 dB
8	5.2 dB	1.4 dB
12	6.0 dB	1.4 dB

The gain shown in the table above can be translated into a linear gain for all number of frequencies in the hopping list. The gain (expressed in interference reduction) for the different number of frequencies can be seen in Figure 204. This gain reflects the 'Hopgain' used in the cost function.

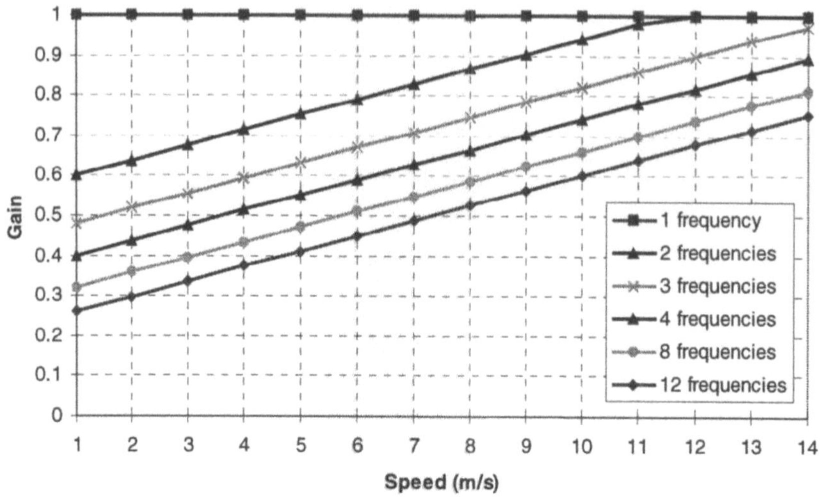

Figure 204. The interference reduction gain from frequency diversity as function of the speed for different number of hopping frequencies.

Including the Interference Diversity Gain

As mentioned previously only random hopping provides a interference diversity gain. This gain can be split into two parts. First of all by hopping between different frequencies the probability of being interfered by another mobile is averaged out on all the frequencies in the cell. This leads to a gain corresponding with the load in the system, e.g. in a system with 50% load 3 dB is achieved by hopping (compared to a non hopping network). However due to the lower load, the coding and interleaving scheme becomes even more effective. A gain referred to as a fractional loading gain is also achieved.

Figure 206 and Figure 207 show the dependency between the *C/I* and the FER at 100 % and 25 % load. Random hopping over 8 frequencies has been used. It can be seen that in the case of 100 % load the *C/I* has to be equal to 8.9 dB for the FER to be between 2 and 5%. At 25% for achieving the same FER, the *C/I* only has to be equal to 7.4 dB. In other words, by going from 100 % to 25 % load a gain of 1.5 dB is achieved, besides of the 6 dB gain from the lower load.

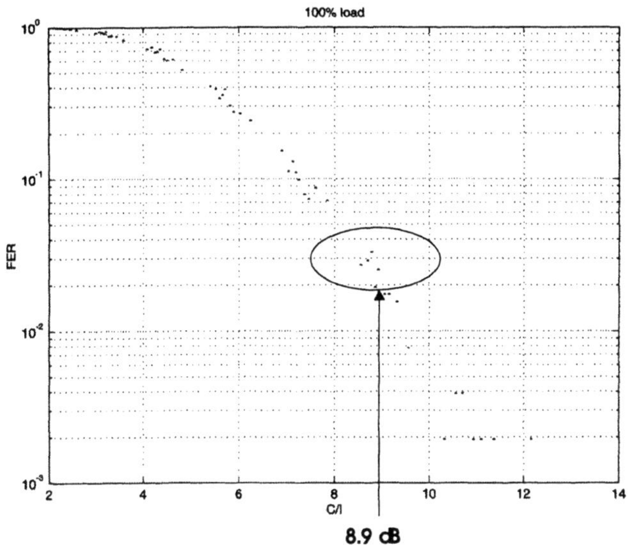

Figure 205. Dependency of *C/I* and FER at 100% load, when hopping over 8 frequencies.

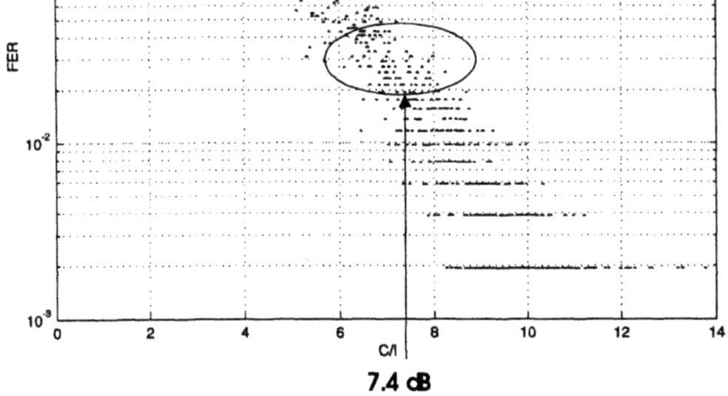

Figure 206. Dependency of *C/I* and FER at 25% load, when hopping over 8 frequencies.

The size of the fractional loading gain is seen in Figure 207 for different number of hopping frequencies. The curves are generated using system level simulations.

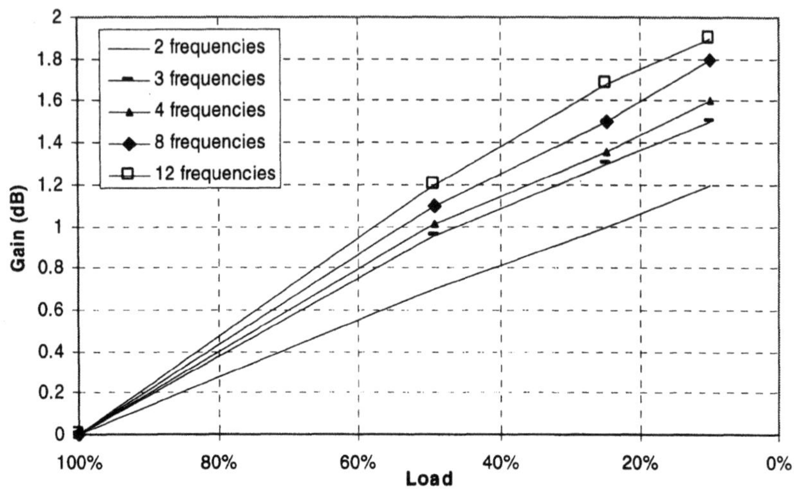

Figure 207. The fractional loading gain from frequency hopping.

This gain can simply be modelled by straight lines depending on the load with different slopes for the different number of hopping. The model can be seen in Figure 208.

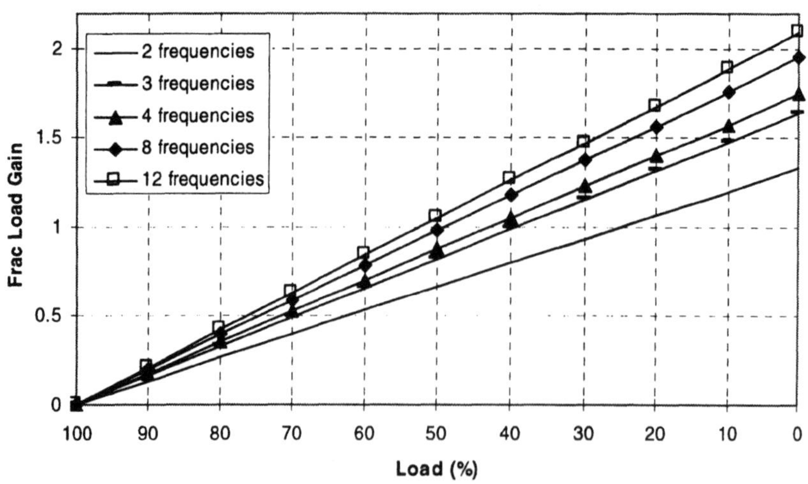

Figure 208. Fractional loading gain model implemented in JETTPlan.

The gain from interference diversity with a load of 100 % depends on the spread in *C/I* on the different frequencies. This is not included in this implementation. In worst case the gain is equal to zero. To illustrate the implementation of the interference diversity gain in JETTPlan, an example is given. The complete interference diversity gain from a hopping network with 8 hopping frequencies in all cells, with 50 % load is 3 dB from only having 50 % load and additionally 1.1 dB from fractional loading. Therefore the total gain is 4.1 dB. In this case the 'Loadgain' in the cost function will be equal to 0.39.

3. PERFORMANCE OF THE FH PLANNING TOOL

This section deals with the performance of JETTPlan. First in Section 3.1 the performance of the search algorithm itself is evaluated. This is done by not splitting the band between BCCH and TCH frequencies. All frequencies are simply assumed to be TCH frequencies and the performance is evaluated. Secondly the spectrum is split into 2 bands and it is shown how much loss this causes. Section 3.2 deals with the tool in frequency hopping mode. The frequency hopping gain is evaluated and the frequency plans with frequency hopping are compared to the ones without frequency hopping. Finally Section 3.3 shows the results of the implementation of a frequency plan made using JETTPlan in the SONOFON network.

3.1 Performance of the Search Algorithm

In this section the performance of the search algorithm is evaluated. The search algorithm was described in Section 2.4.3. Two frequency plans are used to evaluate the search algorithm. They are described in Section 3.1.1, followed by an evaluation of the search algorithm in Section 3.1.2, while the subject of Section 3.1.3 is the trunking loss due to splitting the frequency band between TCH and BCCH frequencies.

3.1.1 Reference Frequency Plans

Two reference frequency plans have been used for comparison with JETTPlan. Experienced SONOFON radio frequency planners have designed both. Exactly the same interference predictions have been used as input for JETTPlan as has been the case for the reference plans. The first of the reference plan were made by hand based on experience and therefore several months were spent optimising the plan. The second reference plan is the best plan the radio planners could come up with using a commercial frequency planning tool followed by manual optimisation. The total cost function value of the two reference plans is seen in Table 47.

Table 47. Cost function of the 2 reference plans.

FREQUENCY PLAN	COST VALUE
Reference plan 1	2654744
Reference plan 2	1230489

The available frequency spectrum consists of 45 frequencies. The geographical area treated has been downtown Copenhagen, which is the area in Denmark with the highest tele-traffic density. The 2 test areas are not completely the same. Reference plan 1 consists of 228 cells with an average of 3.2 TRXs per cell. In the test area of reference plan 2 there are 243 cells with an average of 3.2 TRX per cell. For both test areas the distribution of number of TRXs per cell is quite uniform, i.e. there are not many cells with 5 TRXs or 1 TRX. To avoid border effects, all cells with influence on considered area have been taken into account.

3.1.2 Only TCH Frequencies

The search algorithm is evaluated by setting the number of BCCH frequencies equal to zero and thereby assuming that all frequencies are TCH frequencies. In other words: all frequencies are the same and the entire available spectrum can be used for any frequency. No gain of frequency hopping is taken into account and load is assumed to be 100% for all cells. All this is not realistic, but the goal is to evaluate the search algorithm.

The adjacent channel weight is set to 1,5%, i.e. an adjacent channel is assumed to be filtered, so that its level is 18 dB lower than the level of a co-channel interferer.

In Figure 209 the costs of the frequency plans for reference area 1, made by JETTPlan can be seen. The planning tool was run 3 times, resulting in 3 different frequency plans. Different timers where used. Plan 1 used a timer of 60 seconds, which meant that if not within 60 seconds a new solution is found in a loop, the next loop is started. Plan 2 used a timer of 30 seconds, while for plan 3 the timer was set to only 10 seconds.

A better frequency plan than the reference plan (cost 2654744) is easily found! No good conclusion about the timers can be taken based on these results, since while the plans with short timers go fastest down, but the best plan in all cases are quite alike.

In Figure 210 the results of the planning of the reference area 2 can be seen. Again the plans the tool comes up with are better than the reference plan, but the improvement is not as large as in the reference area 1. It also seems that short timer are a little better than large ones.

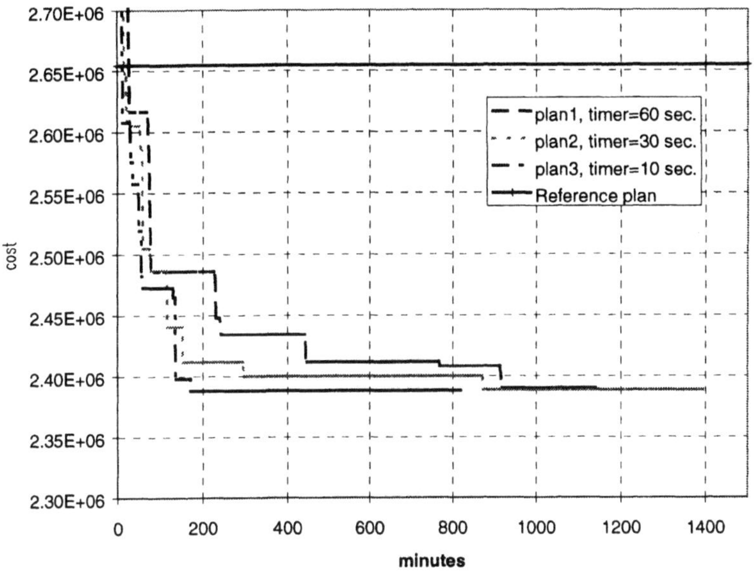

Figure 209. Cost of the plans for reference area 1 as a function of time.

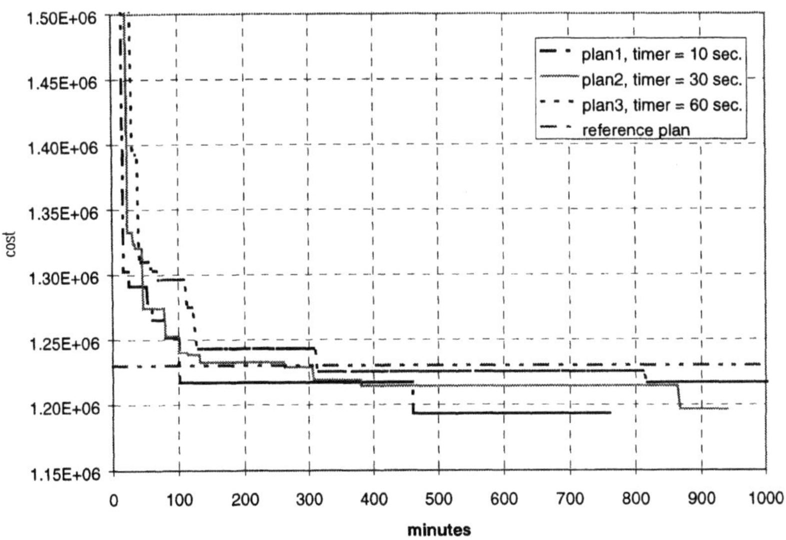

Figure 210. Results of evaluation of the search algorithm with reference area 2.

The best plan for reference area 2 is the plan with cost 1192969 (plan 1, a timer of 10 s.). This plan is now used as input to the tool in cell mode. In Table 48 the improvement of the cost function per complete loop can be seen. One loop means that all cells have had their turn to be optimised. The time per complete loop is also depicted.

Table 48. Improvement of the cost per complete loop of JETTPlan in cell mode.

LOOP NUMBER	COST	TIME (MIN)
0	1192969	0
1	1062455	35
2	1048950	34
3	1046393	34
4	1046294	34

It can be seen that the improvement per loop becomes smaller and smaller. This is as expected, since in the start a lot of improvements can be made, while it becomes harder per loop. The time per loop is about constant (around 34 minutes). It can also be seen that there is a quite an improvement of the cost function. This is because, when the program is run in network mode, a lot of time is spent on improving the worst cells.

In the frequency plan diagram it is tried to push the points to the left. However, there will always be some cells, which are hard to plan and they will keep the line, which is used to set the constraints, to a certain point. Other cells, which may have their points to the left, i.e. which are relatively good and do not contribute much to the total cost function, can be improved. However, they are not improved by the program in network mode, but pulled to the left by the program in cell mode. All this is illustrated by the frequency plan diagrams in Figure 211.

Figure 211 show the frequency diagram of the different plans. It can be seen that the reference plan has no points with an affected traffic over carried traffic ratio higher than 0.1, while in the frequency plan diagram of the plan made by the program in network mode the line used for the constraints settings can be seen. In the frequency plan diagram of the plan made by the program in cell mode, a lot of points have been shifted to the left, compared to the frequency plan diagram after the network mode. A bit surprisingly it can be seen that some points are actually shifted to the right! If looking at the average affected traffic over carried traffic ratios, as seen in Table 49, it can be seen that the plan made by the tool in cell mode gives the best result.

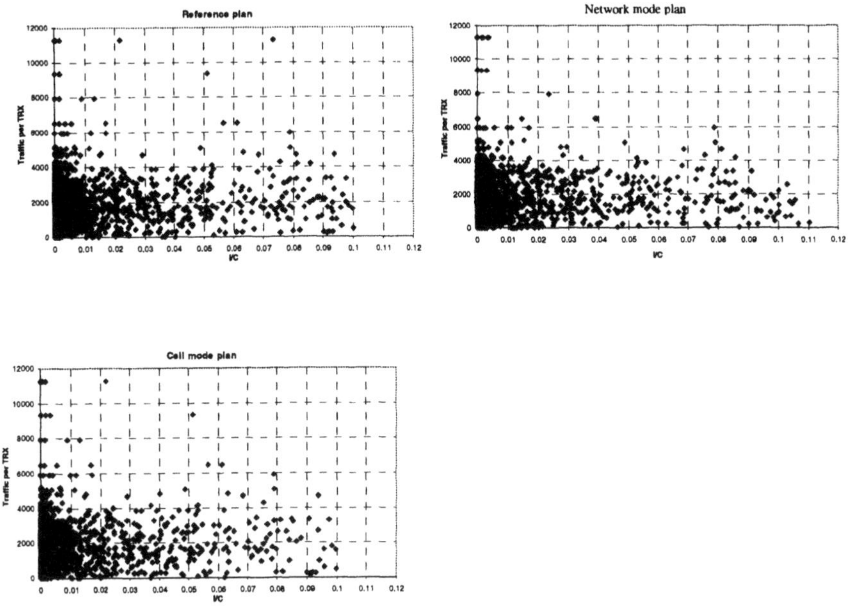

Figure 211. Frequency Plan IC diagrams of the reference plan, the plan made in network mode and in the cell mode.

From this it can be concluded that the limitation of the network mode is that it uses a straight line. That way the program tries to improve only a limited amount of the cells. It also seems to be better to have some cells, which have a high affected traffic over carried traffic ratio, so that there are more freedom degrees for the other cells. It should be noted that all these comments depend on the network topology and traffic distribution in the network treated.

Table 49. Average affected traffic over carried traffic ratio for the different plans.

FREQUENCY PLAN	AVERAGE AFFECTED TRAFFIC OVER CARRIED TRAFFIC RATIO
Reference plan 2	0.006146
Network mode	0.006049
Cell mode	0.005212

The optimum solutions is being found by running the program first in network mode and then use the output as input for the program in cell mode.

3.1.3 Cost of Splitting the Band on Two

As written before it is not preferable from a pure search algorithm point of view to split up the spectrum as is done, since it means a trunking loss. However, there are a lot of good reasons to do it anyway (see Section 2.3). This section shows the effect of band splitting on the cost function.

Reference area 2 is used to evaluate this effect. A frequency band of 45 frequencies has been used. In Figure 212 the cost function can be seen as a function of the number of BCCH frequencies. Different weight factors have been used. A weight factor equal to 1 means that the cost function of the BCCH layer is weighted just as much as the TCH layer. It should be noted that no frequency hopping gain has been taken into account.

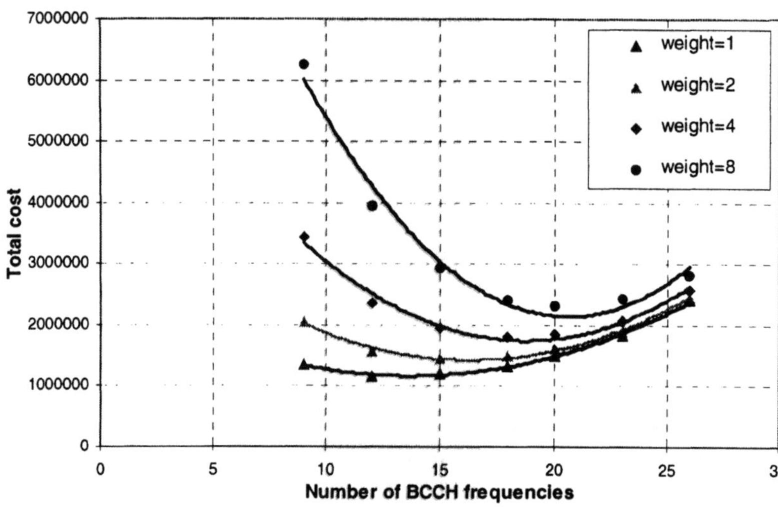

Figure 212. The effect of band splitting on the cost function with different weight factors.

As can be seen in the figure, the optimum number of BCCH frequencies is around 14, when the weight is set to 1. This corresponds very nicely with the fact that there are on average 3.2 TRX's per cell, since 45/3.2 ≈ 14, so on average for each TRX there are 14 frequencies to choose from. Since the BCCH corresponds to one TRX, the BCCH layer uses 14 frequencies in the optimal case.

The weight equal to one is not every realistic though. DTX and power control can be used on the TCH channels and give a gain of respectively 2,5 and 3 dB [66]. The load on the BCCH always is equal to 100 %, while the load on TCH channels is seldom higher than 50 % in a real network. This makes that it is very reasonable to weight the costs of the BCCH network more than the TCH network. If a load of about 50% is assumed on the TCH network and power control and DTX is used, a weight factor of 8 may be quite realistic. In Figure 212 the cost function with weights equal to 1,2,4 and 8 is shown.

It can be seen that the optimum number of BCCH frequencies increases when the weight increases, which is as expected. The optimum when a weight equal to 8 is used lies around 20, which leaves 25 frequencies for the TCH network. This indicates that a more than proportional part of the spectrum should be used for the BCCH network.

It should be noted that this weighting is partly taken into account in JETTPlan, since the TCH frequencies profit from frequency hopping gain and the load is taken into account for the TCH frequencies. The cost function takes care of this, so no extra weights have to put on the different layers.

However if it is desirable to have a better BCCH band than TCH band, so that the quality of access and control information is guaranteed, then weights can be used. Another way of taking this into account is using different interference matrices for the BCCH and TCH band. For example an interference matrix with a threshold of 9 dB is used for the TCH band, while a threshold of 12 dB is being used for the BCCH band.

3.2 Evaluation Method for a Frequency Plan with Frequency Hopping

The previous section shows that the search algorithm works quite well, but the purpose of developing JETTPlan was to be able to plan a network with frequency hopping. This section deals with frequency allocation in a frequency hopping network.

In Section 3.2.1 the reference plans are described, followed by a description of how to visualise the gain of frequency hopping. Results are found in Section 3.2.3.

3.2.1 Reference Plans and Setup

As in the case of non hopping, again a reference area has been used. It is the whole Copenhagen area in Denmark. The area contains 159 cells with an average of 3.1 TRX per cell. However it is chosen to only optimise on the centre part of the 125 cells. Random hopping is used and plans are made for 3 km/h. Two different ways of taking the load of cells into account can be used. First of all the handled traffic in the busy hour can be used, which has the advantage that the most recent traffic data is used. The other option is two use the Erlang B 1% blocking. This means that of each cell the load it taken, which corresponds to having 1 % load on the number of channels of that cell. It is safer, since the traffic can rise to this blocking limit.

The spectrum is split into a part for the BCCH network and a part for the TCH network. Different sizes of bands have been tried to find the optimum. JETTPlan is always first run into network mode, followed by cell mode.

The existing network has been used as reference plan, which has a cost function of 729513 and which does not have separate bands for the BCCH and TCH, but uses the whole spectrum for both kind of channels.

3.2.2 Showing the Gain of Frequency Hopping on a GUI

The output of JETTPlan is simply a frequency plan and a cost function, which indicates how good a plan is. The next step would be to visualise the plan by exporting the frequency plan into a commercially available planning tool with a GUI. Typically such tools will show the worst or average *C/I* per pixel, while the interesting parameter is the FER. Therefore an interface has been made, which turns the different *C/I* values in each pixel into a quality measure, which is closely related to the FER. This is done in the way shown in Figure 213. For every pixel the *C/I* is calculated for each carrier and the combination of these *C/I*'s is mapped to a quality measure closely related to the FER.

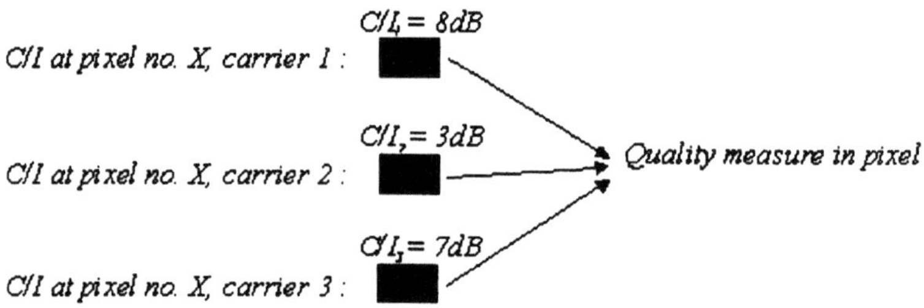

Figure 213. Mapping from *C/I* to a quality measure related to the FER.

The frequency and interference diversity gains have been taken into account in the same way as in the cost function. The frequency diversity gain is found using tables of combinations of the different *C/I* values, while the fractional loading gain is found by taking the load of each cell into account.

3.2.3 Results

In the scenario studied, the available spectrum has been 9 MHz (45 frequencies). This frequency band is split into two parts, one part for the BCCH network and a part for the TCH network. At first JETTPlan was run in network mode. The cost function found for the frequency plans for a different number of BCCH frequencies is seen in Table 50. A speed of 3 km/h has been used. It can be seen that compared to the reference plan, the cost of the plans are a lot lower.

Table 50. Cost of the plans found in the network mode in different situations.

PLAN	NUMBER OF BCCH FREQ.	COST BCCH	COST TCH	COST TOTAL	TIME SPEND
plan98_1	14	240e3	388e3	628e3	16h20m
plan98_2	15	217e3	410e3	627e3	18h30m
plan98_3	16	197e3	451e3	648e3	18h39m
plan98_4	17	179e3	471e3	650e3	12h55m
plan98_5	18	157e3	507e3	664e3	12h40m
plan98_6	19	149e3	542e3	691e3	17h30m
plan98_7	20	137e3	605e3	742e3	19h

It can be seen that when a band gets larger the cost function decreases. The cost function of the network with 15 BCCH frequencies, plan98_2, is lowest. This is a bit strange, since as explained in Section 3.1.3 it could be expected that a greater band for the BCCH frequencies would be preferable. The average cost per frequency is about the same for the BCCH and TCH frequencies in plan98_2, while the TCH frequencies get a gain from interference diversity. The reasons for this are:

1. Low speed and baseband hopping, so both BCCH and TCH get a gain from frequency diversity. So the difference is only the interference diversity gain.
2. There are some fixed cells at the border of the area. Those cells do not use band splitting, so it is very well possible that their BCCH frequency lies in the TCH part of the area, which is planned, and their TCH frequencies can all lie in the BCCH part of the area. The cost function of the two bands is mixed.

Now the plans found above are used in the cell mode of JETTPlan. The results can be seen in Table 51. It can be seen that the cell mode has improved the frequency plans in different layers. The total cost function can be seen in Figure 214.

Table 51. Cost of the plans, found after the cell mode in different situations.

PLAN	NUMBER OF BCCH FREQUENCIES	COST BCCH	COST TCH	COST TOTAL	TIME SPEND
plan98_1	14	232e3	357e3	589e3	38m
plan98_2	15	204e3	372e3	576e3	42m
Plan98_3	16	187e3	410e3	597e3	50m
Plan98_4	17	169e3	431e3	600e3	31m
plan98_5	18	146e3	462e3	608e3	40m
plan98_6	19	134e3	489e3	623e3	50m
plan98_7	20	125e3	524e3	649e3	32m

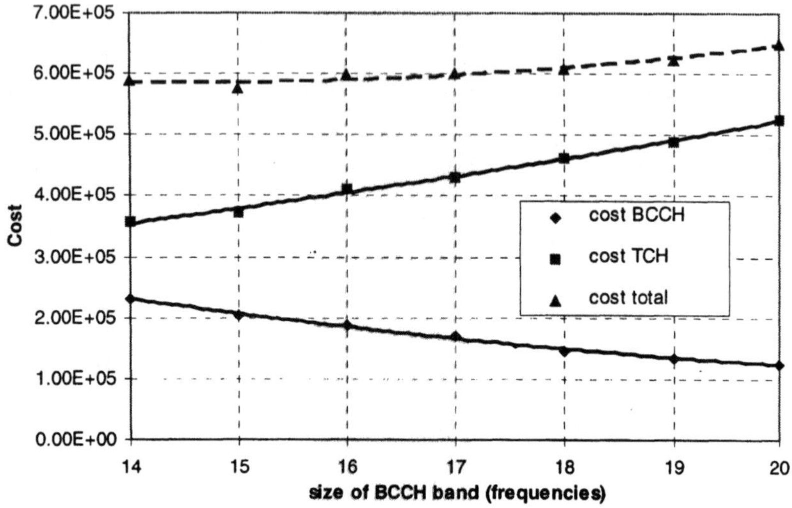

Figure 214. Cost function of both networks, as a function of the size ofthe BCCH band, and the total cost function about 10%, where the largest part comes from the TCH part.

In Figure 215 the frequency plan made by JETTPlan is compared to the reference frequency plan. It shows the cumulative distribution of a quality measure closely related to the FER, here denoted 'pseudo FER'. It can be seen that in both cases 88 % of the area is interference free, while the plan made by JETTPlan gives a slightly better performance in the areas with interference.

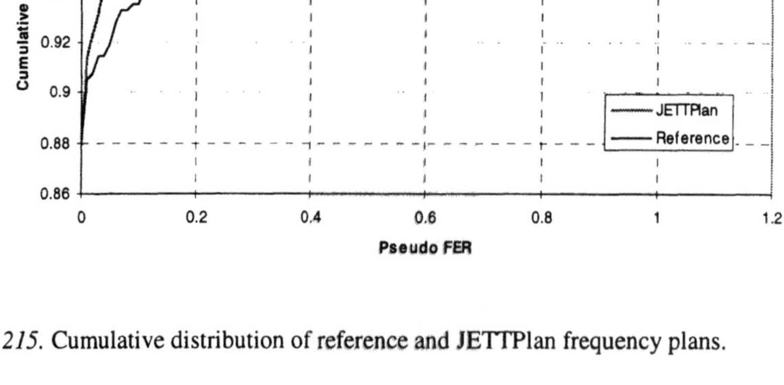

Figure 215. Cumulative distribution of reference and JETTPlan frequency plans.

3.3 Results from Live Network

JETTPlan has been used to make a frequency plan for the SONOFON network in Aalborg, Denmark. About a month after frequency hopping was introduced. In this section this frequency plan shift is described.

The area of the frequency plan shift consists of 34 cells, while 114 cells are taken into account in the interference matrix for the area to take border effects into account. The cost function of the original plan is equal to 67364, if frequency hopping would be used. Not the actual traffic of the cells is taken into account, but the traffic each cell could carry, when the blocking is 1% (1% Erlang-B traffic).

With JETTPlan a frequency plan with a cost equal to 59753 was found, i.e. a gain in cost value of about 13 % was achieved.

The drop rate can be seen in Figure 216. The moments of the frequency plan shift and the introduction of frequency hopping can also be seen. It can be seen that the introduction of the new frequency plan does not seem to effect the drop rate, even though this new frequency plan is designed for a frequency hopping network.

A comparison of the RXQUAL distributions before and after the frequency plan shift indicates no degradation in quality, even though frequency hopping was not switched on from the start.

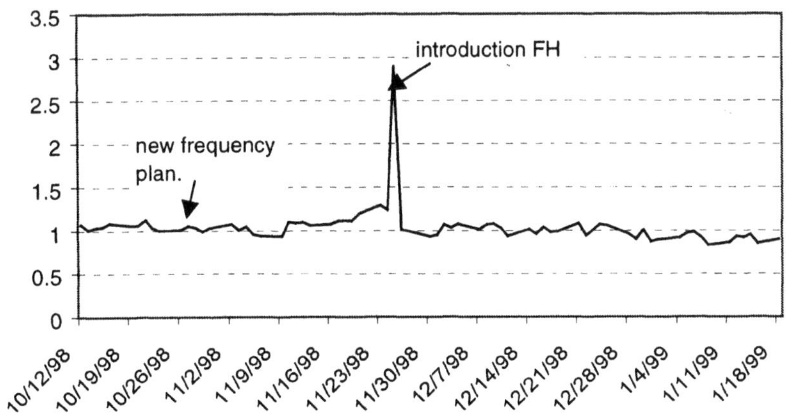

Figure 216. Drop rate before and after the frequency plan shift.

The conclusion is that the frequency plan made by JETTPlan is a success. The network quality is retained, while the frequency plan made by JETTPlan did not need any optimisation, whereas the conventional method of making frequency plan needed a lot of optimisation. The frequency plan used before the shift of the frequency plan were made by hand and optimised during some months. It took 24 hours to make a frequency plan with JETTPlan.

It can be seen that the drop rate increases a lot directly after the frequency hopping is introduced. After some adjustments to the handover algorithm the drop call rate decreased again to the level as before the new frequency plan was introduced. These handover adjustments are like the handover adjustments described in Chapter 8.

After these adjustments the drop rate decreases to about 1 %, where it was about 1.25 % before the introduction of frequency hopping.

4. OTHER PARAMETERS TO BE PLANNED

Not only the frequencies have to be planned in the network, but also several other parameters such as the hopping parameters, training sequences, handover and power control settings are essential. The last two are described in respectively Chapter 7 and 8. The planning of the hopping parameters and training sequences are described in the following 2 subsections.

4.1 Frequency Hopping Parameters

Basically there are two parameters to be planned for each frequency hopping cell, the HSN and MAIO, as described in Chapter 3. There are 64 different HSN leading to 64 different hopping sequences. To ensure random interference the distance between two cells with the same frequencies using the same HSN should be a big as possible, in other words the HSN reuse distance should be kept as high as possible. In the case of grouped frequency planning it is easy, since cells with the same frequencies can easily be recognised.

In the case of non-grouped frequency planning, it is more complex. The best thing to do is here to assign cells, which have a lot of frequencies in common, a different HSN. Even better would be to analyse the existing interference situation of the frequency plan and give the cells, which interfere a lot with each other, a different HSN.

In the case of non-grouped frequency planning it is even possible to use sequential hopping, i.e. use a HSN equal to 0. The interference will come from different cells for every burst since the frequencies are spread around. The advantage of using sequential hopping is that the maximum frequency diversity gain is achieved.

The MAIO parameter, indicating the start in the hopping sequence can be used to avoid adjacent and/or co-channel interference, when RF hopping is used. This is explained with an example. A 3 sector site using 18 frequencies, while only 9 TRXs are distributed equally over the 3 sectors. The 18 frequencies can be used in each of the 3 cells and the HSN number is the same for each of the cells, so their hopping sequence is exactly the same (identical MAI). By setting the MAIO's to the values, shown in Figure 217, no adjacent or co-channel channel interference will occur within this site. A second site could be given the odd MAIO's, so that there is no co-channel interference between the two sites. This needs synchronisation though.

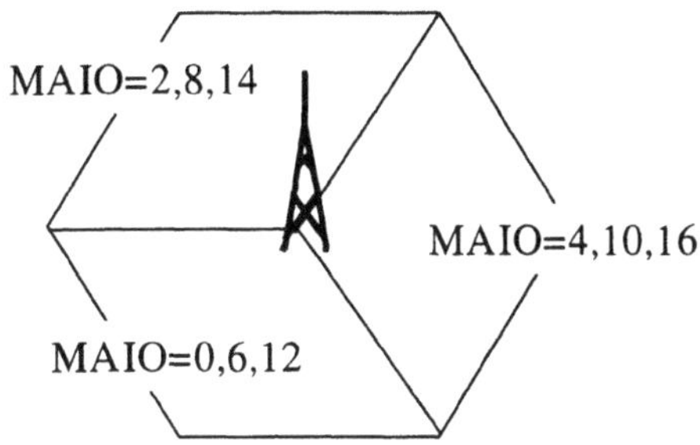

Figure 217. Assigned MAIO's to avoid interference within a site, when the cells have the same hopping sequence.

The optimal MAIO distance between the different sectors in a site, when the cells have the same hopping sequence is equal to:

$$MAIO\_step\_in\_cell = \frac{Number\_of\_frequencies}{Number\_of\_TRXs}$$

$$MAIO\_step\_adj\_cells = \frac{MAIO\_step\_in\_cell}{Number\_of\_\text{sectors}} \quad (3)$$

where *MAIO_step_in_cell* is the MAIO step within one cell (6 in the example) and *MAIO_step_adj_cells* is the step for the adjacent cells 82 in the example).

The MAIO and HSN parameters give other possibilities to optimise the network. So not only the planning of the frequencies is important, but other parameters should be taken into consideration as well.

4.2 Training Sequences

In GSM training sequences have to be assigned to the cells as well. There are 8 different training sequences [64]. In Chapter 4 link simulation results of test conditions were shown. These test conditions can be seen in Figure 218, where C is the carrier and I is the interferer. The interferer is just transmitting random bits, while the carrier (of the desired user) is transmitting random bits with a training sequence (TS) in the middle. In reality however an interferer also will have a training sequence. This can be a different or the same training sequence.

Figure 218. The bursts used for the test conditions of the interferer (I) and the carrier (C).

The influence of the training sequences is studied. Figure 219 shows the FER as function of the *C/I* in the case of test conditions and in case of the interferer having the same or a different training sequence. The desired user is using training sequence 0, while the interferer is using training sequence 0 or 1. 100 % synchronisation is assumed, so interfering and desired user burst arrive at exactly the same time. In the simulations the typical urban profile has been used, while the speed is equal to 3 km/h. Furthermore the results are for random hopping over 8 frequencies and it is assumed that there is no noise.

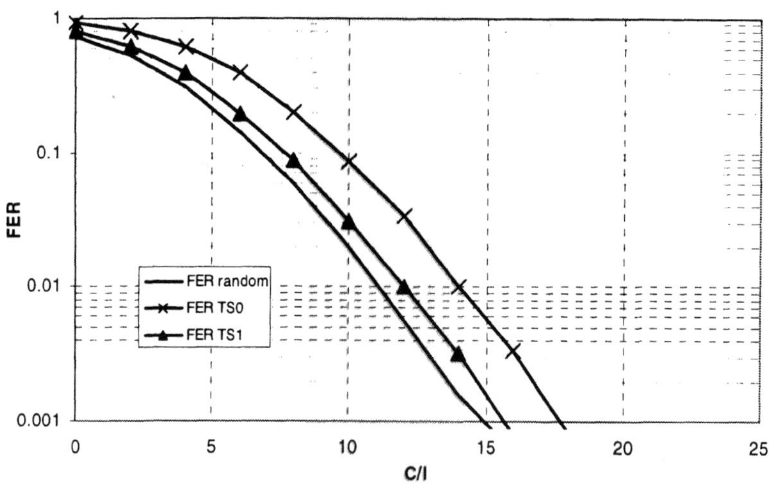

Figure 219. FER as function of the *C/I* in the three different cases.

From the results it can be seen that random bits in the interferer give the best results, followed by the use of another training sequence in the interferer. Using the same training sequence in the interfering burst clearly gives the worst results.

If looked at a 2% FER it can be seen that 3 dB can be lost when using the same training sequence. This is a worst case loss since the interferer and desired burst are 100% correlated.

Figure 220 show the situation, when the interferer is using the same training sequence as the desired user, but they are not 100% synchronised. The FER as function of the difference of arrival time (in bits) between the two signals is shown. So +10 bits shift means that the interfering burst arrives 10 bit times later at the receiver. The Interference is constant, so no interference free areas occur.

It can clearly be seen that if the interfering burst arrives within 5 bit from the desired burst a huge degradation occurs (corresponding to about 3 dB).

In synchronised networks this means that there is a problem, since 1 bit time corresponds to about 1100 meters, so all interfering signals within 5.5 km are having an offset of less than 5 bit. In city areas there are many cells within this range, so 8 training sequences may not be enough to solve this problem.

In a non-synchronised network it is different. On average only in 10/156.25 = 6.4% of the cases the offset will be less than 5 bits. So the average degradation is 0.25 dB. However there is a problem in the downlink since the timing is fixed in that case, i.e. the offset in timing between interferer and desired user is constant. If this offset is small and the same training sequence is used, some performance loss will occur.

The conclusion is that training sequences should be carefully planned to avoid strong interferers using the same training sequence as the desired user.

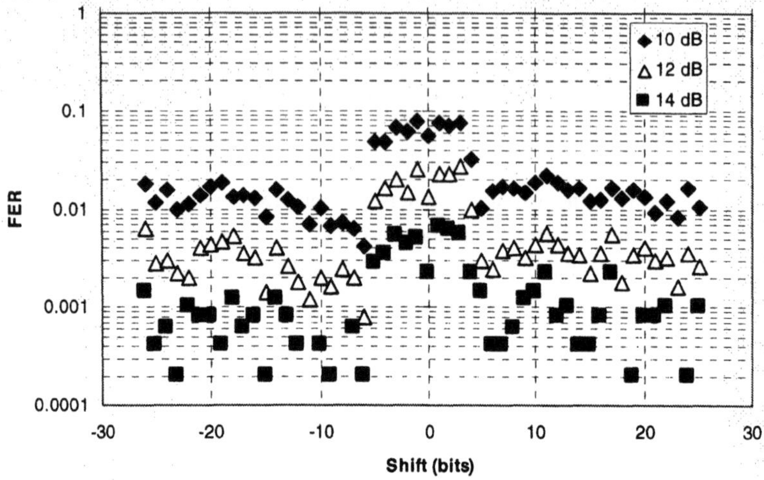

Figure 220. FER as a function of the shift between interferer and desired user for a *C/I* equal to 10, 12 and 14 dB.

5. CONCLUSIONS AND IMPROVEMENTS

This section summarises the results of this chapter. During the development of JETTPlan a lot of different ideas came up. Some of them are implemented while others are not. Some of the ideas that have not been implemented can be found in this section under 'Future Improvements'.

5.1 Summary

Frequency planning for a frequency hopping network differs from frequency planning for non hopping networks. In a non hopping network, the frequencies can be planned individually, but in a frequency hopping network the combined effect of all frequencies has to be taken into account. A frequency planning tool, doing this and including the gains from frequency hopping, has been developed

This frequency planning tool, which falls in the class of heuristic algorithms, is called JETTPlan. It is a search tree, which is optimised for frequency planning. This can be seen in Figure 221. First a random branch is chosen and a solution is found (point A1), then for a certain time this branch is optimised, which gives the solution at point A2. After expiration of the timer a new random branch is chosen and the process continues. Now for example point B or C is found. The searching is stopped, when a predefined timer expires or if the user accepts a solution.

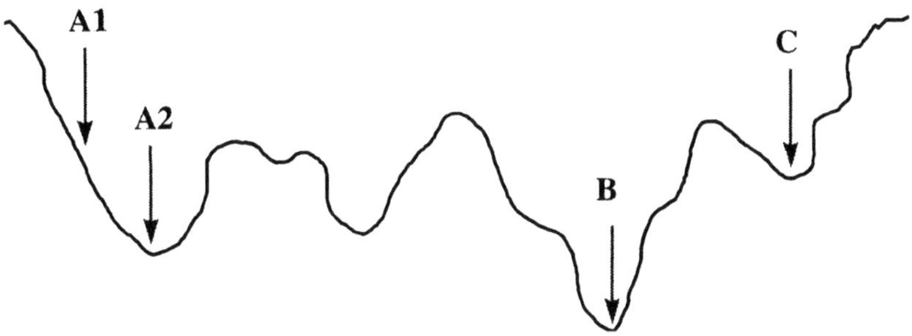

Figure 221. The process treated by the search tree.

The program uses predictions, which indicate how much interference arises from one cell to another, as input. A traditional commercially available planning tool has been used to generate this input, but other input, such as e.g. actual measurements could also have been used.

It was chosen to use band splitting: a special part of the spectrum is dedicated for the BCCH frequencies and the rest can be used for TCH frequencies. In general splitting the band is not a good idea, since there a less combinations to chose from, but the TCH and BCCH frequencies are 2 very different channels and to ensure a good BCCH layer, it is chosen to use band splitting anyway. Results show that more than proportional part of the spectrum should be used for the BCCH layer, if frequency hopping, power control and/or DTX are used. This is due to the fact that frequency hopping gives a higher gain to the TCH frequencies than to the BCCH frequencies. Furthermore DTX and power control can give an extra gain to the TCH frequencies, where the BCCH frequencies do not get any gain.

An automatic constraint settings algorithm is used to speed the search tree up, while a cost function is used to evaluate how good a result is. This cost function represents the amount of interference in the network and includes adjacent and co-channel interference.

The gain from frequency hopping is included by a simple model. The gain is split into a gain from frequency diversity and interference diversity. The frequency diversity gain is speed dependent, while the load of the individual cells gives the interference diversity gain.

When visualising frequency plans, in a non-hopping network it was enough to show the worst *C/I* in every pixel. For frequency hopping networks a new quality measure is introduced, the pseudo FER, to be used instead of worst *C/I*, to visualise the quality of a frequency plan. This pseudo FER is closely related by the FER since effect from hopping over several frequencies with different *C/I* is taken into account. With this new measure it is possible to show the quality of a frequency hopping network.

The frequency plans made by JETTPlan are compared to frequency plans made by commercial tools and by manual planning. The results showed that the plans made by JETTPlan are at least as good.

A frequency plan made by JETTPlan also has been implemented in the SONOFON network in Aalborg, Denmark. The quality in the network remained as it was in the old plan. The old plan however was made by hand over several months, while the new plan was made automatically within 24 hours.

Not only frequencies have to be planned, but also a couple of other parameters, like the hopping parameters and training sequences have to be planned. It is described that these parameters can be used to improve the performance even further.

5.2 Future Improvements

Although the planning program gives good results, a lot of improvements have been though of. Some of them are described in this section.

5.2.1 Input Improvements

The output of the tool depends on the input. If the input is not good, the output can not be good. So if the input can be improved, the output can get better as well. As described, the way the typical interference matrix is made gives us some limitations. The absolute level of the *C/I* values was not taken into account, but they were just compared to a threshold. This means that JETTPlan just tries to minimise the area, which is having a *C/I* worse than this threshold, i.e. 9 dB. JETTPlan is not trying to minimise the level of interference in the area, since it can not see if the *C/I* is 8 dB or -8 dB. It would be an improvement to be able to take the level of every pixel into account, when making the interference matrix. Weighting the amount of affected traffic by how much the traffic is affected can for example do this. The gains from frequency hopping can easily be taken into account this way. In the current situation the gains from frequency hopping just decrease the amount of affected traffic. So even though the affected traffic is suffering from a *C/I* equal to –8 dB, a gain of 3 dB simply halves the amount of affected traffic like it now has a *C/I* which is better than the threshold 9 dB. By taking the level into account, the new amount of affected traffic can be calculated after a gain is applied.

Overlap of interference coming from 2 or more different cells is not possible to model either. If two cells affect the same traffic, the program simply sees it as the amount of affected traffic being twice as big. Taking the absolute level of interference into account can solve this problem. The traffic affected in a cell by two other cells is simply summing the amount of affected traffic weighted by the level of affection. Another improvement to the input would be to take the real traffic distribution into account, but here the practical problem occurs, that this distribution is not known very accurately.

Of course the main improvement would be a good prediction model, but that would mean higher costs, since more accurate clutter data had to be known. A very simple and practical way of avoiding this problem would be to use the measurements made by the mobile stations into account. A method of doing this was presented in [201]. The interference matrix is simply calculated from RXQUAL measurements made by the mobiles. The great advantage of this method is that measurements are used, so no prediction errors are made. Another advantage is that the measurements are made on the places, where the mobile stations actually are, so the traffic distribution is included for free!

5.2.2 Band Splitting Improvement

This improvement was already mentioned in Section 2.3. Instead of just splitting the frequency band into 2 parts for the TCH and BCCH respectively, interleaving can be used. This means that the BCCH frequencies are interleaved between the TCH frequencies. This way adjacent interference between BCCH frequencies is avoided. However adjacent interference from TCH channels can occur in this situation, so the quality of the BCCH band becomes slightly load dependent.

5.2.3 RF Hopping Enhancements

JETTPlan is able to deal with RF hopping, but it is not possible to optimise the length of the MA list, the hopping sequences with the tool. Different lengths can be tried out and the cost function can be compared afterwards, but a tool, which would find the optimal length, would be very valuable. To add it to the existing tool, is a major task, since it adds a dimension: not only the frequencies can be varied, but also the amount of frequencies per cell.

5.2.4 Constraints Improvements

With the automatic constraint settings, as explained in Section 2.4.3, it is possible that a situation occur in which a cell with not much traffic, i.e. a cell in the bottom of a frequency plan diagram, has it frequencies planned so that the affected traffic over carried traffic ratio is very high. This is not desirable, even though it does not influence much traffic. Therefore a certain minimum quality per cell should be guaranteed, i.e. no cells should have an affected traffic over carried traffic ratio worse than a certain minimum.

In this automatic constraints settings a straight line is being used. No other line forms are tried, but it might very well be that another shape works better.

REFERENCES

[1] **Aalto R,** *DX-200 IUO handover - functional descriptions*, Nokia internal document, 1996.

[2] **Aalto R.,** *DX-200 RF Power control and handover - functional descriptions*, Nokia internal document, March 1996.

[3] **Aarts E. and J. Korst**, *Simulated annealing and Boltzmann Machines*, Wiley, 1989.

[4] **Aein, J.M.,** *Power balancing in systems employing frequency reuse*, COMSAT Tech. Rev. Vol. 3, No. 2, Fall 1973.

[5] **Almgren M, H. Andersson and K. Wallstedt,** *Capacity enhancements in a TDMA system*, IEEE Proc of VTC, 1993, pp 277-280.

[6] **Almgren M., H. Andersson and K. Wallstedt.** *Power Control in a Cellular System.* IEEE Proc. VTC, 1994, pp 833-837.

[7] **Amin S.**, *Simulated Jumping*, to appear in Journal of operational research.

[8] **Andersin M., Z. Rosberg and J. Zander**, *Distributed Discrete Power Control in Cellular PCS. TRITA-IT R 94-23, ISSN 1103-534X, ISRN KTH/IT/R-94/23-SE,* Radio Commun. Systems Lab., KTH, Sweden, Oct. 1994 (Submitted to Wireless Personal Communications).

[9] **Andersin M., Z. Rosberg and J. Zander.** *Soft Admission in Cellular PCS with Constrained Power Control and Noise*, In Proc. 5th WINLAB Workshop, Apr. 1995.

[10] **Argyropoulos Y , T. Jeanes and W-H. F. Leung,** *A comparative study of several capacity enhancement techniques in GSM,* IEEE Proc. VTC, 1997, pp. 1847-1851.

[11] **Arregui A., J. Dunlop**, *Distributed Call Admission Control in Partially Loaded TDMA Systems*, IEEE Proc of VTC'98 Ottawa, pp.1361-1365.

[12] **Athanasiadou G.E., A.R. Nix and J.P. McGeehan**, *A Ray Tracing Algorithm for Microcellular and Indoor Propagation Modelling*, ICAP'95, IEE Conference Publication No.407, pp. 231-235.

[13] **Austin M.D. and G.L. Stuber** *Velocity Adaptive Handover Algorithms for Microcellular Systems*, IEEE Proc. ICUPC '93, Ottawa, Canada, Oct 12-15, 1993, pp. 793-797.

[14] **Axén, H.,** *Power Control in Cellular Mobile Telephone Systems*, Diploma Thesis, Ericsson Radio Systems, Internal report, BT/SU 90:1708, 1990.

[15] **Axén, H.**, *Uplink C/I as Control Parameter for Mobile Station Power Control,* Ericsson Radio Systems, Internal report, BT/ST 90:2377, 1990.

[16] **Baier A. and K. Bandelow,** *Traffic engineering and realistic network capacity in cellular radio networks with inhomogeneous traffic distribution,* IEEE Proc. VTC, 1997, pp. 780-784.

[17] **Bajwa A.,** *Determining the gains to be made by optimising frequency re-use patterns and implementing frequency hopping and discontinuous transmission.* In Proc. of IIR Conf.: Maximising Capacity in Mobile Networks, No. 17, London, Nov. 1996.

[18] **Bambos N. and G. J. Pottie.** *On Power Control in High Capacity Cellular Radio Networks.* Proc. 3rd WINLAB Workshop, Oct. 19-20, 1992.

[19] **Bambos N., S. C. Chen and G. J. Pottie,** *Channel Access Algorithms with Active Link Protection for Wireless Communication Networks with Power Control.* Technical Report, UCLA-ENG-95-114, Department of Electrical Engineering, UCLA, 1995.

[20] **Beming P. and M. Frodigh,** *Admission Control in Frequency Hopping GSM Systems,* IEEE Proc. VTC 1997, Phoenix, pp. 1282-1286.

[21] **Benveniste M.** *Cell Selection in Two-Tier Microcellular/Macrocellular Systems,* IEEE Proc. Globecom '95, Singapore, Nov. 1995, pp 1532-1536.

[22] **Bock, F., Ebstein, B.,** *Assignment of transmitter power control by linear programming,* IEEE Trans Electromag. Compt, EMC-6, 2, 36,1964.

[23] **Box F.,** *A heuristic technique for assigning frequencies to mobile radio nets,* IEEE Trans. Veh. Tech., Vol. VT-27, May 1977, pp. 57-64.

[24] **Brockmeyer E., H.L. Halstrøm and A. Jensen,** *The life and works of A.K.Erlang.* Copenhagen 1948.

[25] **Brockmeyer E.,** *Simple overflow problem in the theory of telephone traffic,* Teleteknik. No. 5, 1954.

[26] **Brøndum J.,** *GSM Data Receiver Structures; a Comparison,* Internal AUC report R93-1002, Nov. 1993.

[27] **Brown P.G., C.C. Constantinou,** *Prediction of Radiowave Propagation in Urban Microcell environments using Ray Tracing Methods,* ICAP'95, IEE Conference Publication No.407, pp. 373-376.

[28] **Carlson A.B.,** *Communication systems,* 3. Edition, McGraw-Hill, International Student Edition, 1986.

[29] Carneheim C., S.O. Jonsson, M. Ljungberg, M. Madsfors and J. Näslund, ***FH-GSM Frequency Hopping GSM,*** **IEEE Proc. VTC'94, Stockholm, pp. 1155-1159.**

[30] **Chang K.N. and S. Kim,** *Call blocking performance of new dynamic channel assignment scheme in microcellular radio networks,* IEEE Proc. PIMRC, pp. 198-202, 1996.

[31] **Chia S.T.S.,** *Radio and system design for a dense urban personal communications network,* Electronics and Communication Engineering Journal, IEE, Aug. 1996.

[32] **Chu T-P. and S.S. Rappaport,** *Overlapping coverage with reuse partitioning in microcellular communication systems,* IEEE Proc. PIMRC, 1995, pp. 557-561.

[33] **Chuah C.N. and R.D. Yates,** *Evaluation of a Minimum Power Handoff Algorithm,* IEEE Proc PIMRC'95 Toronto, pp. 814-818.

[34] **Chuah C.N., R.D. Yates and D.J. Goodman,** *Integrated Dynamic Radio Resource Management,* IEEE Proc. VTC'95, pp 584-588.

[35] **Chuang J.C.**, *Operation and performance of a self-organizing frequency assignment method for TDMA portable radio,* IEEE Proc. GLOBECOM'90, pp 1548-1552.

[36] **Clarke H.,** *A statistical theory of mobile radio reception*, BSTJ, No. 47, pp. 957-1000, 1968.

[37] **ComOpt AB,** *CellOpt - User Documentation*, June 1997.

[38] **COST-207,** *Digital land mobile radio communication,* Final report of EURO COST project, No. 207, EIR 12160 EN, ISBM: 92-825-9946-9, 1988.

[39] **Cox D.C. and D.O. Reudink,** *Dynamic channel assignment in High Capacity Mobile Communication Systems,* Bell. System. Tech. Journal, Vol. 50, No. 6, 1971.

[40] **Cox D.C. and D.O. Reudink,** *Increasing channel occupancy in large scale mobile radio systems: dynamic channel reassignment,* IEEE Trans. Veh. Technol. VT-22, 1973.

[41] **Cox D.C., D.O.Reudink,** *A comparison of some Channel Assignment Strategies in Large Scale Mobile Communications Systems,* IEEE Trans. On. Comm., VOL COM-20, NO.2, Feb. 1972.

[42] **Crossley D., P. Stevens, D. Owens, D. Skinner and C. Haines,** *Frequency hopping results from live DCS1800 network,* Proc. EPMCC 1995, Italy, pp. 103-106.

[43] **Dahlin J.**, *Ericsson's multiple re-use pattern for DCS 1800*, Mobile communications international, Nov. 1996.

[44] **Daikoku K. and H. Ohdate,** *Optimal channel reuse in cellular land mobile radio systems,* IEEE Trans. Veh. Technol., vol. VT-32, pp. 217-224, Aug. 1983.

[45] **Dam H.,** *Smart Antennas in the GSM System,* M.Sc.E.E. Thesis Report, Aalborg University, June 1995.

[46] **Dasilva J.S., B. Arroyo, B. Barani, D. Ikonomou et al.,** *European Third-Generation Mobile Systems*, IEEE Comm. Mag., Oct. 1996, pp 68-78

[47] **Del re E., R. Fantacci and G. Giambene**, *Handover and Dynamic Channel Allocation Techniques in Mobile Cellular Networks*, IEEE Trans. Veh. Tech. VOL 40, No.2, May 1995, pp 229-237.

[48] **Delisle, G.Y. and others,** *Propagation Loss Prediction: A comparative Study With Application to the Mobile Radio Channel*, IEEE Trans. Veh Tech., Vol VT-34, 1985, pp. 86-96.

[49] **Dell'Anna M., B.B. Mohebbi, A.E. Brand and A.H. Aghvami,** *A study of polarisation diversity for a microcellular environment,* In Proc. of IEEE PIMRC, Sept 1997, pp. 575-579.

[50] **Divies R., M. Bensebti, M.A. Beach and J.P. McGeehan,** *Wireless propagation measurements in indoor multipath environments at 1.7 GHz and 60 GHz for small cell systems,* in Proc. IEEE VTS Conf., pp. 589-593., St. Louis, MO, USA, 1989.

[51] **Dunlop J. et al.**, *Estimation of the Performance of Link Adaptation in Mobile Radio*, in Proc IEEE VTC'95, Chicago, 1994, pp 326-330.

[52] **Duque-Anton M., D. Kunz and B. Ruber**, *Channel assignment for cellular radio using simulated annealing*, IEEE Trans. on Veh. Tech., Vol. 42, No. 1, Feb. 1993, pp. 14-21.

[53] **Ebert H.,** *Fejlkontrolkodning*, Course note for M.Sc.E.E. telecommunications students at Aalborg University, (in Danish), Aalborg University, Feb. 1987.

[54] **Eggers P.C.F., J. Toftgård and A.M. Oprea,** *Antenna systems for base station diversity in urban small and micro cells,* IEEE Journal on selected areas in communications, Vol. 11, No. 7, Sept. 1993, pp 1046-1057.

[55] **Eisenblätter A.**, *Combinatorial Lower Bounds for Co Channel Interference in GSM Networks*, COST(98) TD104, Duisburg, Sept. 1998.

[56] **Eklundh B.**, *Channel Utilization and Blocking Probability in a Cellular Mobile Telephone System with Directed Retry*, IEEE Trans. Commun., Vol. 34, No. 4, pp.329-337, Apr. 1986.

[57] **Engels S.,** *Measuring Quality of Service in Dual Band Networks*, Conference on Maximising Capacity in Mobile Networks, Cannes, France, Dec. 1998.

[58] **Erätuuli P. and E. Bonek,** *Diversity Arrangements for Internal Handset Antennas*, In Proc. of IEEE PIMRC, Sept. 97, Finland, pp. 589-593.

[59] **Erlang A.K.** *Løsning af nogle problemer fra sandsynlighedsregningen af betydning for de automatiske telefoncentraler.* Elektroteknikeren, Vol. 13, 1917, p. 5, (in Danish).

[60] **ETSI TC-SMG**, *GSM 03.64: Overall Description of the GPRS Air Interface*, version 5.1.0, 1995.

[61] **ETSI TC-SMG**, *GSM 05.01: European digital cellular telecommunication system (Phase 2) – Physical layer on the radio path, General description*, March 1995.

[62] **ETSI TC-SMG**, *GSM 05.02: European digital cellular telecommunication system (Phase 2) - Multiplexing and Multiple Access on the Radio Path*, Sept. 1994.

[63] **ETSI TC-SMG**, *GSM 05.03: European digital cellular telecommunication system (Phase 2) - Channel Coding*, Sep. 1994.

[64] **ETSI TC-SMG**, *GSM 05.05: European digital cellular telecommunication system (Phase 2) - Radio Transmission and Reception*, March 1995.

[65] **ETSI TC-SMG**, *GSM 05.08: European digital cellular telecommunication system (Phase 2) - Radio Sub-System Link Control*, March 1996.

[66] **European Radiocommunications Office (ERO)**, Final Document for the ERO Workshop on Traffic Loading and Efficient Spectrum Use, 1998.

[67] **Fenton L.F.,** *The sum of log-normal probability distributions in scatter transmission systems,* IRE Trans. Commun. Syst., Vol. CS-8, Mar. 1960, pp. 57-67.

[68] **Foschini G.J.** *A Simple Distributed Autonomous Power Control Algorithm and its Convergence.* IEEE Trans. on Veh. Tech., Vol. 42, No. 4, 1993, pp 641-646.

[69] **Fujii, T., Sakamoto, M.,** *Reduction of Cochannel Interference in Cellular Systems by Intra-zone Channel Reassigment and Adaptive Transmitter Power Control*, Proc IEEE Veh Tech Conf, VTC-88, 1988, pp. 668-672.

[70] **Furuskär A., M. Frodigh, H. Olofsson and J. Sköld**, *System Performance of EDGE, a Proposal for Enhanced Data Rates in Existing Digital Cellular Systems,* IEEE Proc VTC'98, Ottawa, pp. 1284-1289.

[71] **Gardiner J. and B. West** , *Personal communication systems and technologies*, Artech House, London, 1995.

[72] **Ghaheri Niri S. and R. Tafazolli,** *Cordless - Cellular Inter System Handover Optimisation*, IEEE Proc. VTC'98, Ottawa, pp.653-657.

[73] **Giancristofaro D., F. Ruggieri and F. Santucci,** *Queuing of handover requests in micro-cell network architectures,* Proc. IEEE VTC, pp. 1846-1849, 1994.

[74] **Giortzis A.I. and L.F. Turner**, *A mathematical programming approach to the channel assignment problem in radio networks*, Proc. IEEE VTC., Atlanta, April 1996, pp. 736-740.

[75] **Glover F., E. Taillard and D. de Werra**, *A user's guide to Tabu search, Annals of operations research*, Vol. 41, 1993, pp. 3-28.

[76] **Goldsmith A.J., L.J. Greenstein and G.J. Foschini,** *Error statistics of real time power measurement in cellular channels with multipath and shadowing*, IEEE Proc. VTC '93, New Jersey, pp 108-111.

[77] **Goodhead H., D. Griggs, J. Maipath and J. Samuelsson.** *Examining the Key Features of Next Generation Automatic Frequency Planning Tools*, In Proc. of IIR Conf. on Advanced Radio Frequency Planning, Sept. 1997, London.

[78] **Gotzner U., A. Gamst, R. Rathgeber**, *Spatial traffic distribution in cellular networks*, IEEE Proc. VTC'98, Ottawa, pp.1994-1998.

[79] **Grandhi S.A., J. Zander and R. Yates.** *Constrained Power Control.* TRITA-IT R 94-06, Dept. of Teleinformation, Radio Commun. System Lab., June 1994.

[80] **Grandhi S.A., R. Vijayan and D. J. Goodman.** *A Distributed Algorithm for Power Control in Cellular Radio Systems.* 13th Annual Allerton Conf. on Commun., Contr. and Computing, Monticello, Illinois, Sept. 1992.

[81] **Grandhi S.A., R. Vijayan, D. J. Goodman and J. Zander.** *Centralized Power Control in Cellular Radio Systems.* IEEE Trans. on Veh. Tech., Vol. 42, No. 4, 1993.

[82] **Grimlund O. and B. Gudmundson,** *Handoff Strategies in Microcellular Systems*, IEEE Proc. VTC'91, St.Louis, pp.505-510.

[83] **Grimlund O., J.E. Berg and B. Gudmundson**, *Handover Performance in Microcellular Systems*, Proc. MRC 1991, pp.126-136.

[84] **Gudmundson M.,** *Analysis of Handover Algorithms* Proc. Vehicular Tech. Conf. '91, St. Louis, MO, May 19-22,1991, pp.537-42.

[85] **Gudmundson M.,** *Correlation model for Shadow fading in Mobile Radio Systems*, Electronic Letters, Vol 27., No.23, Nov. 1991, pp 2145-2146.

[86] **Guérin R.**, *Queueing-Blocking System with Two Arrival Streams and Guard Channels*, IEEE Trans. on Comm, VOL. 36, No. 2, February 1988, pp 153-163.

[87] **Halpern S.W.,** *Reuse partitioning in cellular systems,* IEEE Proc. VTC. 1983, Toronto, Ontario, Canada, May 25-27, pp. 322-327.

[88] **Hanly S.V.,** *An Algorithm for Combined Cell-site Selection and Power Control to Maximize Cellular Spread Spectrum Capacity.* IEEE Journal on Selected Areas, special issue on fundamentals of networking, 1995, pp 1332-1340.

[89] **Hanly S.V.,** *Information Capacity of Radio Networks.* Ph.D. Thesis, Cambridge University, Aug. 1993.

[90] **Hata M.,** *Empirical Formula for Propagation Loss in Land Mobile Radio Services*, IEEE Trans. Veh. Tech., VOL VT-29, No.3, Aug. 1980.

[91] **Haykin S.,** *Communication Systems,* John Wiley & Sons, Second Edition, ISSN 0271-6046.

[92] **Heeralall S. and C.J. Hughes,** *High capacity cellular patterns for land mobile radio systems using directional antennas,* In IEE Proc., Vol. 136, pp. 75-80, No.1, 1989.

[93] **Heinonen M.,** *Current markets, regulation and Finnish accounts for 3rd generation mobile networks,* Nordic Mobile Conference, Feb. 1999.

[94] **Hills D.,** *Maximizing capacity to manage the rapid growth in subscribers: Identifying and evaluating the different solutions,* In Proc. IIR Conf.: Maximizing Capacity in Mobile Networks, No. 17, London, Nov., 1996.

[95] **Ho K.P., and J.M.Kahn,** *Combined source-channel coding using channel-optimized quantizer and multi-carrier modulation,* In IEEE ICC Proc., pp. 1323-1327, 1996.

[96] **Hong D. and S.S. Rappaport,** *Traffic Model and Performance Analysis for Cellular Mobile Radio Telephone Systems with Prioritized and Nonprioritized Handoff Procedures,* IEEE Trans. Veh. Tech., VOL VT-35, No.3., Aug. 1986, pp 77-92.

[97] **Huang W., and M. Nakagawa,** *Channel coding with MPSK/Orthogonal Multi-carrier Transmission,* IEEE Proc. ICC , pp. 1313-1317, 1996.

[98] **Huo D.,** *Simulating slow fading by means of one dimensional stochastic process,* IEEE Proc. VTC '96.

[99] **Ibrahim M.F. and J.D.Parsons,** *Signal strength prediction in Build Up areas: Part 2 Signal Variability,* In Proc. of IEE VOL-130, Pt. F, No. 5, pp. 385-391.

[100] **Iera A., S. Marano, A. Molinaro and G. Serra**, *Effective radio resource assignment in microcellular systems with cell overlapping: The reserve channels scheme.* IEEE Proc. VTC 1997, pp. 964-968.

[101] **Inglis A.F.** *Electronic Communications Handbook.* McGraw-Hill Inc., ISBN 0-07-031711-9, 1988.

[102] **Ivanov K. and G. Spring** *Mobile Speed Sensitive Handover in a Mxed Cell Environmnet,* IEEE Proc. VTC '95, Chicago, July 25-28, 1995, pp. 892-896.

[103] **Iversen V.B.,** *Data- og Teletrafikteori; Lærebog* (In Danish), Inst. for Telekommunikation, DTU, Jan. 1997.

[104] **Jakes Jr. W.C.,** *Microwave mobile communications,* John Wiley & Sons, New York, 1974, ISBN: 0-471-43720-4.

[105] **Johansen J. and B.Vejlgaard**, *Capacity Analysis of a Frequency Hopping GSM System.* M.Sc.E.E. Thesis Report, Aalborg University, June 1995.

[106] **Kahwa T.J. and N.D. Georganas,** *A hybrid channel assignment scheme in large-scale cellular-structured mobile communication system,* IEEE Trans. Comm., Vol. Com-26, No. 4, 1978.

[107] **Kapoor V., G. Edwards, R.Sankar,** *Handoff Criteria for Personal Communication Networks,* Proc. ICC'94, New Orleans, LA, May 1-5, 1994, pp 1297-1301.

[108] **Kawabata K., T. Nakamura T. and E. Fukuda** *Estimating Velocity Using Diversity Reception* IEEE Proc. VTC '94, Stockholm, Sweden, 1994, pp. 371-374.

[109] **Kegel A. and J.C. Arnbak,** *Communicatie-systemen b.o.,* TU Delft, vakgroep Telecommunicatie en Verkeersbegeleidingssystemen, 1992.

[110] **Klingenbrunn T.,** *Capacity in a WLL network based on GSM,* CPK Internal Report, June 1998.

[111] **Kloch C.,** *Influence of surface irregularities on propagation prediction models used in mobile communications,* Ph.D. Thesis, Aalborg University, CPK, Nov. 1997.

[112] **Knudsen J.B. and A.S. Nielsen**, *Soft Output Viterbi data receiver for Digital Mobile Communications*, M.Sc.E.E. thesis, AUC June 1992 (in Danish).

[113] **Kolonits A.**, *Determining the full potential of cell sectorisation and cell splitting,* In Proc. IIR Conf.: Maximising Capacity in Mobile Networks, No. 17, London, Nov., 1996.

[114] **Kronestedt F. and M. Frodigh**, *Frequency planning strategies for frequency hopping GSM,* IEEE Proc. VTC '97, pp. 1862-1866.

[115] **Kuhn, A. W.T. Haggerty, U. Grage, M. Keller, C. Reyering, F. Arndt and W. Scholtholt** , *Validation of the feature frequency hopping in a live GSM network,* IEEE Proc. VTC '96, pp. 321-325.

[116] **Kumar P.S., R. D. Yates and J. Holtzman.** *Power Control Optimal Based on Bit Error Rate (BER) Measurements.* Technical Report, WINLAB, Rutgers University, NJ, USA, 1995.

[117] **Kunz D.**, *Channel assignment for cellular radio using neural networks,* In IEEE Trans. Veh. Tech., Vol. 40, No.1, Feb. 1991, pp.188-193.

[118] **Kunz D.**, *Suboptimum solutions obtained by the Hopfield-Tank Neural Network Algorithm, Biological.* Cybernetics 1991, Vol. 65, pp. 129-133.

[119] **Lachat E., Wagen J.F. et al.,** *Analysis of downlink power control in a GSM system,* IEEE Proc. VTC '98, pp. 2306-2310.

[120] **Lagrange X. and P.Godlewski,** *Performance of a Hierarchical Cellular Network with Mobility-Dependent Handover Strategies,* IEEE Proc. of VTC'96, pp 1868-1872.

[121] **Larsson P., H. Olofsson**, *Performance Evaluation of Different Frequency Reuse Patterns and Channel Coding Schemes in GPRS,* IEEE Proc VTC'98, Ottawa, pp. 139-143.

[122] **Lee W.C.Y.,** *Mobile Cellular Telecommunication Systems,* McGraw-Hill Book Company. ISBN: 0-07-037030-3, 1989.

[123] **Lee W.C.Y.,** *Mobile Communication Engineering,* McGraw-Hill Book Company, ISBM: 0-07-037039, 1982.

[124] **Lin S. and D.J. Costello,** *Error control coding: Fundamentals and applications,* Prentice-Hall 1983, ISBN 0-13-283796-X.

[125] **Ling Y., J. Elling, T.T. Nielsen**., *Capacity of Intelligent Underlay and Overlay Network.* In Proc. of IEE Colloquium on Advanced TDMA Techniques and Applications, London, Oct. 1996, pp. 23-31.

[126] **Lochtie G.D., C.A. van Eijl, M.J. Mehler**, *Comparison of energy minimising algorithms for channel assignment in mobile radio networks,* IEEE Proc. of PIMRC'97, Helsinki, pp. 786-790.

[127] **Lochtie G.D., M.J. Mehler**, *Subspace approach to channel assignment in mobile communications*, IEE Proceedings Communications, VOL.142, No.3, 1995, pp.179-185.

[128] **Lopez R.,** *Performance predictions for cellular switched-beam intelligent antenna systems,* In IEEE Comm. Magazine., Oct. 1996 pp 152-154.

[129] **MacDonald V.H.,** *Advanced mobile phone service; the cellular concept.,* Bell System. Tech. J. Vol. 58, pp. 15-41, No. 1, 1979.

[130] **Madfors M., K. Wallstedt, S. Magnusson, H. Olofsson, P. Backman and S. Engström**, *High Capacity with Limited Spectrum in Cellular Systems*, IEEE Comm. Magazine, August 1997, pp. 38-45.

[131] **Manabe T., K. Taira, K. Sato, T. Ihara, Y. Kasashima and K. Yamaki,** *Multipath measurements at 60 GHz for indoor wireless communication systems,* IEEE Proc. VTC '94, pp. 905-909.

[132] **McFarlance D.A.,** *European personal communication using micro cells,* In Proc. IEE colloquium on micro cellular mobile radio, London, pp. 2/1-2/4, Feb. 1989.

[133] **McGeehan J.P. and K.W. Yates,** *High-capacity 60 GHz micro cellular mobile radio systems,* Telecommunications, Vol. 20, pp. 58-64, 92, No. 9, 1986.

[134] **Mehrotra A.,** *Cellular radio – analog and digital systems,* Artech House, London, 1994, ISBN: 0-89006-731-7.

[135] **Mehrotra A.,** *Cellular Radio Performance Engineering,* Artech House, London, 1994, ISBN: 0-89006-748-1.

[136] **Mehrotra A.,** *GSM System Engineering,* Artech House, Boston, 1997, ISBN:0-89006-860-7.

[137] **Melero J. and T. Halonen,** *Frequency hopping, system performance analysis,* Conf. on frequency hopping and reuse in mobile networks, Amsterdam, 1999.

[138] **Meyerhoff, H.J.**, *Method for Computing the Optimum Power Balance in Mulitbeam Satellite,* COMSAT Tech Rev, Vol 4, No 1, Spring 1974.

[139] **Michaelsen P.H.,** *Course material for Capacity Training Course,* 1999.

[140] **Michealsen P.H.,** *CAPACITY GSM Simulator – output file structure,* CPK internal document, Feb. 1999.

[141] **Mitra D.** *An Asynchronous Distributed Algorithm for Power Control in Cellular Radio Systems.* Proc. 4th WINLAB Workshop, Oct. 19-20, 1993.

[142] **Mogensen P.**, *GSM Base-Station Antenna Diversity using Soft Decision Combining on Up-link and Delayed-signal Transmission in Down-link,* IEEE Proc. VTC, 1993, pp. 611-616.

[143] **Mogensen P.**, *GSM Link Level Studies,* Internal CPK Document, 1992

[144] **Mogensen P., Pedersen K.I.** *Some Aspects of CDMA Technology,* UMTS Workshop, IDA, Aalborg, March 1998.

[145] **Mogensen P.E. and J. Wigard,** *On antenna and frequency diversity in GSM related systems (GSM-900, DCS-1800 and PCS-1900),* IEEE Proc. PIMRC '96, Taipei, pp 1272-1276.

[146] **Mogensen P.E.,** *Determining the full potential and effectiveness of smart antennas,* In Proc. IIR Conf.: Designing and optimising mobile networks, London, Mar. 1997.

[147] **Mogensen P.E.,** *GSM base-station antenna diversity using soft decision combining on up-link and delayed-signal transmission on down-link,* IEEE Proc. VTC '93, pp. 611-616.

[148] **Mogensen P.E., K.I. Pedersen, P. Leth-Espensen, B. Fleury, F. Frederiksen K. Olesen and S.L. Larsen,** *Preliminary measurement results from an adaptive antenna array testbed for GSM/UMTS,* IEEE Proc. VTC '97, pp. 1592-1596.

[149] **Mogensen, P., J. Wigard and F. Frederiksen,** *Performance of Slow Frequency Hopping in GSM (Link-level).* In COST 231, Sept. 1995.

[150] **Møller H.K. and M. Toft**, *Synchronization and optimization of GSM Baseband Receiver*, M.Sc.E.E. thesis, AUC, June 1995.

[151] **Mouly M. and M.B. Pautet**, *The GSM System for Mobile Communications*, ISBN: 2-9507190-0-7, 1992.

[152] **MSI Plc.**, *PlaNet version 2.7 Software documentation*, June 1997.

[153] **Munoz-Rodriquez and others**, *Neural Supported Hand Off Methodology in Micro Cellular Systems*, IEEE Proc VTC '92, Denver, May 10-13, 1992, pp.431-434.

[154] **Nagata Y. and Y. Akaiwa,** *Analysis for spectrum in single cell trunked and cellular mobile radio,* IEEE Trans. Veh. Technol., vol. VT-35, pp. 100-113, Aug. 1987.

[155] **Nagatsu, T., T. Tsuruhara and M. Sakamoto,** *Transmitter Power Control for Cellular Land Mobile Radio*, IEEE Proc. Globecom '83, pp. 1430-1434.

[156] **Nakagami M.,** *The m-distribution – a general formula of intensity distribution of rapid fading, Statistical Methods in Wave Propagation.* W.C. Hoffman Ed. Oxford: Pergamon, pp. 3-36, 1960.

[157] **Narasimhan R. and D.C.Cox,** *A Handoff Algorithm for Wireless Systems Using Pattern Recognition*, IEEE Proc. PIMRC'98, Boston, pp. 335 –339.

[158] **Näslund J., C. Carneheim, C. Johansson, S-O. Jonsson, M. Ljungberg, M. Madfors and J. Skjöld,** *An Evolution of GSM*, IEEE Proc. VTC'94, Stockholm, pp.248-352.

[159] **Nettleton R.W and H. Alavi**, *Downstream Power Control for Spread-Spectrum Cellular Mobile Radio System*, IEEE Proc. Globecom '82.

[160] **Nettleton, R.W and H Alavi,** *Power Control for Spread Spectrum Cellular Mobile Radio System,* IEEE Proc VTC '83, pp. 242-246.

[161] **Nettleton, R.W.**, *Traffic Theory and Interference Management for a Spread Spectrum Cellular Radio System,* Proc ICC-80, Seattle, WA 1980.

[162] **Ng S.,** *Practical implementation of intelligent frequency hopping*, Conf. on frequency hopping and reuse in mobile networks, Amsterdam, 1999.

[163] **Nielsen T.T. and J. Wigard,** *A Network Performance Study of MAIO-management,* Internal CPK document, Aalborg University. Feb. 1998.

[164] **Nielsen T.T. and J. Wigard,** *A Performance Study of the GSM Power Control Algorithm,* Internal CPK document, Aalborg University. March 1998

[165] **Nielsen T.T.,** *Implementation of slow associated control channel (SACCH) in GSM link tool*, Center for PersonKommunikation internal document, May 1997.

[166] **Nielsen T.T., J. Wigard and P. Mogensen,** *On the Capacity of a GSM Frequency Hopping network with Intelligent Underlayer-Overlayer,* IEEE Proc VTC'97, Phoenix, USA, pp. 1867-1871.

[167] **Nielsen T.T., J. Wigard and S. Skjærris,** *Study of the Influence on Quality Estimations in GSM,* Sonofon internal document, June 1998.

[168] **Nielsen T.T., J. Wigard and S. Skjærris,** *Introducing Downlink Power Control in Random Frequency Hopping GSM Network,* Sonofon internal document, Spring 1998.

[169] **Nielsen T.T., J. Wigard and S. Skjærris,** *Introducing Random Frequency Hopping in the SONOFON Network*, Internal Sonofon Report, Fall 1997.

[170] **Nielsen T.T., J. Wigard, P.H. Michaelsen and P.E. Mogensen,** *Resource Allocation in a Frequency Hopping PCS1900/GSM/DCS1800 Type of Network,* to appear in IEEE Proc VTC '99, Houston.

[171] **Nielsen, T.T. and J. Wigard,** *Frequency hopping and re-use in high penetration mobile market,* Nordic Mobile 99 Conference, Copenhagen, Denmark, Feb. 1999.

[172] **Nielsen, T.T. and J. Wigard,** *Frequency planning and optimisation in a frequency hopping network,* Conference on Frequency Hopping and re-use in mobile networks, Amsterdam, Mar. 1999.

[173] **Nielsen, T.T. and J. Wigard,** *How to do Frequency Planning in the Frequency Hopping parts of the Sonofon Network,* IIR Conference on Frequency hopping and frequency re-use, London, Sept. 1998.

[174] **Nielsen, T.T. and J. Wigard,** *Overcoming the challenges of measuring the network quality of service (QoS) and capacity gain when using frequency hopping,* Conference on Maximising Capacity in Mobile Networks, Cannes, France, Dec. 1998.

[175] **Nielsen, T.T. and Y. Ling,** *Combining IUO and frequency hopping in the Sonofon network.* Sonofon internal document, June 1996.

[176] **Nielsen, T.T., J. Wigard, P.H. Michaelsen and P. Mogensen,** *Slow Frequency Hopping Solutions for GSM Networks of Small Bandwidth,* IEEE Proc VTC '98, Ottawa, Canada, pp. 1321-1325.

[177] **Nielsen, T.T., J. Wigard, S. Skjærris, C.O. Jensen and J. Elling,** *Enhancing Network Quality using Baseband Frequency Hopping, Downlink Power Control and DTX in a Live GSM Network,* IEEE Proc. PIMRC '98, Boston, pp 905 -909.

[178] **Nielsen, T.T.,** *Procedure and results of verification of IUO traffic model, fall 1996.* Sonofon internal document, Oct. 1996.

[179] **Nokia Telecommunications,** *NPS/X Functional Software Description.*

[180] **Oh S. and D. Tcha,** *Prioritized Channel Assignment in a Cellular Radio Network,* IEEE Trans. on Comm., VOL. 40, No. 7, july 1992, pp 1259-1269.

[181] **Okada M., S. Hara and N. Moringana,** *A novel Concatenated Block Coded Modulation Scheme in Rayleigh Fading Channel,* IEEE Proc. VTC '94, pp. 967-971.

[182] **Okumura Y., E. Ohmoni, T. Kawato and K. Fukuda,** *Field Strength and Its Variability in VHF and UHF Land-Mobile Radio Service,* Review of the Electrical Communication Laboratory, Vol. 16, Numbers 9-10, Sept.-Oct., 1968.

[183] **Olofsson H., J. Näslund and J. Sköld,** *Interference diversity gain in frequency hopping GSM,* IEEE Proc VTC'95, pp. 102-106, 1995.

[184] **Olofsson H., M. Almgren, M. Höök,** *Transmitter Diversity with Antenna Hopping for Wireless Communication Systems,* IEEE Proc. VTC '97, pp. 1743-1747.

[185] **Olofsson H., S. Magnusson and M. Almgren** *A Concept of Dynamic Neighbor Cell List Planning in a Cellular System,* IEEE Proc PIMRC'96, Taipei, pp.138-142.

[186] **Ostling P.E.,** *Implications of Cell Planning on Handoff Performance in Manhattan Environments,* IEEE Proc. PIMRC'94, Den Haag, Netherlands, pp 625 -629.

[187] **Papavassiliou S., L. Tassiulas and P. Tandon,** *Meeting QoS requirements in a cellular network with reuse partitioning.* IEEE Journal Select. Areas in Commun., vol. 12, No. 8, Oct. 1994, pp 1389 -1400.

[188] **Parsons J.D.,** *The Mobile Radio Propagation Channel*, Pentech Press Ltd., 1992.

[189] **Partanen M.,** *Test report for MS co-channel performance in SFH environment,* Nokia internal document, 1997.

[190] **Pollini G.P.,** *Trends in Handover Design*, IEEE Communications Magazine, March 1996, pp 82-90.

[191] **Prasad R. and A. Kegel,** *Improved Assessment of interference Limits in Cellular Radio Performance,* IEEE Trans. On Veh. Tech., Vol. 40, No. 2, 1991, pp 412 -419.

[192] **Prasad R. and J.C. Arnbak,** *Comments on "Analysis for Spectrum Efficiency in Single Cell Trunked and Cellular Mobile Radio",* IEEE Trans. On Veh. Tech., Vol. 37, No. 4, Nov. 1988, pp 220 –222.

[193] **Proakis J.D.,** *Digital Communications,* 2. Ed., 1993, ISBN: 0-07-100269-3.

[194] **R.S. Krup,** *Stabilization of Alternate Routing Networks*, ICC, Philidelphia, June 1982, 31.2.1-31.2.5.

[195] **Ramsdale P.A. and W.B. Harrold** *Techniques for Cellular Networks Incorparating Microcells*, IEEE Proc PIMRC'92, Boston, pp. 169-173.

[196] **Rapp Y.,** *Planning of Junction Network in a Multi-exchange Area,* Ericsson Technics No. 1, 1964.

[197] **Rappaport S.S. ,** *The Multiple-call Hand-off problem in High-capacity cellular communications systems*, IEEE Proc. VTC'90, pp 287-294.

[198] **Rappaport S.S.,** *Modeling the Hand-off Problem in Personal Communications Networks*, IEEE Proc. VTC'91, pp. 517-523.

[199] **Raymond P.**, *Performance analysis of cellular networks,* IEEE Trans. On Communications, VOL.39, NO.12 Dec. 1991, pp. 1787-1793.

[200] **Riordan J.**, *Derivation of moments of overflow traffic (Appendix 1 in 'Theories for Toll Trafic Engineering in the U.S.A.', by R.I.Wilkinson, 1956, pp. 507-514).*

[201] **Ritzen B.**, *Automatic Frequency Planning based on measurements performed by the system and mobiles*, Slides from Frequency Hopping & Re-use in Mobile Networks, 14-16 september 1998, London.

[202] **Robertson P. and Wörz T,** *A novel bandwidth efficient coding scheme employing turbo codes*, IEEE Proc. ICC'96, pp. 962-967.

[203] **Rodriguez D.M. and R.A.Herrera,** *DTX evaluation in Nakagami – Nakagami and Rician – Nakagami channels*, IEEE Proc. IEEE VTC '94, pp. 1475-1479.

[204] **Rosenberg, A.N,** *Simulation of Power Control and Voice-Channel Selection in Cellular Systems*, IEEE Proc. VTC '85, pp. 12-15.

[205] **Ross S.M.** *Introduction to probability and statistics for engineers and scientists,* John Wiley and Sons, 1987.

[206] **Sallberg K., B. Stavenow and B. Eklundh ,** *Hybrid channel assignment and reuse partitioning in a cellular mobile telephone system*, IEEE Proc. 1987.

[207] **Salvalaggio T.,** *On the application of reuse partitioning*. IEEE Proc. VTC '88, pp. 182-185.

[208] **Schiml A.** *Interference Analysis in the GSM-System*, IEEE Proc. VTC '94, Stockholm, Sweden, pp. 686-690.

[209] **Schramm P. et al.**, *Radio Interface Performance of EDGE, a Proposal for Enhanced Data Rates in Existing Digital Cellular Systems*, IEEE Proc VTC'98, Ottawa, pp. 1064-1068.

[210] **Schwartz M.,** *Telecommunication Networks: Protocols, Modeling and Analysis*, Addison-Wesley Publishing Company, November 1988.

[211] **Schwartz S.C. and Y.S. Yeh,** *On the distribution function and moments of sums with log-normal components,* Bell Syst. Tech. J., vol. 61, pp. 1441-1462, Sept. 1982.

[212] **Schwartz S.C. and Y.S. Yeh,** *The sum of log-normal probability distributions in scatter transmission systems,* IRE Trans. Commun. Syst., Vol. CS-8, Mar. 1960, pp. 57-67.

[213] **Senerath G., A. Abu-Dayya and R. Matyas**, *Adaptive Handoff Algorithms Using Absolute and Relative Thresholds for Cellular Mobile Communication Systems*, IEEE Proc. VTC'98, Ottawa, pp. 1603-1607.

[214] **Senerath G.N. and D. Everitt**, *Performance of Handover Priority and Queueing Systems under Different Handover Request Strategies for Microcellular Mobile Communication Systems,* IEEE Proc. VTC '95, Chicago, pp. 897-901.

[215] **Serizawa, M., Goodman, D.J.**, *Instability and Deadlock of Distributed Dynamic Channel Allocation*, IEEE Proc. VTC '93, pp 528 –531.

[216] **Silventoinen M.I., Ranta P.A. and Raitola M.,** *Fast power control for GSM HBS using training sequences*, IEEE Proc. VTC '97, Phoenix.

[217] **Skjôld J., B. Gudmundson and J. Fârjh,** *Performance and characteristics of GSM-based PCS,* IEEE Proc. VTC '95, pp.743-748.

[218] **Smith D.H., S.M. Allen, S.Hurley and W.J. Watkins**, *Frequency Assignment: Methods and Algorithms*, NATO symposium: frequency assignment, Aalborg, Denmark, oct. 98.

[219] **Soest van J.**, *Elementaire Statistiek*, Delftse Uitgevers Maatschappij, 1992.

[220] **Sørensen T.B.** *Correlation Model for Slow Fading in a Small Urban Macro Cell,* IEEE Proc. PIMRC'98, pp 1161-1165.

[221] **Steele R. and V.K. Prabhu,** *Mobile radio cell structure for high user density and large data rates*, IEEE Proc., pp. 396-404, No. 5, Aug.1985.

[222] **Strang G.**, *Linear algebra and its applications*, third edition, Massachusetts Institute of Technology, USA, 1986.

[223] **Suzuki H.,** *A statistical model for urban radio propagation,* IEEE Trans. Comm., COM-25, No.7, pp. 673-680, 1977.

[224] **Tschirks W.**, *Effects of transmission power control on the cochannel interference in cellular radio networks*, Electrotechnik und Informationstechnik, vol. 106,no. 5, 1989

[225] **Turin G.L., F.D. Clapp, T.L. Johnston, S.B. Fine and D. Lavry,** *A statistical model of urban multipath propagation,* IEEE Trans. Veh. Techn. VT-28, No.3, pp. 182 - 203, 1979.

[226] **Turkmani A.M.D., A. Demery and J.D. Parsons,** *Measurement and Modelling of Wideband Mobile Radio Channels at 900 MHz,* In IEE Proc. Commun. Speech &Vision, 138(5), 1991, pp 447-457.

[227] **Ulukus S., R.D. Yates,** *Power control using stochastic measurements*, WINLAB publication, In 34th Annual Allerton Conference on Comm. Oct. 1996.

[228] **Vijayan R., and J.M. Holtzman,** *A Model for Analyzing Handoff Algorithms*, IEEE Trans. on Veh. Tech., VOL. 42, NO.3, August 1993, pp 351 -356.

[229] **Walfisch J. and H.L. Bertoni**, *A Theoretical model of UHF Propagation in Urban Environments*, IEEE Trans. on Ant. and Prop., vol 36, no.12, pp. 1788-1796, Dec. 1988.

[230] **Wallstedt K. et al.**, *Micro Cellular Performance in a TDMA System*, IEEE Proc. VTC'93, 1993, pp 293 -296.

[231] **Wang L., K. Chawla and L.J.Greenstein**, *Performance studies of narrow beam trisector Cellular Systems*, IEEE Proc VTC'98, Ottawa, pp. 724-728.

[232] **Watanabe F., T. Buot, T. Iwama and M. Mizuno,** *Load sharing sector cells in cellular systems*, IEEE Proc PIMRC '95, pp. 547-551.

[233] **Whitehead,** Signal-level-based Dynamic Power Control for Cochannel interference management, IEEE Proc. VTC '93, pp 499-502.

[234] **Wigard J. and P. Mogensen** , *A Simple Mapping from C/I to FER and BER for a GSM type of interface,* IEEE Proc. PIMRC '96, Taipei, pp 78 -82.

[235] **Wigard J. and P. Mogensen** , *Capacity of a GSM Network with Fractional Loading and Random Frequency Hopping*, IEEE Proc. PIMRC '96, Taipei, pp 723 -727.

[236] **Wigard J., and T.T. Nielsen,** *Examining quality measurements and the influence of DTX and power control in a frequency hopping network,* Conference on Frequency Hopping and re-use in mobile networks, Amsterdam, March. 1999.

[237] **Wigard J.,** *GSM Link Simulation Tool v. 3.0,* Center for PersonKommunikation, internal document, April 1996.

[238] **Wigard J., P. Mogensen, J. Johansen and B. Vejlgaard**, *Capacity of a GSM Network with Fractional Loading and Random Frequency Hopping,* IEEE Proc. PIMRC 1996, Taipei, pp. 723-727.

[239] **Wigard J., T.T. Nielsen, P.E. Mogensen and P.H. Michaelsen,** *Frequency Planning for Frequency Hopping Networks,* COST 259, Duisburg, Germany, Sept. 1998.

[240] **Wigard J., T.T. Nielsen, P.H. Michaelsen and P.E. Mogensen,** *Dynamic handover parameters for a PCS1900/GSM/DCS1800 Type of Network,* to appear in IEEE Proc VTC '99, Houston.

[241] **Wigard J., T.T. Nielsen, S. Skjærris and P.E. Mogensen,** *On the influence of Discontinuous Transmission in a PCS1900/GSM/DCS1800 Type of Network,* to appear in IEEE Proc VTC '99, Houston.

[242] **Wigard, J., T.T. Nielsen and P. Mogensen,** *Improved Intelligent Underlay-Overlay Combined with Frequency Hopping in GSM,* IEEE Proc. PIMRC '97, Helsinki, pp. 376-380.

[243] **Wigard, J., T.T. Nielsen, P.H. Michaelsen and P. Mogensen,** *BER and FER Prediction of Control and Traffic Channels for a GSM type of interface*, IEEE Proc.VTC '98, pp. 1588-1592.

[244] **Wilkinson I.**, *Theories for Toll Traffic Engineering in the USA,* Bell System Tech. Journal, Vol. 35, No. 2, 1956.

[245] **Wille V., H. Multimaki and S. Irons**, *A Practical Approach to 'Channel Borrowing' for Microcells in GSM systems* IEEE Proc. VTC'98, Ottawa, pp.144-148.

[246] **Wille V., S.M. Irons and King,** *Capacity increase in cellular radio networks using intelligent underlay overlay,* 10th Int. IEE Conf. on Antenna and Propagation, April 1997.

[247] **Yates R. and C. Y. Huang.** *Optimal Uplink Control and Base Station Assignment.* IEEE Proc. ICUPC '94, San Diego, CA, 247-251.

[248] **Yoshihisa et al. Okumura**, *Field Strength and its Varability in VHF and UHF Land-Mobile Service*, IEEE Press, New York. ISBN: 0-87942-174-6., 1984

[249] **Zako B. et al.,** *The GSM radio link performance with space diversity and slow frequency hopping,* IEEE Proc. VTC '93, pp. 480-482.

[250] **Zander J.,** *Distributed Cochannel Interference Control in Cellular Radio Systems.* IEEE Trans. on Veh. Tech., Vol. 41, No. 3, 1992, pp 305 -311.

[251] **Zander J.**, *Optimum Global Transmitter Power Control in Cellular Radio Systems,* IEEE Proc. PIMRC '91, London, pp. 24 –27.

[252] **Zander J.,** *Performance of optimum transmitter power control in cellular radio systems*, IEEE trans. on Veh. Technology, vol.41, no. 1 Feb 1992, pp 57 -62.

[253] **Zander J.,** *Transmitter Power Control for Co-channel Interference Management in Cellular Radio Systems.* Proc. 4th WINLAB Workshop, Oct. 19-20, 1993.

[254] **Zeng Q., K. Mukumoto and A. Fukuda**, *Performance Analysis of Two-level Priority Reservation Handoff Scheme in Mobile Cellular Radio Systems*, IEEE Proc. VTC'97, Phoenix, pp. 974-978.

[255] **Zhang N. and J.M. Holtzman**, *Analysis of a CDMA Soft Handoff Algorithm*, IEEE Proc of PIMRC'95 Toronto, pp. 819-823.

[256] **Zhang N. and J.M. Holtzman**, *Analysis of Handoff Algorithms using both Absolute and Realtive Measurements*, IEEE Proc. VTC '94, Stockholm pp. 82-86.

[257] **Zonoozi M. and P. Dassanayake,**, *Shadow Fading in Mobile Radio Channel*, IEEE Proc. PIMRC '96, Taipei., pp 291 –295.

Index

MIX
Papier aus verantwortungsvollen Quellen
Paper from responsible sources
FSC® C105338

If you have any concerns about our products,
you can contact us on
ProductSafety@springernature.com

In case Publisher is established outside the EU,
the EU authorized representative is:
Springer Nature Customer Service Center GmbH
Europaplatz 3, 69115 Heidelberg, Germany

Printed by Libri Plureos GmbH
in Hamburg, Germany